QUANTITATIVE BIOLOGY
FROM MOLECULAR TO CELLULAR SYSTEMS

CHAPMAN & HALL/CRC
Mathematical and Computational Biology Series

Aims and scope:

This series aims to capture new developments and summarize what is known over the entire spectrum of mathematical and computational biology and medicine. It seeks to encourage the integration of mathematical, statistical, and computational methods into biology by publishing a broad range of textbooks, reference works, and handbooks. The titles included in the series are meant to appeal to students, researchers, and professionals in the mathematical, statistical and computational sciences, fundamental biology and bioengineering, as well as interdisciplinary researchers involved in the field. The inclusion of concrete examples and applications, and programming techniques and examples, is highly encouraged.

Series Editors

N. F. Britton
Department of Mathematical Sciences
University of Bath

Xihong Lin
Department of Biostatistics
Harvard University

Hershel M. Safer
School of Computer Science
Tel Aviv University

Maria Victoria Schneider
European Bioinformatics Institute

Mona Singh
Department of Computer Science
Princeton University

Anna Tramontano
Department of Biochemical Sciences
University of Rome La Sapienza

Proposals for the series should be submitted to one of the series editors above or directly to:
CRC Press, Taylor & Francis Group
4th, Floor, Albert House
1-4 Singer Street
London EC2A 4BQ
UK

Published Titles

Published Titles (continued)

Chapman & Hall/CRC Mathematical and Computational Biology Series

QUANTITATIVE BIOLOGY
FROM MOLECULAR TO CELLULAR SYSTEMS

EDITED BY
MICHAEL E. WALL

CRC Press
Taylor & Francis Group
Boca Raton   London   New York

CRC Press is an imprint of the
Taylor & Francis Group, an **informa** business

A CHAPMAN & HALL BOOK

The Los Alamos unclassified release number to the Preface is LA-UR-11-12291, and the release number to Chapter 10 is LA-UR-11-00459. Los Alamos National Laboratory strongly supports academic freedom and a researcher's right to publish; however, the Laboratory as an institution does not endorse the viewpoint of a publication or guarantee its technical correctness.

Cover Image: Bacteria. Image captured using fluorescence microscopy. Gerd Guenther, Duesseldorf, Germany. Honorable Mention, 2011 Olympus BioScapes Digital Imaging Competition®.

CRC Press
Taylor & Francis Group
6000 Broken Sound Parkway NW, Suite 300
Boca Raton, FL 33487-2742

© 2013 by Taylor & Francis Group, LLC
CRC Press is an imprint of Taylor & Francis Group, an Informa business

No claim to original U.S. Government works

Printed in the United States of America on acid-free paper
Version Date: 20120529

International Standard Book Number: 978-1-4398-2722-2 (Hardback)

Library of Congress Cataloging-in-Publication Data

Quantitative biology : from molecular to cellular systems / editor, Michael E. Wall.
        p. cm. -- (Chapman & Hall/CRC mathematical and computational biology series)
     Includes bibliographical references and index.
     ISBN 978-1-4398-2722-2 (alk. paper)
     1. Molecular biology--Mathematical models. 2. Cytology--Mathematical models. 3. Biological systems. I. Wall, Michael E., 1967-

     QH506.Q36 2012
     572.8'2--dc23   2012006899

Visit the Taylor & Francis Web site at
http://www.taylorandfrancis.com

and the CRC Press Web site at
http://www.crcpress.com

For Jack and Simone, and their generation

Contents

Preface

To see a World in a Grain of Sand
And a Heaven in a Wild Flower,
Hold Infinity in the palm of your hand
And Eternity in an hour.

WILLIAM BLAKE, *Auguries of Innocence*

Quantitative methods are bringing about a revolution in modern molecular and cellular biology. Groundbreaking technical advances are fueling a rapid expansion in our ability to observe, as seen in multidisciplinary studies that integrate theory, computation, experimental assays, and the control of microenvironments. Although a relatively small number of risk-takers and visionaries have led most of the advances, many of the ideas and methods are maturing and are ready to be adopted by a wider group of researchers. This book is intended to provide them with a solid foundation for starting work in this field.

These pages provide an introduction to the fundamentals of the emerging science of modern quantitative biology and what it is telling us about molecular and cellular behavior. The chapters, which span diverse topics and were contributed by recognized experts, are organized into three sections. Chapters 1–4 (Fundamental Concepts) cover bold ideas that inspire many questions and approaches of modern quantitative biology. There you'll find perspectives on evolution dynamics (Robert H. Austin), system design principles (Michael A. Savageau), chance and memory (Theodore J. Perkins, Andrea Y. Weiße, and Peter S. Swain), and information processing in biology (Ilya Nemenman). Chapters 5–10 (Methods) describe recently developed or improved techniques that are transforming biological research. You'll read about experimental methods for studying single-molecule biochemistry (Jeffrey A. Hanson and Haw Yang), small-angle scattering of biomolecules (Cy M. Jeffries and Jill Trewhella), subcellular localization of proteins (Jeffrey J. Saucerman and Jin Zhang), and single-cell behavior (Philippe Cluzel). You'll also find theoretical methods for synthetic biology (Ying-Ja Chen, Kevin Clancy, and Christopher A. Voigt) and modeling random variations among cells (Brian Munsky). Chapters 11–13 (Molecular and Cellular Systems) focus on specific biological systems where modern quantitative biology methods are making an impact. At the molecular level is protein kinase (Susan S. Taylor and Alexandr P. Kornev), at the regulatory system level is the genetic switch of phage lambda (John W. Little and Adam P. Arkin), and at the cellular level is *Escherichia coli* chemotaxis (Howard C. Berg). Chapters 1–10 also incorporate case studies of systems for which new concepts or methods are increasing our understanding.

The chapters stress focused studies of a single system, whether cell, protein, molecular network, or other system. Such studies already have revealed many marvels of biological behavior and have suggested general principles, such as evolutionary optimization of information processing and other measures of performance, that might apply much more broadly in biology. Quantitative biology often seeks to discover, in Blake's words, "a world in a grain of sand." It is this spirit that the book aims to capture.

The absence of a relevant topic shouldn't be seen as a sign of diminished importance or lack of relevance for the book. For example, recent advances in molecular dynamics simulations will accelerate the integration of biomolecular structural biology (a subject critical to early quantitative biology research and the development of the entire discipline of molecular biology) with modern quantitative biology. Recent important quantitative advances in developmental biology are also missing from this volume. The field is large, and there are many reasons a topic might not be included.

As an exception, some readers might wonder at the absence of *high-throughput* methods. After all, they are responsible for an explosive growth in the availability of genome sequence data, which is crucial to many quantitative methods. However, because much has already been written about high-throughput methods in systems biology, it is not necessary to cover the topic here.

This book also reflects the excitement among those in the quantitative biology community, a large number of whom have participated in the Annual q-bio Conference (http://q-bio.org). Similar to the long-established Cold Spring Harbor Symposia on Quantitative Biology, the q-bio Conference, now entering its sixth year, has been a lively forum for presenting many new ideas, methods, and findings as well as complete stories where new understanding is achieved through integration of modeling and experiments. The conference and the accompanying summer school are helping to create a community of researchers that is realizing the potential of advances in quantitative biology. In this book's pages, the q-bio community reaches out to a wider group and shows some of the amazing things that can now be done. In reading this book, you'll gain some understanding and I hope be excited (or even bothered!) by some of the perspectives. Perhaps you'll be inspired to ask a bold and provocative question yourself, bring a new tool into your lab, or subject a new system to quantitative scrutiny.

Lastly, I have no doubt that in time the impact of modern quantitative biology on medicine will be profound. For example, quantitative studies of protein kinase structure and function already have led to development of imatinib, an important cancer drug. Much of the quantitative biology community, however, is driven by the need to advance fundamental understanding as opposed to biomedical applications. Not everything described in this book will be equally important to human health, and it is too early to say where the most important contributions will come from. That is why I will be donating my royalties from this book to the Cystic Fibrosis Foundation. Thus, even if the work found in this volume doesn't directly help people live longer, healthier lives, the book sales will improve the future of young people who are living with this heartbreaking disease.

Additional material is available from the CRC Web site: http://www.crcpress.com/product/isbn/9781439827222

Michael E. Wall

Acknowledgments

Many thanks to all of the authors for their contributions and to Rebecca and Jack for their inspiration and patience.

Much of the editing of this book took place in the quiet halls of the Meem Library at St. Johns College and in the Southwest Room of the Main Branch of the Santa Fe Public Library.

The Editor

Michael E. Wall is a scientist at Los Alamos National Laboratory. He has a Ph.D. in physics from Princeton University and did postdoctoral training in biochemistry. His work has addressed problems in protein dynamics and gene regulation, among other things. He lives with his wife and son, mostly in Santa Fe, New Mexico, and occasionally in Dunedin, New Zealand.

Contributors

Adam P. Arkin
University of California, Berkeley
    and Lawrence Berkeley National
    Laboratory
Berkeley, California

Robert H. Austin
Princeton University
Princeton, New Jersey

Howard C. Berg
Harvard University
Cambridge, Massachusetts

Ying-Ja Chen
Massachusetts Institute
    of Technology
Cambridge, Massachusetts

Kevin Clancy
Life Technologies Corporation
Carlsbad, California

Philippe Cluzel
Harvard University
Cambridge, Massachusetts

Jeffrey A. Hanson
University of California, Berkeley
Berkeley, California

Cy M. Jeffries
University of Sydney and
The Australian Nuclear Science
    and Technology Organisation,
    Lucas Heights
Australia

Alexandr P. Kornev
University of California, San Diego
La Jolla, California

John W. Little
University of Arizona
Tucson, Arizona

Brian Munsky
Los Alamos National Laboratory
Los Alamos, New Mexico

Ilya Nemenman
Emory University
Atlanta, Georgia

Theodore J. Perkins
McGill University
Montreal, Québec
Canada

Jeffrey J. Saucerman
University of Virginia
Charlottesville, Virginia

Michael A. Savageau
University of California, Davis
Davis, California

Peter S. Swain
University of Edinburgh
Scotland

Susan S. Taylor
University of California, San Diego
La Jolla, California

Jill Trewhella
University of Sydney
Sydney, Australia

Christopher A. Voigt
Massachusetts Institute
   of Technology
Cambridge, Massachusetts

Andrea Y. Weiße
University of Edinburgh
Edinburgh, Scotland

Haw Yang
Princeton University
Princeton, New Jersey

Jin Zhang
Johns Hopkins University
Baltimore, Maryland

1. Free Energies, Landscapes, and Fitness in Evolution Dynamics

Robert H. Austin

1.1   Introduction: Is Evolution Theory in Trouble?

Charles Darwin transformed biology; however, he provided no analytical template, and we still have experimentally a great deal to learn about the dynamics of evolution, particularly under the response to stress [1]. Fortunately we can analyze the dynamics, and here we provide an analytical framework. We have been engaged in a program for several years using microfabricated environments to probe adaptation and evolution dynamics [2–4]. This chapter presents some of the results that have come from these attempts to accelerate evolution in a deterministic manner. The work is perhaps a little controversial, as typically it is believed that evolution proceeds slowly over many generations and in a random manner. However, we do not believe this is necessarily the case, and we provide some quantitative reasons we hold this opinion.

We will discuss how, on a complex landscape of stress (which corresponds to a free energy landscape), the common bacterium *E. coli* can evolve very high resistance to an antibiotic in about 10 hours, which is remarkably only at most 20 generations. Furthermore, the resistance has a surprisingly parsimonious genetic profile: only four single nucleotide polymorphisms (SNPs). How is this possible? How did a bacterium with the presumably low mutation rate μ of only 10^{-9} SNPs/basepair-replication so quickly find the magic four

Quantitative Biology: From Molecular to Cellular Systems edited by Michael E. Wall © 2012 CRC Press / Taylor & Francis Group, LLC. ISBN: 978-1-4398-2722-2

Chapter 1

SNPs needed to replicate in the presence of such high minimum inhibitory concentration (MIC) excesses of Cipro? In this chapter we will try to show that this is not magic. The consequences of this work have ramifications for other areas of biology and for human health, in particular cancer. We have published a recent review discussing these ramifications [5], in which perhaps the underpinnings can be found.

The title of this section, "Is Evolution Theory in Trouble," brings up a subject beloved by physicists at least of a certain age: we like to poke holes in the current dogmas, and finding "problems" is great fun even if it is a bad career move (in biology, that is). A certified fundamental unsolved problem in physics is: why do the fundamental physical constants have the values they do? One of the more radical ideas put forward is that there is no deeper meaning to these numbers: there are many universes with different numbers, but no one to ask the question [6]. We just happen to be in a particular universe where the physical constants have values conducive to form life and eventually evolve organisms who ask such a question.

This idea of a landscape of different universes actually came from biology (evolutionary theory, in fact) and was first applied to cosmology by Lee Smolin [7]. It is interesting that such a fundamental idea in cosmology came from biology by studies of evolution dynamics. Biology inherently deals with landscapes because the biological entities, whether they are molecules, cells, organisms, or ecologies, are inherently heterogeneous and complex in their interactions with their environment: landscape theory provides a way to visualize the movement of a biological entity in this multidimensional space. This chapter attempts to show how a landscape perspective can help visualize the dynamics of biological systems moving in many dimensional spaces. As an experimentalist, I suspect that the theorists will find this very child-like, but perhaps the basic ideas will come through.

We all intuitively understand what a landscape is: we move in three dimensions but are constrained to follow the contours of the surface of the earth, a surface determined by "something else" but that we are forced to move upon. This landscape is a simple potential energy landscape. Gradients determined by the potential energy landscape set the direction of the gravitational forces acting on a particle and can be used to determine the path of the particle on the surface.

However, a potential energy landscape with a single particle moving on it really is no great conceptual breakthrough. Let us broaden our concept of what is moving on the surface to include more than one particle, and let the particles interact with each other in some way. Then the dynamics of the collective motion within the potential energy landscape become a more complex question. The total energy of the system is collectively a function of the potential energy, the number of particles, and the way that the particles interact. If this were a Newtonian system of N particles, one would be tempted to write down N differential equations and try to solve them deterministically. Of course, as N becomes much bigger than 2 the task becomes hopeless, and the dynamics of the equations for each particle hold little intuitive value. The hope, however, is that the idea of movement on a collective landscape can provide a basic understanding of evolution dynamics.

We recently published a paper showing that by construction of a fitness landscape and other concepts explained in this section it is possible to greatly accelerate the evolution dynamics of the emergence of resistance of a genotoxic antibiotic (Cipro) in bacteria [8]. Hopefully what is written here will explain some aspects of this work in more depth.

1.2   Free Energies Are Crisply Defined in Physics

Stubbornly, although evolution concerns fitness landscapes, our discussion begins with junior-level physics to make sure we are not just throwing jargon around in the hopes that biologists don't notice we can't do simple physics problems. Sadly, *free energy* must first be defined. The concept of free energy came out of statistical mechanics and thermodynamics as the idea of conservation of energy was generalized to include cases where it seemed to be violated due to the presence of many internal degrees of freedom. Any standard statistical physics textbook, such as Kittel [9], can be used to elucidate the next sections if readers have forgotten their undergraduate thermodynamics course. For example, we are taught in freshman physics that the (scalar) center of mass mechanical energy, K, of a particle is conserved as long as there are no "hidden" degrees of freedom that can steal away the mechanical energy, and thus all the internal energy in the system can be extracted to do work: the free energy is equal to the internal energy. However, it is easy to hide center of mass kinetic energy because it is a scalar quantity and has no direction associated with it. The vector quantity momentum, $\vec{P}$ (or the angular momentum $\vec{L}$), is rather harder to hide since it has both magnitude and direction. The classic demonstration of this is to shoot a bullet into a block of wood at various impact parameters, b, to the block center of mass and then to determine how high the block rises into the air. The block always flies into the air to the same height h no matter where it was hit, but this constant height is not due to conservation of the incoming center of mass kinetic energy of the bullet, K_{bullet}. Rather, because of conservation of linear momentum $\vec{P}$, which can't be hidden, the block height rise h is easily calculated, even though the spin rate, ω, of the block varies depending on the impact parameter, b. Since physics is supposed to be an exact and quantitative science, it is disturbing that the kinetic energy is not conserved. A nice presentation of this can be found by the great Richard Feynman [10].

So, it would appear that conservation of kinetic energy is false from the block experiment since the work done on the system does not result in 100% extractable internal energy: center of mass kinetic energy is lost. Physicists believe that in fact the total kinetic energy is conserved and can be quantitatively shown to be conserved if all the various places the kinetic energy could have been stashed away in all the other degrees of freedom can be accounted for. Trying to do this is the province of statistical physics, which is an exact analytical science only if you stick to the area of little point particles bouncing around in a box. However, the key idea behind free energy is that the amount of internal energy, E, that can be extracted to do work, W, is not E because the many internal degrees of freedom can be described only using a probability distribution. The entropy, S, is defined as a measure of the size of the phase space occupied by this probability distribution, and as E decreases some of it eventually must be used to squeeze the system into a smaller phase space with lower S; the penalty for doing this is more severe at lower temperatures. Quantitatively, thermodynamics gives the equation for the free energy, A, which is the amount of work that can be extracted from the internal energy:

$$A = E - TS \tag{1.1}$$

So this makes good physical sense. Of course, we don't know yet how to calculate S, and in fact it is not at all trivial to do so. For biological systems, it is probably impossible.

Chapter 1

Let's first review how we can calculate the extractable energy from a system given the inherent disorder contained within it for the simple systems physics can handle so well.

The partition function, Z, of a system is the fundamental statistical function from which all thermodynamic parameters such as entropy can in principle be calculated, and for the present purposes it is simply the sum over all the Boltzmann-weighted probabilities of the microscopic states of the system. The partition function deeply shows how the energy, E, and the entropy, S, of a system are intertwined so that all the mechanical energy, E, cannot be taken out of a system with S left unchanged, nor can all the S be removed from a system leaving E the same from pure statistical considerations.

$$Z = \sum_i e^{-\beta E_i} \tag{1.2}$$

where $\beta = \dfrac{1}{k_B T}$. For a monatomic ideal gas, the mechanical energy is the average translational energy $<E>$ (there is no potential energy in the system), and the entropy S is a measure of the number of states that can be occupied given a fixed amount of internal energy, pure statistical counting. The average internal energy $<E>$ is given in terms of the partition function:

$$<E> = -\frac{\partial Z}{\partial \beta} \tag{1.3}$$

The entropy, S, of the ideal gas is equal to the Boltzmann constant, k_B, times the logarithm of the total number of possible states that can be occupied. In terms of the partition function, Z, and the average internal energy per degree of freedom $<E>$, S is

$$S = k_B \log Z + k_B \beta <E> \tag{1.4}$$

The Helmholtz free energy, A, related to how much center of mass mechanical energy, which carries no entropy, can be extracted from a system within the constraints of the second law of thermodynamics, is

$$A = <E> - TS = \frac{-\log Z}{\beta} \tag{1.5}$$

Equation (1.5) is a little deceptive if it is simply memorized. The relationship $A = <E> - TS$ comes from the concept of a partition function Z that sums over all probabilities for a fixed total energy of the ways of distributing the energy in different states. All of E cannot simply be pulled out of a system with S remaining intact, because Equation (1.4) shows that S and E are fundamentally connected. If you don't understand this you can stop reading right now and go do some biology, because we both will be wasting each other's time.

The temperature, T, is probably the trickiest thing to connect to evolution dynamics. The thermal temperature, T, previously used, which connects entropy with the average translational energy of an atom, will distinctly not also be used here. E for biology will be defined in terms of the average fitness, s, an individual has, and then the evolution temperature, T_e, connects the genomic disorder with the average fitness.

For an ideal gas, a free-energy landscape would be the surface defined as $A(P,T)$ for fixed V and N as P and T are varied. Of course, for an ideal gas this surface is extremely boring and has a global minimum where the Helmholtz free energy is minimized, namely, where $PV = \dfrac{N}{\beta}$.

At this point the reader may ask: this is all great thermal physics, but wasn't the mission to explain evolution dynamics? First, I claim if you don't understand fundamental statistical physics you can't understand evolution dynamics at a deep level. Second, of course to move into evolution biology we will have to lift a couple of the restrictions we have labored under. The most important one is that the number of particles in the volume V is fixed. This is clearly wrong for a biological system; after all, biological agents can reproduce, changing the number of particles. That can't be found in a thermal physics textbook. In thermodynamics, this is handled by allowing for diffusive contact of a reservoir with another one so that atoms can move back and forth between the reservoirs. The definition of free energy then needs to be broadened to include the possibility that the number of particles can change, especially if we want to talk about evolution. Also, typically in biology the pressure and temperature are fixed, but the volume and number of particles can vary: that is, the system can do work and can reproduce. That sounds more like biology. Much more of a problem is that biological entities do not replicate with absolute fidelity; they change (mutate), so they are hardly identical. This actually is a huge problem.

If biological entities were identical but could change in number there would be no real problem in principle. In that case, the minimized free energy in equilibrium conditions would not be the Helmholtz free energy, A, in Equation (1.5), since for the Helmholtz case the volume was held fixed and no work could be done by the system. In the case of allowing the volume to change, the free energy that is minimized is not A but rather the Gibbs free energy, G:

$$G = E + PV - TS \tag{1.6}$$

where the PV term has units of energy and allows both extraction of internal energy via entropy flow (heat) and physical work. The partition function expression for the Gibbs free energy is unfortunately quite a bit less elegant than the expression for the Helmholtz free energy:

$$G = -\frac{Z}{\beta} + \frac{V}{\beta}\left[\frac{\partial \ln Z}{\partial V}\right]_{N,T} \tag{1.7}$$

A free-energy landscape is not necessary to understand how a system of noninteracting atoms moves on the pressure and temperature surface at fixed volume to minimize the Helmholtz free energy. A van der Waals gas has just two more variables: the finite volume, b, taken up by each atom; and the mean pairwise attractive interparticle potential energy, a. This is really the simplest thing that can be done to move toward a real system. Suppose the particles are allowed interact with each other and thus store internal amounts of potential energy as well as kinetic energy. The statistical analysis becomes much harder, but it also becomes much more interesting. For example, if we move from an ideal gas of point particles to the van der Waals system where the particles have a finite volume and pairwise attractive interparticle forces, we get phase transitions in the system at critical temperatures. In an ideal gas there are no phase transitions, the gas is a gas at any temperature, and while it might be intellectually satisfying it is a dead end in terms of vastly more interesting new phenomena.

In the van der Waals system, the free energy becomes now crucially important in mapping the states of the system, but it is difficult to deal with. If this simple system gives so much trouble, shouldn't we worry that quantitative biology at a basic level may face far more difficult tasks? At the simplest level in the van der Waals system, a mean-field approach can be used. The problem with this is that all true heterogeneity in the system gets averaged out, which might be catastrophic in terms of missing fundamentally important biological phenomena. In the van der Waals system, the mean interaction energy, a, is really an average over all of space of a potential function, $\varphi(r)$:

$$\int_b^\infty dV \varphi(r)n = -2na \tag{1.8}$$

where n is the average density of the particles, and a is the average radius.

We can move here to the equation of state for the van der Waals system by calculating the free energy analytically, but it isn't too illuminating as a generalized model of interacting systems except that the equation of state has coexistence lines below a critical temperature and thus captures a feature of phase transitions, but rather obscurely. It may seem at this point as if we are wandering farther and farther away from biology, but actually we are moving toward an aspect of biology that many physicists try to actively avoid: namely, that biological entities, like cells, not only take up space but also actually interact, giving rise to highly collective phenomena. The collective aspect of biological interactions, like for the atoms in the van der Walls equation, is both messy and liberating because it opens up the realm of complex biological phenomena. The fear is that quantitative descriptions can no longer be possible, but this is a risk that must be taken to avoid mere hand waving.

To connect the dots with evolution theory, it is helpful to look at Landau's [11] insights on how to combine mean field theory with the free-energy formulation of thermodynamics and the idea of a changing free-energy surface. At the most basic level, Landau surmised that in general the free energy for a system consisting of interacting particles will have a local minimum corresponding to the state of stability and can be expanded as an even function of some order parameter around that stable point. To put a biological spin on this, imagine that the free energy is the negative of the fitness of a particular genome in an environment and that the order parameter is some sort of generalized change in the genome. This would be boring, except that Landau showed that such a simple model could describe basic aspects of phase transitions in thermodynamic systems, and the hope is that it extends to generalized instabilities in genomic systems under stress.

Staying on terra firma at present, we note that the generalized expansion of the Gibbs free energy, $G(T, \zeta)$, of a system can be written in terms of the order parameter, ζ, and the temperature, T, as

$$G(T,\zeta) \sim g_o + \frac{1}{2}g_2(T-T_c)\zeta^2 + \frac{1}{4}g_4\zeta^4 + \frac{1}{6}g_6\zeta^6 + \ldots \tag{1.9}$$

where the second-order term g_2 is "put in by hand" as a function of temperature. There is a clear physical meaning and need for this negative sign: for physics, a negative sign means that higher-order terms lead to a net attraction for particles in the system, just as in the van der Waals case net attraction between nearest neighbors leads to phase changes. The negative terms are needed to make things physically interesting.

Such a generalized free energy with attractive (negative) signs can display either continuous (second-order) or discontinuous (first-order) phase transitions as a function of T. This is of course a very simple, 1-D free energy landscape, but it has surprisingly deep physics connects, which we cannot go into since our mission is to map this to biological fitness landscapes. The abruptness of biological population changes, such as the concept of *punctuated equilibrium* popularized by Stephen J. Gould [12], can be seen as an example of questioning whether a population transition is a first-order or second-order phase transition in genome space.

The two cases of first- and second-order phase transitions are easily seen in the Landau formulation. For example, if we ignore the sixth-order term g_6 and higher in Equation (1.9) and assume that $g_4(T)$ is everywhere positive, then the stable points where the free energy is a minimum as a function of the order parameter are given by

$$\frac{\partial G(T,\zeta)}{\partial \zeta} = 0 = g_2(T - T_c)\zeta + g_4\zeta^3 \tag{1.10}$$

For $T > T_c$ the minimum is at $\zeta = 0$, while for $T < T_c$ the value of ζ_c gradually increases:

$$\zeta_c^2 = \frac{g_2(T_c - T)}{g_4} \tag{1.11}$$

Thus, the equilibrium value, ζ_c, is zero above T_c and smoothly moves from zero below T_c, a second-order phase transition with no discontinuity in the order parameter, ζ, as the critical temperature, T_c, is passed. However, if g_4 is negative and the g_6 term is retained, the change in the free-energy surface with temperature is qualitatively different:

$$\frac{\partial G(T,\zeta)}{\partial \zeta} = 0 = g_2(T - T_c)\zeta - |g_4|\zeta^3 + g_6\zeta^5 \tag{1.12}$$

Again, there are two solutions: $\zeta = 0$ above the critical temperature, but now at $T = T_c$ the order parameter suddenly jumps from zero to $\zeta_c^2 = \frac{g_4}{g_6}$. This sudden jump is indicative of a first-order phase transition, or what Gould called punctuated equilibrium, although Gould, like Darwin, used no mathematics and so invited endless debate. Figure 1.1 shows the differences in the shapes of these free-energy landscapes as a function of temperature.

Note that for a first-order phase transition there is a barrier between the high temperature minimum of the order parameter at $\zeta = 0$ and the nonzero value at $\zeta_c^2 = \frac{g_4}{g_6}$. The system at this point is *frustrated*: there are two values of the order parameter with

Chapter 1

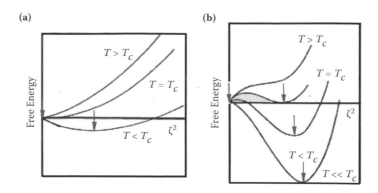

FIGURE 1.1   (a) The Landau free energy versus the order parameter ζ^2 as a function of temperature for a second-order transition. The minimum of the free energy moves smoothly from 0 to positive values as the temperature is lower than T_c. (b) The Landau free energy versus the order parameter ζ^2 as a function of temperature for a first-order transition. The minimum of the free energy moves abruptly from 0 to positive values as the temperature is lower than T_c, and at T_c there are two equivalent values. The minimum in the free energy is denoted by arrows, and the barrier separating the two minima at T_c is shaded. Adapted from Kittel and Kroemer [9].

the same minimum value of free energy. This bistability then results in a delocalization of the order parameter, although the time scale at which the order parameter moves between the two values is not well defined at present.

Well, the reader once again may note that this is all well and good regarding free-energy landscapes and how they represent phase transitions, but where is the biology, aside from analogy? What does *temperature* mean here? Where do these numbers g_n come from in the free energy? Alas, for the remainder of this chapter we must no longer be so formal but instead try to see how the dynamics of a simple interacting system of particles can map over into a case of biological agents, well aware that the rigors of statistical mechanics will be left behind. Perhaps the reader can help us make the links between formalism and biology as we proceed.

1.3   Fitness Landscapes and Evolution

Now we want to move into fitness landscapes and connect to evolution, having immersed ourselves in physics so we aren't complete phonies [13]. An excellent introduction to conventional energy landscapes without the evolution twist can be found in *Energy Landscapes* by David Wales [14] or *Rugged Free Energy Landscapes* by Wolfhard Janke [15]. Up to this point we have used the thermodynamic concept of free energy, which is well defined by Equation (1.5). We now seek stable points in population numbers, N, rather than in free energies, A. In biological systems, what is critical is the rate at which an organism reproduces in an environment. The faster it reproduces, the more fit the organism is said to be. All other things being equal, a mutation that increases the reproduction rate of an organism can be said to give a fitness advantage, s, to that organism:

$$\frac{dN}{dt} = +R_o[1+s]N \tag{1.13}$$

Thus, at least for open-ended exponential growth a fitness increase gives rise to an exponential increase in the number of offspring, in the same way that increasing the energy of a state decreases the probability of occupation of that state exponentially. Because of this sign change, biologists speak about maximizing fitness in the same way that physicists speak about minimizing a free energy. To cater to the biologists, we will continue with the positive sign for fitness enhancement and leave it to the reader to recast things in terms of free energies if desired. At the most naïve level, called the Bob Park level of naiveté regarding biological systems, we can set $N(g_s)$ as the occupation level of a genomic configuration, g_s, which leads to a fitness enhancement, s, and we can set the fitness free energy as the log of the ratio of the population levels N_{g_s} and N_{g0}. Thus, fitness free energy, A_s, can be defined as

$$A_s = T_e \log\left[\frac{N(g_s)}{N_{g0}}\right] \tag{1.14}$$

Written in this way, there seems to be a connection between A_s and the Helmholtz free energy, A, given by Equation (1.5) (with an inverted sign), but it is by no means transparent how to map the critical idea of a partition function, Z, into the growth rates or the entropy, S, or the temperature, T. We first simply point out that just as the log of some function related to occupation levels is taken through the Boltzmann relationship to find the free energy in Equation (1.5), the log of some population number N is taken to get the fitness, which also becomes in effect a free energy because there is a communal aspect to the fitness of an organism: we cannot ignore the presence of other organisms. The evolution temperature, T_e, then must be, in analogy to the thermal temperature T, a measure of the genomic heterogeneity of the collection of individuals. We can say, in analogy to thermodynamics, that

$$\frac{1}{T_e} \sim \frac{\partial N}{\partial s} \tag{1.15}$$

That is, the more rapidly the number of individuals changes with a change in fitness, the lower is the system's T_e. A *hot* ecology then cares little about the relative fitness of individuals, whereas a *cold* ecology cares a great deal.

A free-energy landscape provides a way of calculating occupation levels of many states with different free energies. If each cell grew independently, it would not be necessary to know about the number of cells, and we could simply work with fitness "energies" rather than free energies, whereas a fitness landscape provides a way of calculating rates of population numbers. In thermodynamics it is necessary to account for the number of particles in the system because the energy is shared between the particles, but in biological systems the number of organisms need to be accounted for because the resources of the environment must also be shared with the individuals. The analogies to the thermodynamic systems discussed earlier seem clear, but making things quantitative is not clear.

In particular, the time and length scales of evolution dynamics are a puzzle in large populations of complex cells with large genomes and are not well understood in spite of what we are to tell our school boards fending off the creationists. The problem is that simple theories of evolution give rise to very slow dynamics. This became clear as evolution theories tried to move from Darwin's descriptive anecdotes to a firmer mathematical foundation. Near the celebration of the 150th Anniversary of Darwin's publication of the *Origin of the Species*, it is fair to ask: what took so long for us to figure this out, and

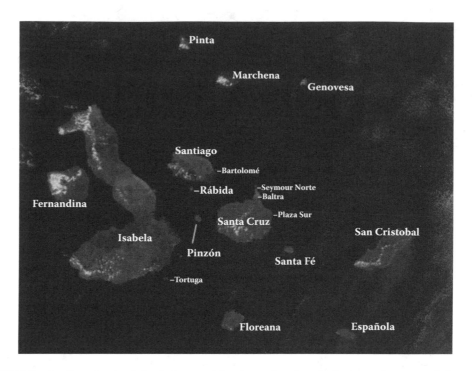

FIGURE 1.2 Satellite photo of the Galapagos Islands overlayed with the Spanish names of the visible main islands. Based on the public domain NASA satellite photo Image:Galapagos-satellite-2002.jpg.

why are we still debating it? The answer lies in the Galapagos Islands, which are quite a unique set of microhabitats weakly coupled with each other yet remote from the South American continent. Figure 1.2 shows a satellite view of the islands. Note that they are small, relatively close to each other, and isolated from any large land masses.

Although much has been made of the different beaks of Darwin's finches, recent work by the Grants [16] on the Galapagos Islands revealed something far more startling then the slow evolution of beak sizes. The Grants caught and banded thousands of finches and traced their elaborate lineage, enabling them to document the changes that individual species make, primarily to their beaks, in reaction to the environment. During prolonged drought, for instance, beaks may become longer and sharper to reach the tiniest of seeds. Here is the problem: we are talking about thousands of birds, not millions. We are talking about beaks that change during a drought period, not thousands of years. How could evolution proceed so quickly?

The problem can be put into simple mathematical form. Darwinian evolution is typically believed to be due to random mutation of N genes, which occurs at some very small rate u of approximately 10^{-9} mutations/basepair-generation. The fitness, s, is also believed to be very small when dealing with a highly optimized genome. Thus, the number of mutations/generation, ΔN, that are fixed or are selected to enter the genome is very small:

$$\Delta N \sim usN \tag{1.16}$$

A further problem, due to Haldane in [17], is the problem of trying to juggle the optimization of many genes simultaneously. Some rather tricky math yields an estimate of about 300 generations to fix a gene under mild selection pressure, and given that genes interact with one another [18] optimizing the fitness of N' genes becomes a logistic nightmare: Haldane's dilemma. A related issue to Haldane's dilemma is Muller's ratchet [19], which discusses how in the absence of recombinant sex the genomes accumulate deleterious mutations in an irreversible manner (hence the ratchet) and lose fitness. Interestingly, the larger the genome the stronger the ratchet effect: small genomes (such as in the mitochondria) cannot degrade further. Presumably the nonrecombining mammalian Y male chromosome is degrading under the action of Muller's ratchet.

Sewall Wright realized that the mathematics of genomic evolution lead to either clonal isolation at a local fitness peak if the mutation rate, u, was too small and the selection pressure, s, was too large (so no evolution of complex organisms) or would ratchet down to a poor fitness valley if u becomes too large and particularly if there is no recombinant exchange mechanism. Within this scenario, mutation rates, u, lead to either nonadaptive clonal populations or unstable populations ratcheting down in fitness. To quote Wright [20, p. 3]: "The problem of evolution as I see it is that of a mechanism by which the species may continually find its way from lower to higher peaks in such a field." Wright proposed instead that "in order that this may occur, there must be some trial and error mechanism on a grand scale by which the species may explore the region surrounding the small portion of the field which it occupies." In effect, he was saying that the deliberate creation of a heterogenous distribution of genomes across a large range of local fitness peaks and valleys was necessary for the evolution of a species (really a quasi-species for the genome is deliberately heterogeneous) to higher fitness peaks. The classic drawing by Wright from his landmark 1932 paper [20, p. 4] is presented here (Figure 1.3), for it presents so clearly the concepts we have been trying to talk about.

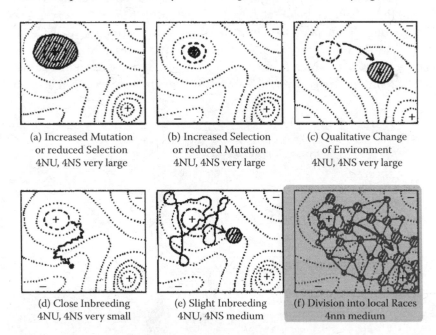

FIGURE 1.3 The fitness landscape as a function of population size N, mutation rate u and selection pressure s. The lower right hand corner in box f represents Wright's enhance evolution via a heterogenous genomic distribution of quasi-species.

The key idea here is that by breaking a population of N individuals (genomes) into a set of quasi-species consisting of $n_i \ll N$ closely related genomes, and letting these quasi-species interbreed weakly using the matrix m_{ij}, which connects quasi-species i with quasi-species j, nature can avoid the traps of Muller's ratchet and realize "some trial and error mechanism on a grand scale by which the species may explore the region surrounding the small portion of the field which it occupies [20, p. 4]."

1.4 Spin Glasses, Genomic Dynamics, and Genomic Heterogeneity

The spin glass model has played a very important role in the development of understanding complexity in many systems consisting of weakly interacting parts. The spin glass is a very simple idea. At its simplest, it is just a collection of spins on a lattice. The i-th spin on the lattice is labeled S_i and for the simplest spin object can point either up ($S_i = +1$) or down ($S_i = -1$). The spins interact, just as individuals respond to other individuals in a prisoner's dilemma game through the reward matrix R_{ij} that determines payoffs in games or the interbreeding matrix m_{ij} that couples genomic contents together in a fitness landscape. In the case of a spin glass, two spins i and j interact through the matrix, J_{ij}, the coupling matrix. At its simplest, J_{ij} can be nonzero only for adjacent spins on the lattice and can have only values +1 (a ferromagnetic interaction, because the two spins want to line up parallel) or –1 (an antiferromagnetic interaction, since the two spins will then want to be antiparallel).

What makes this spin system interesting (and why it is called a glass) is the emerging concept of frustration, shown in Figure 1.4. The simplest nontrivial lattice is a triangle with three spins at each corner. Let $J_{12} = 1$, $J_{23} = 1$, and $J_{13} = -1$. Figure 1.4 shows that spin 3 has two degenerate places to points of equal energy. Because such a lattice is built of more and more points, this frustration leads to many low-lying energy states and a rough energy landscape, which freezes the spin system into a glass with many local spin islands as the temperature is lowered. Such a spin glass picture has been used to describe

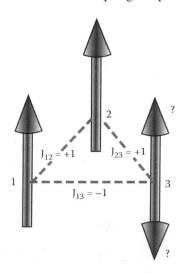

FIGURE 1.4 A simple frustrated spin lattice. Spin 3 has 2 degenerate spin values.

the main conformational states that a protein freezes into at low temperature [21,22]. A spin glass can also be used to make an analogy to the heterogeneous genomic state that a metapopulation of quasi-species can form.

Spin glass analogies to genomic diversity, although only analogies, have some interesting consequences since the spin glass model is inherently a "digital" form of reconstructing a complex system and naturally leads to concepts such as information, distance between states, entropy of states, and the concept of speciation. First, it is possible to define the distance between two genomes by aligning the genomes σ and σ′ and mapping backward the mutational paths to a common ancestor. This is of course a fundamental operation in genomics, and it is not necessary to speak of spin glasses. However, the spin glass analogy mates smoothly into concepts that came out of errors in coding transmissions [23] and allows quantitative measures of the non-Euclidian distance, $d_{\sigma,\sigma'}$, between the two genomes using concepts from the Hamming distance in digital transmissions:

$$d_{\sigma,\sigma'} \sim \Sigma_i |\sigma_i - \sigma_i'| \tag{1.17}$$

Within this spin glass analogy to genomic distance, the axes of Sewall Wright's Figure 1.3 [20] can be calculated, in principle some estimation of the fitness can be attempted for a given genome, and hence the fitness landscape could move from a concept to a reality, probably for bacteria and maybe someday for mammalian cells.

Further, the Hamming distance, d, in Equation (1.17) leads naturally to the biological concept of quasi-species through a powerful and by no means obvious relationship of the distances among three genomes A, B, and C. Normally, if these distances were regular Euclidian distances measured with a ruler the *triangle inequality* would be assumed to hold:

$$d_{AC} \leq d_{AB} + d_{BC} \tag{1.18}$$

For genomic distances, however, a stronger relationship can hold [24]:

$$d_{AC} \leq Max[d_{AB}, d_{BC}] \tag{1.19}$$

where the notation $Max[d_{AB}, d_{BC}]$ just means to take the bigger of the two quantities. The consequences of this are quite profound; the field is called ultrametric spaces [25]. The most amazing statement is that genomes in an ultrametric space must naturally fall into disjoint quasi-species—that is, that a given genome cannot be contained within two classes of genomes connected by two different distances. It belongs to one class or the other, not both.

The final, and most important, lesson that comes from a spin glass model is the answer to a simple question: suppose you apply a weak magnetic field, so now the spins want to point up. How long does it take for the spins to point up? In genomics, the question is: suppose you make the environment fitter for a particular genome σ_i. How long, in τ, does it take for the fittest genome to "fix" in a population of N heterogeneous but closely related individuals (a quasi-species)? If the genomes did not interact, the answer is very simple: the system exponentially relaxes to the new fittest genotype, and the time constant, τ, is independent of the population size.

Chapter 1

But things change dramatically when the spins interact through the J_{ij} coupling we talked about, or in the case of metapopulations the coupling m_{ij} between the metapopulations on the fitness landscape. Because of this coupling, and the frustration we talked about in Figure 1.4, the new fitter mutation propagates much more slowly through the population through the rough fitness landscape. The time dependence is now a power law rather than an exponential. That is, the number of original, less fit individuals, $N(t)$, decays very slowly at long times:

$$N(t) \sim N(0)\left(1 - \frac{t}{\tau}^{-\alpha}\right)$$

(1.20)

where α is some number less than 1, and the characteristic time, τ, scales as

$$\tau \sim \frac{\log(N(0))}{s}$$

(1.21)

This is a very strange result. It says: the larger the population, the longer it takes to fix a mutation. Large populations buffer evolution rates, and this is connected to the power-law decays. Thus, evolution can proceed more rapidly if a population, N, is broken into subpopulations with n members, and they interbreed at rate m. This, then, is nature's grand experiment, and it is ongoing in us. The hope is that spin glasses may provide insights to the dynamics of evolution, while thermodynamics as previously modified might provide a measure of the final equilibrium occupation of the genomic states and the relative numbers of individuals. Further, the spin glass model may help us understand how a system behaves if cooled rapidly enough, using the definition of ecological temperature, T_e, presented in Equation (1.15).

1.5   Landscape Dynamics of an Ecology under Stress

Shark cartilage is one of the more common "alternative medicines" for cancer. Why? Well, mortality rates have not changed, and people facing reemergence are desperate. There exists somewhat of an urban legend that sharks do not get cancers. This is no doubt wrong, but they may have lower incidence rates than *Homo sapiens*. According to the reasoning, there must be something magical in a shark that prevents cancer, and desperate people buy various parts of sharks in the vain hope there is a magic bullet there. There is not.

From what we have learned so far, we can say something else about sharks from an evolutionary perspective: they are an evolutionary dead end. Judging from the fossil record, they have evolved very little in 300 million years and have not attempted to scale the fitness landscape peaks that the mammals eventually have. Wright [20] spoke of "some trial and error mechanism on a grand scale by which the species may explore the region surrounding the small portion of the field which it occupies"; apparently the shark is not doing this anymore. We can ask based on what we have developed here: is cancer an inevitable consequence of rapid evolution—and in that sense not a disease at all but an inevitable consequence of rapid evolution? Is cancer then inextricably connected with high evolution rates and high stress conditions and not susceptible to being "cured"?

Let us assume that in any local habitat or ecology there exists a distribution of genomes—some of high fitness and some of low fitness. The low-fitness genomes are under stress but contain the seeds for evolution. *Stress* in this section is defined as something that either directly generates genomic damage (e.g., ionizing radiation chemicals like those that directly attack DNA, viruses) or something that prevents replication of the genome (e.g., blockage of DNA polymerases, blocking of the topological enzymes required for chromosome replication). All of these events are stress inducers and, left unchallenged, will result in the extinction of the quasi-species.

This is the business end of the grand experiment in exploring local fitness peaks and ultimately in generating resistance to stress. The system must evolve in response to the stress, and it must do this by deliberately generating genetic diversity to explore the fitness landscape ... or not. There is a name for such a collective mutational response to stress in bacteria: the SOS response [26]. We discussed earlier in this chapter the concept of a phase transition in thermodynamics and how, under the appropriate conditions of temperature, T, and interaction terms, g_n, in the Landau model for free energies, that a system can spontaneously transit to a new order parameter. The SOS response is analogous to an entropy generating term in the system, or, in terms of our evolution temperature, T_e, it makes the system hot in terms of occupation of genomic states far from the optimum value of fitness. Figure 1.5 shows schematically the network transition that occurs, moving from ensuring faithful genomic replication to highly error-prone DNA

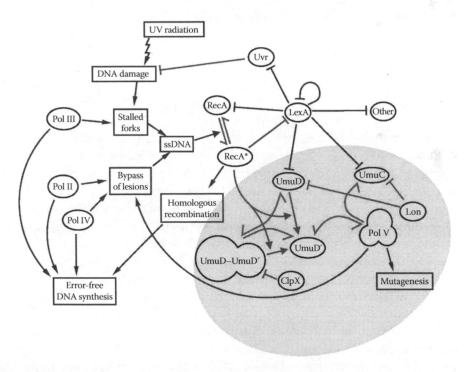

FIGURE 1.5 Schematic representation of the SOS network in *E. coli*. The lines indicating interactions represent underlying processes such as transcriptional regulation, active degradation, proteolytic cleavage, and complex formation. The shading highlights the proteins involved in mutagenesis, centered on the Pol V DNA polymerase. The mutation rate in the shaded region is at least 100 times greater than in the normal region. (From [27]. With permission.)

replication. The basic mechanism of the SOS response in bacteria generating genome diversity may be the main key to the shocking rise of resistance to antibiotics in bacteria at a rapid time scale [28]. By heating the system in the evolution sense above the critical temperature, a much wider genomic landscape can be explored.

It is possible for something like the SOS response to initiate a genomic meltdown by overheating the system in the evolutionary sense. Previous sections have discussed that smaller populations, n_i, result in more rapid fixation of high fitness mutations. Further, the effect of the activation of SOS or mutator genes is to increase the mutation rate u. The combination of increasing mutations and decreasing quasi-species size as the heterogeneity of the genomic landscape increasingly fragments can lead to a genomic meltdown [29,30], which can be viewed as explosive evolution but necessarily of a fitness increase at this point. Referring to Wright's [20] fitness landscape, the mutational meltdown can take the form of Panel a or Panel e of Figure 1.3 depending on a delicate combination of parameters. With great generality, we can say that if n^* is the critical population size at meltdown and u^* is the critical mutational rate at meltdown, then

$$\frac{u^*}{n^*} \sim 1 \tag{1.22}$$

That is, the effect of severe stress in activating mutator genes and reducing population size can lead to a mutational meltdown of the genome, spreading the genome over a large range of fitness values in an uncontrolled manner.

1.6　An Introduction to Some Experiments

We have been engaged in a program for several years to use microfabrication to probe adaptation and evolution dynamics [2–4]. The landscape model we are attempting to present here suggests the idea that certain areas of the fitness landscape, called *Goldilocks points,* are best for rapid evolution. Remember that Goldilocks broke into a temporarily empty house owned by bears and sampled the three bears' porridges for the "just-right" combination of taste, fit, and comfort [31]. Like her need for the just-right parameters, evolution proceeds most rapidly when there is the just-right combination of a large number of mutants and rapid fixation of the mutants [32]. These landscape concepts make it possible to fix resistance to the powerful antibiotic ciprofloxacin (Cipro) in wild-type *E. coli* rapidly through a combination of extremely high population gradients, which generate rapid fixation, convolved with the just-right level of antibiotic, which generates a large number of mutants and the motility of the organism. Although evolution occurs in well-stirred chemostats without such Goldilocks conditions [33], natural environments are rarely well stirred in nature, as Darwin realized on the Galapagos Islands [34]. For complex environments such as the Galapagos Islands, spatial population gradients and movement of mutants along these population gradients can be as important as genomic heterogeneity in setting the speed of evolution. The design of our microecology is unique in that it provides two overlapping gradients: one an emergent and self-generated bacterial population gradient due to food restriction; and the other a mutagenic antibiotic gradient. Further, it exploits the motility of the bacteria moving across these gradients to drive the rate of resistance to Cipro to extraordinarily high rates.

Intuitively, the need for a combination of population gradients and motility of the mutants for rapid evolution of resistance comes from the following considerations: (1) if a mutation occurs that increases the fitness of a mutant, the mutation can fix if the mutant moves with velocity, $\vec{v}$, down the population gradient into regions of lower population density [17,35]. This stepwise movement of motile mutant bacteria via successive mutations into regions of higher stress is accelerated if the basal mutation rate, u^*, is very high and the population gradient is very steep $-\nabla N$. Figure 1.6a shows that a rare mutation that confers resistance may not be able to fixation if the local wild-type population is very high but that a resistant mutation in a region of low population can result in rapid fixation. Figure 1.6b shows that by creating a gradient in nutrients that generates a gradient in population density, the mutants will move away from the highly populated regions to the lower ones, if the organisms are motile. Figure 1.6b shows a Goldilocks point where the gradients are highest and evolution should be most rapid.

Several parameters are relevant to enhancement of the evolution of resistance in mutagenic stress gradients: (1) the N_h, population size in a high-stress microhabitat; (2) the stress-induced mutation rate, u^*; (3) the net velocity of bacteria, $\vec{v}$, along gradient; (4) the time before entering stationary phase τ; (5) the doubling time, $t_{1/2}$, of the mutant bacteria in the presence of the antibiotic; (6) the fitness advantage of the resistant bacteria, s. The time, T_e, for the adaptive mutants to reach a frequency that assures escape from stochastic loss [36] can be estimated to be

$$\frac{1}{T_e} = N_h u^* + (-\nabla N_h \cdot \vec{v})\tau u^* \tag{1.23}$$

The time to fix T_f for a mutant is expected to scale as [37]

$$T_f \sim \frac{\log(N_h \times s)}{s} \times t_{1/2} \tag{1.24}$$

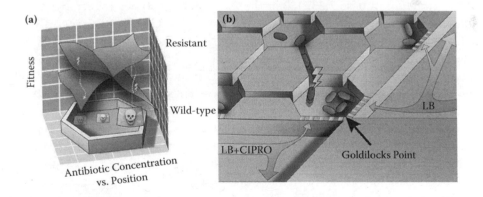

FIGURE 1.6 (See color insert.)   (a) The blue surface and the red surface are the fitness of the wild-type and resistant bacteria, respectively, versus space in a combination of antibiotic gradient and population gradients. Mutation from the wild-type to the resistant genome is represented by vertical transitions between the two fitness surfaces. (b) The basic design of our microecology creates high-population gradients using constriction of nutrient flow via nanoslits in the presence of an antibiotic gradient. Here the apex of the device is shown, where the gradients are highest. Channels allow movement of motile mutant bacteria along the population gradient. The nutrients and nutrient + Cipro streams are circulated by syringe pumps as shown.

Chapter 1

1.7   Changes in the Fitness Landscape with Evolution

We refer the reader to [4] for some details of how we conducted bacterial evolution experiments on a stress landscape. In reference to our idea of the temperature of an evolution landscape, Equation (1.15), we created a very cold landscape. That is, we used Cipro, a highly mutagenic antibiotic with an MIC of only 50 nM. At the MIC, the growth rate of wild-type bacteria approaches zero. In the experiments we carried out we ran Cipro concentrations across the landscape from approximately 0 to 10 µM, so that at the high end we have antibiotic concentrations of approximately 200 × the MIC. Such a huge concentration places an extremely strong selection pressure on the bacteria, and in our terminology we have a very cold evolution temperature in many parts of the landscape.

Ordinarily if the evolution temperature is cold everywhere, any evolution is frozen out because of the extremely deep free energy valleys that confine the genomic distribution (we relax here to the sign convention of the physicist rather than the biologist, and let the free energy minimize rather than maximize the fitness). However, because there is a large gradient in the Cipro concentrations across the device the evolution temperature is not everywhere uniform. Under these conditions, as we have tried to explain already, we can expect an evolution flux driven by the evolution temperature gradient, the equivalent of an out-of-equilibrium thermodynamic system. Such a genomic flux means physically that we can expect rapid evolution under these conditions.

Figure 1.7 shows that this rapid evolution dynamic occurs exactly under this high evolution temperature gradient along the fitness landscape. In the presence of a high gradient along the fitness landscape due to the presence of the mutagenic antibiotic, and given the motility of the *E. coli* bacteria used in this experiment we were able to observe the emergence of resistance of the bacteria to Cipro in less than 10 hours, which is remarkable. We found four SNPs that dominated the mutations, and they were found in three (DG-1, DG-2, DG-3) of four experiments:

1. T→C mutation in base 2,337,183 of the *E. coli* K12 genome. This causes a missense mutation at amino acid 87 from Aspartic Acid (D) to Glycine (G) in *gyrA*. Structural alignment of the protein sequences of the *E. coli* and *S. aureus* gyrase A subunits revealed that amino acid 87 (Aspartic Acid) in *E. coli* aligned to amino acid 88 (Glutamic Acid) in *S. aureus*. Aspartic acid and Glutamic acid differ by a single carbon, and these two residues are often substituted in similar proteins. Reconstructing this mutation in the known X-ray structure of *S. aureus* gyrase A [38] reveals that Cipro inhibits *gyrA* function by sitting in the active site of the enzyme. More specifically, the Ciprofloxacin intercalates between bases in the nicked strand of DNA. We can also see that the mutated amino acid (red) sits very close to Cipro. Thus, this SNP is very likely highly functional.

2. A missense A→T in base 3,933,247 in a region coding for the *rbsA* gene which is a component of the ribose ABC transporter complex and has been previously reported to export other antibiotics (Erythromycin, Tylosin, and Macrolide) [39]. Thus, this SNP is also likely to be highly functional.

3. A pair of SNPs (1,617,461: A→C and 1,617,460:C→G) found in the coding sequence for *marR*. *marR*'s normal function is to repress the multiple antibiotic resistance (*mar*) operon [40]. These mutations occur consecutively and create an alanine to

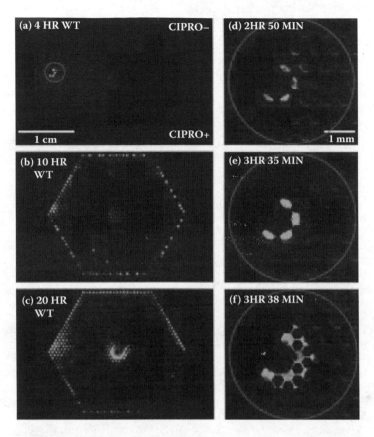

FIGURE 1.7 (See color insert.)   (a) Ignition of resistance to Cirpro at the Goldilocks point 4 hours after inoculation. (b) Spread of resistant bacteria around the periphery of the microecology at 10 hours after inoculation. (c) Return of Cipro-resistant bacteria to the center after 20 hours. The dynamics of the resistant bacteria at the Goldilocks point is shown with higher time resolution in (d)–(f) during the ignition of resistance over a 1 hour period.

serine mutation at codon 105 of *marR* (A105S). Alanine is a nonpolar base, whereas serine is a polar base. It is possible that this mutation alters the ability of *E. coli* to regulate the expression of antibiotic resistance genes. Beside the high Cipro resistant isolates, we also got low-resistant isolates in experiment DG-4.

We believe that we have put forth here the basic foundations for understanding the quantitatively rapid rate of evolution observed in this experiment, but a rigorous calculation will be very difficult and is beyond the scope of this chapter.

We close by noting that some people are upset to see evolution proceed at such a rate. However, it should not be viewed as a mystery. We start with a million bacteria, and using our measured Cipro mutation rate of 10^{-7} per bp/generation at the MIC (data not shown), each progeny in a replication will have approximately one SNP, and in a million bacteria basically every possible SNP in the genome exists somewhere. It is then a matter of devising a microenvironment that allows fixing of a mutant within a few generations, which we have done. It is unlikely that the four SNPs emerge all at once; in all likelihood they emerge in a sequential manner. We think that the rapid emergence of resistance to

Cipro demonstrated here using the powerful tools of microenvironment design reveal the importance of understanding the foundations of the dynamics of evolution. The rapid emergence of bacterial resistance to antibiotics has become one of the most important issues in human health. The design principles that we discuss here should provide a powerful template for exploring the rates at which resistance can occur in very complex fitness landscapes, including the rapid evolution of resistance to genotoxic chemotherapy in cancer. While a great deal of work remains to be done to characterize the dynamics of bacterial resistance shown here, the implications of this work we hope are clear.

References

[1] Michel, B. After 30 years of study, the bacterial SOS response still surprises us. *PLoS Biol.* **3**, 1174–1176 (2005).

[2] Keymer, J. E., Galajda, P., Muldoon, C., Park, S., & Austin, R. H. Bacterial metapopulations in nanofabricated landscapes. *Proc. Natl. Acad. Sci. USA* **103**, 17290–17295 (2006).

[3] Keymer, J. E., Galajda, P., Lambert, G., Liao, D. & Austin, R. H. Computation of mutual fitness by competing bacteria. *Proc. Natl. Acad. Sci. USA* **105**, 20269–20273 (2008).

[4] Austin, R. H., Tung, C.-K., Lambert, G., Liao, D., & Gong, X. An introduction to micro-ecology patches. *Chem. Soc. Rev.* **39**, 1049–1059 (2010).

[5] Lambert, G. et al. An analogy between the evolution of drug resistance in bacterial communities and malignant tissues. *Nat. Rev. Cancer* **11**, 375–382 (2011).

[6] Davies, P. C. W. Multiverse cosmological models. *Modern Phys. Lett. A* **19**, 727–743 (2004).

[7] Smolin, L. Cosmological natural selection as the explanation for the complexity of the universe. *Physica A* **340**, 705–713 (2004).

[8] Zhang, Q. C. et al. Acceleration of emergence of bacterial antibiotic resistance in connected microenvironments. *Science* **333**, 1764–1767 (2011).

[9] Kittel, C. & Kroemer, H. *Thermal Physics,* 2d ed. (W. H. Freeman, New York, 1980).

[10] Feynman, R. P. *Six Easy Pieces: Essentials of Physics Explained by Its Most Brilliant Teacher* (Helix Books, Cambridge, MA, 1963).

[11] Yukhnovskii, I. R. *Phase Transitions of the Second Order—Collective Variables Method* (World Scientific, Singapore, 1987).

[12] Gould., S. J. *Punctuated Equilibrium* (Harvard University Press, Cambridge, MA, 2007).

[13] Salinger, J. *The Catcher in the Rye* (Little, Brown and Company, New York, 1951).

[14] Wales, D. *Energy Landscapes* (Cambridge University Press, Cambridge, UK, 2003).

[15] Janke, W. *Rugged Free Energy Landscapes: Common Computational Approaches to Spin Glasses, Structural Glasses and Biological Macromolecules* (Springer, Berlin, 2010).

[16] Grant, B. R. & Grant, P. R. What Darwin's finches can teach us about the evolutionary origin and regulation of biodiversity. *Bioscience* **53**, 965–975 (2003).

[17] Dronamraju, K. R. Haldane's dilemma and its relevance today. *Current Science* **70**, 1059–1061 (1996).

[18] Briollais, L. et al. Methodological issues in detecting gene-gene interactions in breast cancer susceptibility: a population-based study in Ontario. *BMC Medicine* **5**, 22 (2007).

[19] Moran, N. A. Accelerated evolution and Muller's rachet in endosymbiotic bacteria. *Proc. Natl. Acad. Sci. USA* **93**, 2873–2878 (1996).

[20] Wright, S. The roles of mutation, inbreeding, crossbreeding and selection in evolution. *Proceedings of the VI International Congress of Genetics* **1**, 356–366 (1932).

[21] Bryngelson, J. D. & Wolynes, P. G. Spin-glasses and the statistical-mechanics of protein folding. *Proc. Natl. Acad. Sci. USA* **84**, 7524–7528 (1987).

[22] Dill, K. A., Ozkan, S. B., Shell, M. S., & Weikl, T. R. The protein folding problem. *Ann. Rev. Biophys.* **37**, 289–316 (2008).

[23] Hamming, R. W. Error detecting and error correcting codes. *Bell System Tech. J.* **29**, 147–160 (1950).

[24] Gusfield, D. *Algorithms on Strings, Trees, and Sequences: Computer Science and Computational Biology* (Cambridge University Press, New York, 1997).

[25] Rammal, R., Toulouse, G., & Virasoro, M. A. Ultrametricity for physicists. *Rev. Modern Phys.* **58**, 765–788 (1986).

[26] McKenzie, G. J., Harris, R. S., Lee, P. L., & Rosenberg, S. M. The SOS response regulates adaptive mutation. *Proc. Natl. Acad. Sci. USA* **97**, 6646–6651 (2000).

[27] Krishna, S., Maslov, S., & Sneppen, K. UV-induced mutagenesis in *Escherichia coli* SOS response: A quantitative model. *PLoS Computational Biology* **3**, 451–462 (2007).

[28] Cirz, R. T. et al. Inhibition of mutation and combating the evolution of antibiotic resistance. *PLoS Biol.* **3**, 1024–1033 (2005).

[29] Lynch, M., Burger, R., Butcher, D., & Gabriel, W. The mutational meltdown in asexual populations. *J. Heredity* **84**, 339–344 (1993).

[30] Bagnoli, F. & Bezzi, M. Eigen's error threshold and mutational meltdown in a quasispecies model. *Int. J. Modern Phys. C* **9**, 999–1005 (1998).

[31] Opie, I. & Opie, P. *The Classic Fairy Tales* (Oxford University Press, Oxford, 1974).

[32] Pulliam, H. R. & Danielson, B. J. Sources, sinks, and habitat selection—a landscape perspective on population-dynamics. *Am. Nat.* **137**, S50–S66 (1991).

[33] Barrick, J. E. et al. Genome evolution and adaptation in a long-term experiment with *Escherichia coli*. *Nature* **461**, U1243–U1274 (2009).

[34] Keynes, R. D. *Charles Darwin's Beagle Diary* (Cambridge University Press, Cambridge, UK, 1988).

[35] Hallatschek, O. & Nelson, D. R. Population genetics and range expansions. *Physics Today* **62**, 42–47 (2009).

[36] Gillepsie, J. H. *The Causes of Molecular Evolution* (Oxford University Press, Oxford, 1991).

[37] Karasov, T., Messer, P. W., & Petrov, D. A. Evidence that adaptation in *Drosophila* is not limited by mutation at single sites. *PLoS Genet.* **6**, (2010).

[38] Klaus, W., Ross, A., Gsell, B., & Senn, H. Letter to the editor: Backbone resonance assignment of the N-terminal 24 kda fragment of the gyrase B subunit from *S. aureus* complexed with novobiocin. *J. Biomol. NMR* **16**, 357–358 (2000).

[39] Fath, M. J. & Kolter, R. ABC transporters—bacterial exporters. *Microbiol. Rev.* **57**, 995–1017 (1993).

[40] Miller, P. F. & Sulavik, M. C. Overlaps and parallels in the regulation of intrinsic multiple-antibiotic resistance in *Escherichia coli*. *Mol. Microbiol.* **21**, 441–448 (1996).

Chapter 1

2. System Design Principles

Michael A. Savageau

2.1   Introduction

Biological systems involve complex networks of reactions and gene circuits that exhibit enormous variation in design. For years the most convenient explanation for these variations was historical accident. According to this view, evolution is a haphazard process in which many different designs are generated by chance; there are many ways to accomplish

Quantitative Biology: From Molecular to Cellular Systems edited by Michael E. Wall © 2012 CRC Press / Taylor & Francis Group, LLC. ISBN: 978-1-4398-2722-2

Chapter 2

the same thing, so no further meaning can be attached to such different but equivalent designs. In recent years we have found a more satisfying explanation based on design principles for at least certain aspects of biological systems. A *System design principle* is a rule that characterizes some biological feature exhibited by a class of systems such that discovery of the rule allows not only for the understanding of known instances but also the prediction of new instances within the class. For cases that involve subtle relationships among several elements of the design, it is expected that rules will be expressed effectively in only quantitative as opposed to qualitative form. Although the issue of design also arises at the level of the elements themselves (e.g., in the field of protein engineering), the term *system design* is used in this chapter to emphasize the larger-scale issues of system organization, variously referred to as *topology, architecture*, or *structure*. The central importance of signal transduction and gene regulation in modern cell biology provides strong motivation to search in these fertile areas for more of these underlying system design principles. Although there is expanding interest in biological design principles (for an excellent engaging account see Alon 2006), the search for quantitative design principles is in its infancy, and undoubtedly many such design principles remain to be discovered.

After providing some general background for context, the mathematical methods used to represent, analyze, and compare alternative designs are briefly outlined. Mathematically controlled comparison, which has been used successfully to identify a number of quantitative system design principles, is emphasized. Subsequent sections treat specific design principles, operating at different levels of organization, that make predictions supported by experimental evidence. The chapter concludes with a few general remarks.

2.2   Design Principles

In a sense, all biologists can claim that they are interested in the integrated behavior of biological systems, but it is possible to distinguish between different modes of understanding that are complementary and overlapping (Savageau 1991). Phenomenological description deals with the intact system, but it is not a very fundamental kind of understanding. Component isolation and characterization are more fundamental, but at the same time it is difficult to relate them to the behavior of the intact system. An approach that attempts to bridge the gap between the component and the system levels is selective mutation of components. However, this is still a relatively blunt instrument that often does not discriminate between primary and secondary influences and tends to yield qualitative relationships.

Each of these methods has contributed to our understanding of biological systems. However, one important mode of understanding has been largely absent from the biological realm: the understanding that derives from knowledge of design principles, as exemplified in engineering. By knowing these *rules of organization*, we could, in principle, make a television set from scratch. We would know why it's designed this way and not some other way. This is the kind of understanding that the scientist or engineer had when designing the set in the first place. In my view, it is this latter type of understanding— quantitative understanding of design principles that relate molecular and systemic properties—that will be at the very heart of integrative cell biology in the near future.

The richness of biological interactions and their nonlinear character leads to critical quantitative relationships and design principles that characterize systemic behavior. Only a quantitatively precise language like mathematics has the power to elucidate, and the structure to efficiently represent, such design principles. Rigorous comparisons of alternative designs often lead to the discovery and elucidation of system design principles. The results in turn form the basis for a predictive theory of alternative design.

2.3 Comparative Approach to the Study of System Design

The elucidation of design principles for a class of systems requires (1) a systematically structured formalism to represent alternative designs, (2) methods of analysis capable of predicting behavior, (3) methods for making well-controlled comparisons, and (4) quantitative criteria for making objective comparisons. The following sections provide a brief review of methods for dealing with each of these issues, but with special attention to making well-controlled comparisons by a method called *mathematically controlled comparison*.

2.3.1 Power-Law Formalism

The power-law formalism combines nonlinear elements having a very specific structure (products of power laws) with a linear operator (differentiation) to form a set of ordinary differential equations that are capable of representing any suitably differentiable nonlinear function. In particular, it includes the elementary rate laws of chemical kinetics and the rational function rate laws of biochemical kinetics. This makes it an appropriate formalism for representing alternative designs of biochemical systems.

The two most common representations within the power-law formalism are generalized-mass-action (GMA) systems

$$\frac{dX_i}{dt} = \sum_{k=1}^{r} \alpha_{ik} \prod_{j=1}^{n+m} X_j^{g_{ijk}} - \sum_{k=1}^{r} \beta_{ik} \prod_{j=1}^{n+m} X_j^{h_{ijk}} \qquad i=1,\cdots,n \tag{2.1}$$

and synergistic (S) systems

$$\frac{dX_i}{dt} = \alpha_i \prod_{j=1}^{n+m} X_j^{g_{ij}} - \beta_i \prod_{j=1}^{n+m} X_j^{h_{ij}} \qquad i=1,\cdots,n \tag{2.2}$$

The derivatives of the state variables with respect to time t are given by dX_i/dt. The α and β parameters are rate constants, whereas the g and h parameters are kinetic orders. There are in general n dependent variables, m independent variables, and a maximum of r terms of a given sign. The resulting power-law formalism can be considered a canonical nonlinear representation from at least four different perspectives: fundamental, recast, local, and piecewise (Savageau 2001).

The power-law formalism can be considered a fundamental representation that includes traditional chemical kinetics. Moreover, the GMA representation provides a context for assessing the importance of noninteger kinetic orders in the quantitative characterization of heterogeneous systems (Savageau 1995, 1998). For an example of an experimental study involving enzymatic activity in an artificial membrane preparation see Clop et al. (2008). However, the fundamental representation is typically not the most useful level of representation for comparing biochemical systems consisting of many reactions because it is much too detailed and values for many of the elementary parameters will not be available. Nor is the structure of the GMA equations conducive to general symbolic analysis. Similarly, the power-law formalism in the context of recasting (Savageau and Voit 1987) is not the most useful level of representation for general symbolic analysis. However, the recast representation provides a general framework for representation and comparison of nonlinear systems. It also plays a critical role during construction of the *system design space*, a recently introduced concept that facilitates the identification, enumeration, analysis, and comparison of qualitatively distinct phenotypes exhibited by a model (Savageau et al. 2009).

In contrast to the utility of the previous two perspectives, the local S-system representation within the power-law formalism is most important in that it combines the advantages of analytical power and tractability. In the nonlinear realm of biological systems, it functions much like conventional linearization based on a Taylor expansion. Indeed, Taylor's theorem (in log space) gives a rigorous justification for the local S-system representation and specific error bounds within which it will provide an accurate representation. However, it is typically accurate over a wider range of variation than the corresponding linear representation. Moreover, it has a very desirable structure from the standpoint of general theory and symbolic analysis (Savageau 1969). The piecewise power-law representation (Savageau 2009) is a logical extension of the local power-law representation capable of providing a global characterization for nonlinear systems. It involves the conversion of an intractable nonlinear system of equations into a series of simpler S-system equations. This greatly simplifies the analysis and also captures the essential nonlinear behavior of the original system, as is evident from results involving application of the design space methodology (Savageau and Fasani 2009; Coelho, Salvador, and Savageau 2010; Savageau 2011a, 2011b; Tolla and Savageau 2011). In this methodology, the piecewise and local representations play a critical role along with the recast representation, as previously mentioned.

2.3.2   Methods of Analysis

The regular, systematic structure of the power-law formalism implies that methods developed to solve equations efficiently having this form will be applicable to a wide class of phenomena. This provides a powerful stimulus to search for such methods. The potential of the power-law formalism in this regard has yet to be fully exploited.

The simplicity of the local S-system representation has led to the most extensive development of theory, methodology, and applications within the power-law formalism (Voit 1991). These advances have occurred because it was recognized from the beginning that the steady state analysis of S-systems reduces to conventional linear analysis in a logarithmic space. Hence, it was possible to exploit the powerful methods already developed for linear systems. Steady state logarithmic gain matrices, with elements defined as the relative change in a dependent variable in response to a relative change in an independent variable $L(X_d, X_i) = \partial \log X_d / \partial \log X_i$, provide a complete network analysis of the signals that propagate through the system. Similarly, steady state sensitivity matrices, with elements defined as the relative change in a dependent variable in response to a relative change in a rate constant parameter $S(X_d, \beta_k) = \partial \log X_d / \partial \log \beta_k$, provide the basis for a complete sensitivity analysis that defines the local robustness of a system. The linear structure also permits the use of a well-developed optimization theory such as the simplex method and efficient feasibility tests (Fasani and Savageau 2010).

Analytical solutions for the local dynamic behavior are available, including eigenvalue analysis for characterization of the relaxation times in terms of the exponential parameters and the turnover numbers for the pools of the system. The regular structure allows the conditions for Hopf bifurcation to be expressed as a simple formula involving the kinetic-order parameters (Lewis 1991). The regular form of the S-system equations also lends itself to the development of efficient methods for identifying both structure and parameter values of biochemical systems (e.g., Cho, Cho, and Zhang 2006; Chou and Voit 2009; Gennemark and Wedelin 2009).

2.3.3 Mathematically Controlled Comparison

An adequate theoretical framework for understanding alternative designs of biochemical systems should provide explanations for universal (or nearly universal) designs relative to hypothetical alternatives as well as for existing alternative designs in terms of conditions that might promote their selection or maintenance. Although a general theory for understanding the various well-documented designs has yet to be formulated, the rudiments of such a theory have been developed and applied successfully in a number of instances.

The method, called mathematically controlled comparison, combines aspects from a number of existing methods. Its characteristic features are as follows:

1. The two designs being compared are represented within a common canonical nonlinear formalism.
2. The two designs being compared are restricted to having differences in a single specific process that remains embedded within its natural milieu. This is equivalent to a single mutational difference in an otherwise isogenic background.
3. One of the two designs, often the more complex, is chosen as the reference to which the alternative will be compared.
4. The values for the parameters that characterize the unaltered processes of the alternative are assumed to be strictly identical to the values for the corresponding parameters of the reference system. This equivalence of parameter values from a perspective within the systems is called *internal equivalence*. It provides a means of nullifying or diminishing the influence of the background, which in complex systems is largely unknown. Again, this is analogous to the isogenic control in an experimental comparison.
5. The two systems are required to be as nearly equivalent as possible in their interactions with the outside environment, that is, from a perspective external to the system. This is called *external equivalence*. The one altered process will in general have a different set of values for all of its parameters. This introduces extra degrees of freedom that must be eliminated; otherwise, arbitrary differences will arise in the comparison. The constraints imposed by external equivalence fix the values of the

Chapter 2

parameters for the altered process in such a way that arbitrary differences in system behavior are largely eliminated. Functional differences that remain between the two systems with maximum internal and external equivalence constitute necessary differences. Such differences, which can be eliminated only by reducing the systems to identity, represent inherent functional differences for the designs in question.

6. Once the parameters for both systems have been properly fixed, the systems can be rigorously analyzed by mathematical, statistical, and computational methods to determine their irreducible differences.

7. Finally, comparisons are made on the basis of quantitative criteria for functional effectiveness that have been established a priori.

The step involving external equivalence is critically important for nonlinear systems. It is used to establish the same basal state for the systems being compared, that is, to have the systems operating with the same values for their concentrations and fluxes. Failure to have the systems in the same state can lead to an erroneous understanding of the results from the comparison. For example, an amino acid biosynthetic pathway in a mutant having lost end-product inhibition can exhibit large-scale oscillatory instability, which can lead to the conclusion that the role for this feedback is ensuring stability of the system (Bliss, Painter, and Marr 1982). However, for many (but not all) feedback resistant mutants, the flux through the pathway can be much greater than that in the wild type; this leads to saturation of an enzyme downstream of the pathway (the aminoacyl-tRNA synthetase), which can cause instability of an otherwise stable system. Once there is such an uncontrolled difference in the state of a nonlinear system, such as the gross difference in pathway flux in this case, all bets are off. Secondary effects elsewhere in the model can completely obscure understanding of the differences under investigation.

A mathematically controlled comparison in this case would involve the equivalent of selecting a different feedback-resistant mutant from the large class of such mutants. The parameters of this mutant enzyme would be altered to reflect not only the loss of the recognition site for the end product of the pathway but also a reduced molecular activity such that pathway flux is the same as that in the wild-type organism. That is, one of the constraints imposed by external equivalence to maintain the same basal state of the systems being compared is that pathway flux be the same in the two systems. In such a well-controlled comparison, the downstream enzyme is not saturated and there is no large-scale oscillatory instability.

The method of mathematically controlled comparison has many advantages for making well-controlled comparisons of complex systems. Two key features of this method should be noted. First, because much of the analysis can be carried out symbolically, the results are often independent of the numerical values for particular parameters. This is a marked advantage since all the parameter values of a complex system are not known, and in many cases it would be impractical to obtain them. Second, the method allows the relative optima of alternative designs to be determined without actually having to carry out an optimization (i.e., without having to determine explicit values for the parameters that optimize the performance of a given design). If it can be shown that a given design with an arbitrary set of parameter values is always superior to the alternative design that has been made internally and externally equivalent, whether the set of parameter values represents an optimum for either design, the given design is proven to be superior to the alternative even if the alternative were assigned a parameter set that optimized its performance. This feature is a decided advantage because the difficult procedure of optimizing complex nonlinear systems can be avoided.

A numerical approach to this problem has been developed that combines the method of mathematically controlled comparison with statistical techniques to yield numerical results that are general in a statistical sense (Alves and Savageau 2000). This approach retains some of the generality that makes mathematically controlled comparison so attractive and the same time provides quantitative results that are lacking in the qualitative approach.

A beautiful example illustrating the experimental equivalent of a mathematically controlled comparison in the context of a stochastic system is provided by the work of Çagatay et al. (2009). They examined the implications of alternative gene circuitry controlling the induction of competence for DNA exchange in the bacterium *Bacillus subtilis*. The question addressed was whether it makes a difference if the sequence of influences is positive and then negative or the reverse in a net negative feedback circuit. In engineering the alternative circuits, Çagatay et al. were careful to adjust the strength of promoters, to ensure the same average behavior from the alternative systems, and the numbers of molecules in the system, to tune out differences in the distribution of the noise profile. By means of such techniques, Çagatay et al. were able to establish the functional implications of this subtle difference in gene circuitry, which otherwise would have been completely obscured by other adventitious differences; namely, the differences alter the dynamic distribution of stochastic fluctuations. These dynamic differences alter the timing of competence events and the range of response to the environmental concentrations of DNA, thus revealing a trade-off between temporal precision and physiological response range. Such carefully controlled comparisons are rare, undoubtedly because they are difficult to achieve experimentally, but in many cases essential for discovering system design principles.

2.3.4 Quantitative Criteria for Functional Effectiveness

It is critically important in making well-controlled comparisons to have criteria that can be quantified. The criteria that would be important for a particular system will depend on our knowledge of its biology. In this regard, there is no substitute for an intimate knowledge of the biology. However, even when there is uncertainty concerning its biological role, a number of potentially relevant criteria for the system can be postulated, and the results of the analysis will often suggest functions that the system might be well adapted to perform. In this way we can learn about potential roles for the system.

In addition to criteria that might be specific to a particular class of systems, there is a suite of rather general criteria that apply to many systems. The following are examples of criteria that have frequently proved useful. Local stability, determined by eigenvalues, is expected to be an important criterion for any system that functions in a homeostatic context. Margins of stability, determined by Routh criteria (Dorf and Bishop 1995), are important for ensuring stability despite fluctuations in parameter values. Response time, typically measured in terms of dominant eigenvalues or more empirically in terms of a half-time, is important for any system responding to a rapidly changing environment.

Minimizing changes in the concentration of intermediates in many metabolic systems is important because there is limited solvent capacity in cells, because many of these intermediates are toxic, and because large pools of intermediates compromise the response time of these systems. Relative parameter insensitivity is a local criterion conventionally used to highlight the *robustness* of biological systems. However, more difficult to determine is the ability of a biological system to tolerate large changes in its parameter values. In this regard, it is worth mentioning that the system design space methodology offers a criterion for tolerance to large changes in parameter values (Coelho et al. 2009). Other quantitative criteria include latency, first passage time, and information theoretic quantities (see Chapter 4) such as channel capacity.

The methods outlined in this section have been applied to a number of systems. The following three sections provide representative examples for metabolic pathways, signal transduction cascades, and elementary gene circuits.

2.4 System Design Principles for Feed-Forward Control

The penultimate product of the histidine pathway, histidinol, was discovered to inhibit histidyl t-RNA synthetase, the enzyme that links histidine to its cognate tRNA for use in protein synthesis (Ames and Hartman 1961). Similar results were also discovered in the arginine pathway (Nazario 1967; Sussenbach and Strijkert 1969; Yem and Williams 1971). The function of this feed-forward mechanism remained a paradox for many years. In some cases, an exogenous addition of a pathway intermediate caused an inhibition of growth (Sussenbach and Strijkert 1969), which was believed to be mediated by this mechanism; in other cases, if anything such additions stimulated growth. An early analysis showed that this feed-forward inhibition mechanism alone could not produce the inhibition of growth (Savageau 1972).

2.4.1 Resolution of a Paradox

A rigorous comparison of models with and without this feed-forward from the penultimate product (Figure 2.1) resolved the paradox by revealing a simple design principle involving three factors influencing the distribution of flux among alternative fates for the amino acid (Savageau 1972, 1979): (1) the ratio of the kinetic order for the loss of the feed-forward modifier to the magnitude of the kinetic order for its inhibition of the synthetase; (2) the ratio of the kinetic orders for the alternative fates with respect to the branch-point amino acid; and (3) the ratio of the steady state flux for the alternative fate

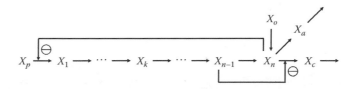

FIGURE 2.1 Model of an amino acid biosynthetic pathway in which the penultimate product X_{n-1} acts as a feed-forward inhibitor of the last reaction converting the branch-point amino acid to the charged aminoacyl-tRNA X_c, and X_n acts as a feedback inhibitor of the first reaction in the sequence. The exogenous supply of the amino acid is X_o, and X_a represents an alternative fate for the use of the amino acid.

to that through the pool of the feed-forward modifier. In the conventional symbols of the power-law formalism:

$$L(V_c, X_{io}) > 0 \quad if \quad \frac{h_{n-1,n-1}}{-g_{c,n-1}} > \frac{g_{an}}{g_{cn}} \frac{V_a}{V_{n-1}}$$

$$L(V_c, X_{io}) < 0 \quad if \quad \frac{h_{n-1,n-1}}{-g_{c,n-1}} < \frac{g_{an}}{g_{cn}} \frac{V_a}{V_{n-1}}$$

(2.3)

where V_j is the steady state flux through pool X_j, X_{io} is an exogenous supply of intermediate $X_i (i \leq n-1)$, and $L(V_c, X_{io})$ is the logarithmic gain in flux defined as $(\partial V_c / \partial X_{io}) \cdot (X_{io} / V_c)$ at the steady state operating point. Following an increase in the exogenous intermediate, the flux to protein synthesis (and hence growth) will decrease if the ratio $h_{n-1,n-1}/|g_{c,n-1}|$ is less than the product of the ratios g_{an}/g_{cn} and V_a/V_{n-1}. In any case, this *choking* behavior is hardly expected to be the physiological function of such a mechanism.

2.4.2 Distribution of Flux at a Branch Point

Further insight into the function of feed-forward control was obtained by a mathematically controlled comparison of the more general case (Savageau and Jacknow 1979) in which the feed-forward modifier had other locations in the pathway (see Figure 2.1, but with the X_k being the feed-forward modifier). The same state is imposed on the alternative systems by letting the system with $k = n-1$ be the reference and adjusting the two parameters $g_{ck} = (h_{kk}/h_{n-1,n-1})g_{c,n-1}$ and $\alpha_c^{(k)} = \alpha_c^{(n-1)}(\alpha_{n-1}/\beta_{n-1})^{(g_{c,n-1}/h_{n-1,n-1})}$ as the position k of the feed-forward modifier is varied.

A comparison of the flux to protein coming from endogenous precursor X_p (or from endogenous intermediates $i \leq k$) in systems with ($1 \leq k \leq n-1$ and $g_{ck} < 0$) or without ($k = 0$ and $g_{ck} = 0$) feed-forward inhibition reveals

$$\frac{L(V_c, X_p)_k}{L(V_c, X_p)_0} = \frac{V_p h_{pp} [h_{kk} g_{cn} + g_{ck} g_{an} (V_a / V_p)]}{V_p h_{pp} h_{kk} g_{cn}} < 1$$

(2.4)

Thus, with feed-forward inhibition, the increment in flux from endogenously provided material is diverted preferentially away from protein synthesis into the alternative fate.

On the other hand, a comparison of the flux to protein coming from exogenous amino acid X_0 (or from endogenous intermediates $k < i \leq n-1$) in systems with and without feed-forward inhibition (and noting that $g_{1n} < 0$ for the feedback inhibition) reveals

$$\frac{L(V_c, X_o)_k}{L(V_c, X_o)_0} = \frac{V_o h_{oo} [h_{kk} g_{cn} + g_{ck} g_{1n}]}{V_o h_{oo} h_{kk} g_{cn}} > 1$$

(2.5)

Now, with feed-forward inhibition, the increment in flux from exogenously provided amino acid is diverted preferentially into protein synthesis and away from the alternative fate.

In most cases, there is no inhibition of growth, and the detailed analysis involving a half-dozen different criteria for functional effectiveness showed that, other than the

Chapter 2

preferential diversion of amino acid flux, feed-forward inhibition from the penultimate product has very little effect on the predicted behavior of the system. However, when it was postulated that other intermediates in the pathway might be the feed-forward modifier ($1 \leq k < n-1$ in Figure 2.1), the analysis showed dramatic effects on the stability of the system (Savageau and Jacknow 1979).

2.4.3 Feed–Forward Inhibition Improves Robustness

An earlier comparison of pathways with and without end-product inhibition, based on half a dozen quantitative criteria for functional effectiveness, used external equivalence to eliminate many of the extraneous differences between such systems and identified significant irreducible differences regarding local robustness (Savageau 2009). For example, the parameter sensitivities of the pathway with this feedback inhibition were less than or equal to those of the pathway without this feedback inhibition, in particular for the first and last enzymes. Moreover, the differences were all of the form

$$\frac{S(V,p_k)^{(B)}}{S(V,p_k)^{(0)}} = \frac{g_{cn}}{g_{cn} - g_{1n}} < 1 \tag{2.6}$$

since $g_{1n} < 0$ and all other parameters of the model are positive.

A comparison of the robustness of feedback inhibited pathways with and without feed-forward inhibition from the first intermediate (Savageau and Jacknow 1979) shows that the sensitivities are less than or equal to those of the pathway with only feedback inhibition, in particular for the second and last enzyme. Again, the differences have a common form

$$\frac{S(V,p_k)^{(B+F)}}{S(V,p_k)^{(B)}} = \frac{g_{21}}{g_{21} - g_{c1}} < 1 \tag{2.7}$$

since $g_{c1} < 0$, and $g_{21} > 0$. These represent irreducible differences because they can be eliminated only by making $g_{c1} = 0$, which also eliminates the feed-forward inhibition and makes the systems identical.

2.4.4 Stabilizing Effect of Feed–Forward Inhibition

Stability of an unbranched biosynthetic pathway subject to end-product inhibition was shown some time ago to be influenced by three features of the design (Savageau 1975): (1) strength of feedback inhibition; (2) balance among the kinetic parameters of the unregulated reactions in the pathway; and (3) pathway length. The analysis of the general case is intractable. However, for the worst-case scenario, in which all the reactions have the same values for their kinetic parameters, stability can be determined by directly examining the roots of the characteristic equation, which results in

$$-g_{1n} < g_{cn} \sec^n(\pi/n) \tag{2.8}$$

For a given strength of inhibition ($g_{1n} < 0$), the threshold of instability is reduced as the length of the pathway n increases. Eventually, for some sufficiently large n the margin of stability vanishes, provided g_{1n} is greater in magnitude than the kinetic order for loss of the end product, g_{cn}, and the system becomes unstable.

If one starts with feedback inhibition in the most unstable n-step pathway on the threshold of instability [$-g_{1n}/g_{cn} = \sec^n(\pi/n)$ from Equation (2.8)] and alters the kinetics of the unregulated reactions so as to make m steps temporally dominant, then the dominant eigenvalue becomes

$$\mathrm{Re}\left[\lambda_{max}^0\right] = \left(-g_{1n}/g_{cn}\right)^{1/m}\cos\left[\pi/m\right] - 1 \tag{2.9}$$

for an effective pathway length $m \le n$. By making use of the threshold condition from Equation (2.8), we can rewrite Equation (2.9) as

$$\mathrm{Re}\left[\lambda_{max}^0\right] = \sec^{(n/m)}(\pi/n)\cos\left[(n/m)\frac{\pi}{n}\right] - 1 \tag{2.10}$$

Again, starting with feedback inhibition in the most unstable n-step pathway on the threshold of instability [$-g_{1n}/g_{cn} = \sec^n(\pi/n)$ from Equation (2.8)] but then adding feed-forward inhibition from the kth intermediate in the pathway, the dominant eigenvalue becomes

$$\mathrm{Re}\left[\lambda_{max}\right] = \sec^{(n-k)}(\pi/n)\cos\left[(n-k)\frac{\pi}{n}\right] - 1 \tag{2.11}$$

A comparison of the results in Equations (2.10) and (2.11) highlights that if $m = n/(n-k)$, then equivalent degrees of stabilization will be produced by each mechanism. For example, an n-step pathway with only three temporally dominant reactions has an effective path length of 3; an n-step pathway with feed-forward inhibition from the modifier two-thirds of the way down the pathway will also have an effective path length of 3. If the modifier is located halfway down the pathway, the effective path length is 2. The optimal position for the feed-forward modifier ($k = 1$) results in an n-step pathway that appears from a kinetic perspective to consist of a single reaction.

2.4.5   Experimental Evidence

The predicted optimal position reveals a special relationship between the first and last enzymes. They share the same reactants and modifiers; namely, a reactant of the first enzyme (X_1) is a modifier of the last enzyme, and a reactant of the last enzyme (X_n) is a modifier of the first enzyme (Figure 2.2). Consider the dynamic response to a disturbance in the initial substrate X_1. In the pathway without feed-forward inhibition, the disturbance will propagate down the entire length of the pathway before there is a response in the end product X_n, which is then fed back to the first enzyme to correct for the disturbance. In the pathway with the optimal position of the feed-forward modifier, the disturbance will be sensed almost immediately by a response in the first

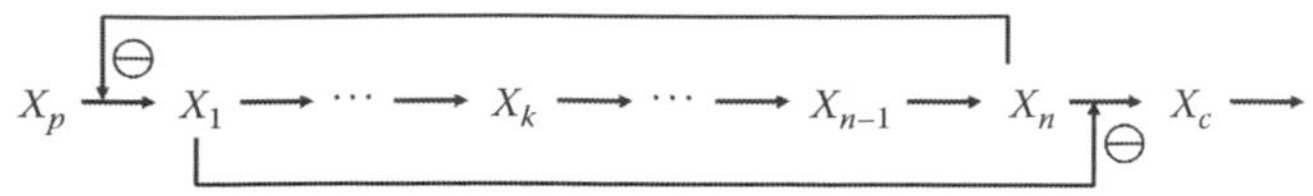

FIGURE 2.2 Model of an amino acid biosynthetic pathway in which the first intermediate X_1 acts as a feed-forward inhibitor of the last reaction converting the amino acid X_n to the charged aminoacyl-tRNA X_c, and X_n acts as a feedback inhibitor of the first reaction in the sequence.

intermediate, which is fed forward to the last enzyme whose substrate response is then fed back to the first enzyme to correct for the disturbance. In this case, much of the intervening biochemistry is bypassed by the kinetic short circuit involving the first and last enzyme. This creates a much faster response time that leads to a greater degree of stability.

If this reduction in time delay around the control loop is the desired goal of feed-forward inhibition, then it can be predicted that having the first and last enzymes in a multienzyme complex would be the ideal, since this arrangement would reduce to a minimum the diffusion time for reactants to find their targets in the bulk phase (Savageau and Jacknow 1979). As soon as the product of the first reaction is produced it would already be on the surface of the last enzyme, and as soon as the substrate of the last enzyme increases it would be sensed on the surface of the first enzyme.

Experimental studies of the isoleucine biosynthetic pathway in *Escherichia coli* revealed that a deletion of the gene encoding the first enzyme of the pathway, threonine deaminase, caused the last enzyme of the pathway, isoleucyl-tRNA synthetase, to become physically unstable (Williams, Whitfield, and Williams 1977). This was the first hint that these two enzymes might bind to each other and promote their stability. However, it could also be that one of the intermediates in the pathway might normally bind to the synthetase and promote its stability, since these intermediates were not being produced in the deletion mutant. Subsequent studies by Singer, Levinthal, and Williams (1984) showed biochemically that the first intermediate in the pathway (α-ketobutyrate) is in fact a feed-forward inhibitor of the synthetase and that the first and last enzymes copurify as a multienzyme complex. Thus, both of the theoretical predictions—first intermediate as the feed-forward modifier and first and last enzyme in a multienzyme complex—have received independent experimental confirmation.

The prediction of complexes among other enzymes for regulatory purposes (Savageau 2009) is supported by experimental data for regulatory enzymes in the branched pathways for synthesis of aromatic (Cotton and Gibson 1965; Gibson and Pittard 1968) and aspartate-family (Patte, Truffa-Bachi, and Cohen 1966; Truffa-Bachi and Cohen 1966) amino acids in *E. coli*. However, complexes among nonregulatory enzymes, which are the great majority, are the most abundant (Jardine et al. 2002) and can be rationalized in terms of channeling intermediates from one enzyme to the next with limited loss by diffusion to the bulk phase (Ovádi 1991).

Given that the first evidence for feed-forward inhibition involved the penultimate product of the pathway, what are we to make of the fact that this position has essentially no functional significance? Since the penultimate product and the product are so similar in chemical structure, it might be that the penultimate product will be an inescapable competitive inhibitor of the synthetase. Indeed, the penultimate product of the histidine biosynthetic pathway, histidinol, is a competitive inhibitor of the histidyl-tRNA synthetase in *Salmonella typhimurium* (Ames and Hartman 1961); however, it also is an inhibitor of the histidyl-tRNA synthetase in humans (Hansen, Vaughan, and Wang 1972), even though humans have no histidine biosynthetic pathway and presumably make no natural histidinol.

It is perhaps not surprising that this form of anticipatory control involving feed-forward-inhibition would be found in biological systems. Feed-forward control has been used in technological settings for more than 200 years, as practiced by the early millwrights, and its advantages in modern control theory are well established. Indeed, a multiplicity of uses for feed-forward control in other biological contexts has been suggested (Mangan and Alon 2003; Wall, Dunlop, and Hlavacek 2005).

2.5 System Design Principles for Two-Component Signal Transduction

Signal transduction may be thought of as the process by which environmental signals are detected, modified, and propagated to ultimately influence expression of cellular genes or other cellular processes. This process typically represents the input side of genetic circuitry, although signal transduction can also be involved in connecting outputs of one gene circuit to inputs of others (Martínez-Antonio et al. 2006). Two-component systems in the prototype form are signal transduction mechanisms consisting of two cognate proteins: sensor kinase and response regulator (Stock, Robinson, and Goudreau 2000; Inouye and Dutta 2003; Mizuno 2005; Mascher, Helmann, and Unden 2006). The sensor kinase responds to signals in the environment by undergoing autophosphorylation and then transfers the activated moiety to the response regulator, and the phosphorylated regulator typically binds DNA targets to influence gene expression.

Two-component systems are among the most ubiquitous signal transduction mechanisms. Several hundred different two-component systems have been identified in bacteria (http://www.genome.ad.jp/kegg/regulation.html). A much larger number can be inferred from genome sequence data, because well-conserved similarities characterize their gene sequence. Approximately 1,000 putative sensors and response regulators were identified in over 100 organisms that had their gene sequence determined several years ago from homology searches among.

2.5.1 Bifunctional and Monofunctional Sensors

The design of prototype two-component systems exhibits a common variation in which the sensor kinase exhibits a second function (Figure 2.3). In this variation, the unphosphorylated sensor acts as a phosphatase that dephosphorylates the response regulator. In these cases it is called a bifunctional sensor, whereas when this second function is lacking it is called a monofunctional sensor (Stock et al. 2000; Alves and Savageau 2003). Changes in the protein levels occur on a timescale that is much slower than that of phosphorylation, phosphotransfer, and dephosphorylation in these systems. Thus, on the timescale of interest here, gene regulation can be ignored and the total amount of sensor protein (S_T) and the total amount of regulator protein (R_T) can be considered conserved quantities (i.e., $S + S^* = S_T = $ constant and $R + R^* = R_T = $ constant).

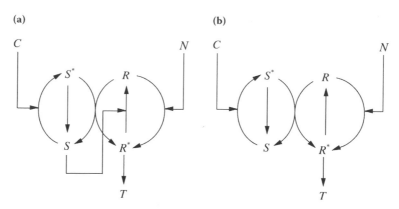

FIGURE 2.3   Alternative designs for a two-component system consisting of a sensor kinase, which is converted from the inactive form S to the active form S^* in response to the cognate input signal C, and a response regulator, which is converted from the inactive form R to the active form R^* by the transfer of phosphate from S^*. The activated response regulator binds to the target T, which is considered the output function. Both S^* and R^* undergo limited spontaneous deactivation. The noncognate input signal N also can stimulate the conversion of the response regulator from R inactive form to the active form R^*. Models in which the inactive form S (a) does (bifunctional sensor), or (b) does not (monofunctional sensor), have a positive influence on the rate of dephosphorylation of R^*.

Three-dimensional structures have been determined for several response regulators and, by homology modeling, numerous putative response regulators. These structures reveal no obvious differences that depend upon whether the cognate sensor proteins are bifunctional or monofunctional. Partial structures have been determined for both types of sensors. In this case, there appear to be differences in a region of the ATP-binding domain known as the ATP lid. Alves and Savageau (2003) used these different structures as templates to predict the 3-D structure of other putative sensor kinases by homology modeling.

Although these results are suggestive and allow predictions to be made (Dalton and Scott 2004), there are known cases in which protein domains with very similar sequence have very different folds and cases in which protein domains with very different sequence have very similar folds. In any case, it is beneficial to know what, if any, functional implications can be found of this variation in design. In our comparisons we must be careful to eliminate as many as possible of the accidental differences that always exist among alternatives. This is done using the method of mathematically controlled comparison and a set of quantitative criteria for functional effectiveness.

2.5.2   Quantitative Criteria for Functional Effectiveness

A signal transduction cascade should have large logarithmic gains to amplify physiological signals and small logarithmic gains that attenuate pathological noise. The cascade should be robust; that is, it should function reproducibly despite perturbations in the values of the parameters that define the structure of the system. The steady state of the system should be stable and have a sufficient margin of stability such that it will not become unstable when subjected to random fluctuations in the parameters of the system. Finally, the system should respond quickly to changes in its environment because otherwise the system is unlikely to be competitive in rapidly changing environments. Perhaps other criteria could be added, but the above are representative of those that have revealed significant functional differences associated with alternative designs in other classes of systems.

2.5.3   Qualitative Functional Differences

A rigorous analytical comparison of the alternative sensor designs uses external equivalence to eliminate many of the extraneous differences between such systems. This can be accomplished in a variety of ways. For example, the alternatives can be made equivalent in their response to the combined cognate and noncognate signals,

$$L(R^*,C)_{Bi} + L(R^*,N)_{Bi} = L(R^*,C)_{Mono} + L(R^*,N)_{Mono} \tag{2.12}$$

and then a comparison of their responses gives the following result, which may be considered a simple design principle for these systems:

$$\frac{L(R^*,C)_{Bi} \,/\, L(R^*,C)_{Mono}}{L(R^*,N)_{Bi} \,/\, L(R^*,N)_{Mono}} = \left(\frac{g_{R^*S^*} - h_{R^*S^*}}{g_{R^*S^*}} \right) > 1 \tag{2.13}$$

The kinetic orders in this case represent the influence of S^* on R^* directly via the phosphotransfer reaction ($g_{R^*S^*} > 0$) and indirectly via the phosphotase reaction ($h_{R^*S^*} < 0$). Thus, for a given difference in response to the noncognate signal $L(R^*,C)$, the difference in responses to the cognate signal will be even greater by the factor $(g_{R^*S^*} - h_{R^*S^*}) \,/\, g_{R^*S^*}$; alternatively, for a given difference in response to the cognate signal $L(R^*,C)$, the difference in responses to the noncognate signal $L(R^*, N)$ will be even less by the factor $g_{R^*S^*} \,/\, (g_{R^*S^*} - h_{R^*S^*})$.

The alternative sensor designs could also be made equivalent in their response to the cognate signal alone or the noncognate signal alone, and then their responses could be compared with the noncognate and cognate signal, respectively. Regardless of how the external equivalence is imposed, the qualitative differences revealed are the same (Alves and Savageau, 2003). These differences regarding discrimination between cognate and noncognate input signals are irreducible. The only way they can be eliminated is by making $h_{R^*S^*} = 0$, which is equivalent to making both systems identical and monofunctional.

The result is a prediction from the simple design principle in Equation (2.11): systems with a bifunctional sensors tend to amplify the cognate signal and attenuate the noncognate signal, whereas those with a monofunctional sensors tend to have a more balanced response to the two inputs. This prediction also can be expressed as bifunctionality tends to diminish pathological cross-talk (expressed quantitatively as the reduction in logarithmic gain between the input signal to one sensor and the output signal from a noncognate response regulator compared with the logarithmic gain between the same input signal and the output signal from the cognate regulator), whereas monofunctionality tends to enhance physiological cross-talk (expressed quantitatively as a balance of these two logarithmic gains).

2.5.4   Experimental Evidence

Kato and Groisman (2004) provided a striking demonstration that a sensor kinase can be converted in effect from bifunctional to monofunctional. They found that a protein connector (PhoP-activated PmrD) blocks the spontaneous dephosphorylation of the response regulator PmrA, along with the dephosphorylation activity of the cognate sensor kinase PmrB, in essence converting it from bifunctional to monofunctional and thereby promoting the integration of signals from a second two-component system PhoP/PhoQ. However, in the numerous examples of cross-talk reported between two-component systems in vitro, or in vivo when components are highly overexpressed, the physiological significance remains questionable.

The analysis of Alves and Savageau (2003) predicts that an advantage of bifunctional sensors is the amplification of cognate versus noncognate signals that reduces the impact of noisy cross-talk. This is an irreducible consequence of bifunctional over monofunctional mechanisms, independent of the numerical values of the parameters. However, the magnitude of the effect will depend on the specific parameter values. Recent experimental studies have found evidence that supports the role of bifunctional sensors in suppressing cross-talk (Siryaporn and Goulian 2008; Groban et al. 2009). The diagram in Figure 2.4 includes the relevant interactions for this discussion.

Siryaporn and Goulian (2008) experimentally examined cross-talk between the EnvZ–OmpR and CpxA–CpxR systems in *Escherichia coli*. These systems were chosen because they exhibit considerable sequence similarity that might be expected to increase the probability of detecting cross-talk and because both are reasonably well characterized. Siryaporn and Goulian used imaging of fluorescently tagged proteins to examine cross-talk in a variety of mutant strains. By deleting the cognate sensor, and hence the bifunctional mechanism, cross-talk increased and also was potentiated by eliminating the response regulator for the noncognate sensor, as this gets rid of the primary receptor that competes for binding of the noncognate sensor.

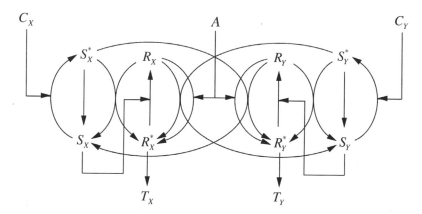

FIGURE 2.4 Potential interactions between two-component systems. The two-component system on the left (X) consists of a sensor kinase, which is converted from the inactive form S_X to the active form S_X^* in response to the cognate input signal C_X, and a response regulator, which is converted from the inactive form R_X to the active form R_X^* by the transfer of phosphate from S_X^*. The activated response regulator binds to the target T_X, which is considered the output function. Both S_X^* and R_X^* undergo limited spontaneous deactivation. The inactive form of the sensor S has a positive influence on the rate of R_X^* dephosphorylation. The two-component system on the right (Y), which can be considered a potential source of noncognate cross-talk, is analogous to the system on the left. Another potential source of noncognate cross-talk is the result of transphosphorylation from acetyl-phosphate A.

Groban et al. (2009) studied the same system experimentally and determined in vitro values for the kinetic parameters of the cognate and noncognate phosphotransfer reactions between the systems. These data were integrated into a mathematical model that successfully predicted the extent of cross-talk in vivo resulting from mutations that eliminated specific components of the system. Groban et al. demonstrated that the bifunctional sensor mechanism reduces cross-talk and also found that cross-talk is enhanced by eliminating other acceptors of the phosphotransfer from the noncognate sensor kinase (i.e., the noncognate response regulator) and other phosphodonors that competed with the noncognate sensor kinase (i.e., the transphosphorylation from acetyl-phosphate). Removing any of the three—unphosphorylated cognate sensor, noncognate response regulator, or acetate kinase A and phosphotransacetylase—still results in substantial suppression of cross-talk. In fact, cross-talk continues to be suppressed following any pair-wise removal of the three mechanisms, and only with the triple knockout of all three mechanisms is significant cross-talk manifested.

These experiments do not involve comparisons with a monofunctional sensor. To mimic the mathematically controlled comparison of Alves and Savageau (2003), not only would the unphosphorylated sensor have to be removed, as was done in these experiments by knockout mutation, but also the level of spontaneous dephosphorylation of the response regulator would have to be adjusted to match that of the system with a putative monofunctional sensor and the alternatives put in the same state, which might be difficult to achieve experimentally.

The system used in the aforementioned experimental studies exhibits a continuous graded response (Batchelor, Silhavy, and Goulian 2004). The possibility of a discontinuous all-or-none response in other two-component systems remains to be explored experimentally; a dead-end complex between the unphosphorylated form of a monofunctional sensor kinase and its response regulator is predicted to promote this type of response (Igoshin, Alves, and Savageau 2008). Moreover, the inclusion of gene regulation coupled with the posttranslational interactions generates a rich repertoire of responses (Ray and Igoshin 2010) that were previously thought to be contradictory, because some groups focused on responses associated only with positive feedback and others focused on responses associated only with negative feedback.

2.6 System Design Principles for Coupled Expression in Gene Circuits

The output of a transcriptional unit is typically an expression cascade of several levels, each with the potential for the fan-out of several effectors: (1) there can be one or several RNAs; (2) each may undergo processing to one or more functional mRNAs; (3) the mRNA may code for one or more proteins; (4) the proteins may have one or more functions; and (5) some are enzymes that catalyze the production of metabolites that may have one of more fates. Gene circuits are created by connecting the outputs of one expression cascade to the inputs of another in patterns that include various forms of feedback and feed-forward to produce coupled expression. The simplest example, called an elementary gene circuit, involves a single effector function and its specific transcriptional regulator.

Chapter 2

2.6.1 Extremes of Coupling

In the first form of coupled gene expression to be characterized experimentally, the regulator and the effector were encoded in separate transcriptional units, and regulator expression did not change while the effector expression was undergoing induction or repression (Figure 2.5a, top). This represents complete uncoupling, which is one extreme form of coupling. Examples were subsequently found in which regulator and effector functions were encoded in the same transcriptional unit so that regulator expression and effector expression varied coordinately during induction or repression (Figure 2.5a, bottom). This represents perfect coupling, which is the other extreme form of coupling.

There are examples of each form of coupling in bacteria, so the question naturally arises as to why. Are these different forms of coupling simply the result of historical accident without functional consequences, or have they been selected because there are advantages to each in different contexts; if the latter, can we discover the underlying design principles?

An analytical comparison of these extreme forms of coupling, based on a half-dozen quantitative criteria for functional effectiveness, used external equivalence to eliminate many of the extraneous differences between such systems and identified significant irreducible differences regarding local robustness and temporal responsiveness. For example, the parameter sensitivities of the circuit with repressor control and perfect coupling

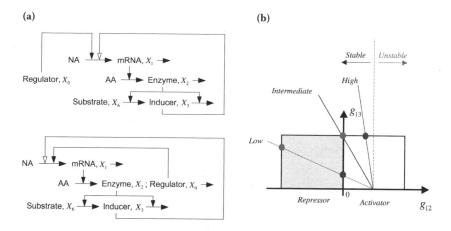

FIGURE 2.5 Model of an elementary gene circuit with an inducible expression cascade for the effector protein that catalyzes the processing of the substrate signal. (a, top) Circuitry with complete uncoupling of regulator and effector expression. (a, bottom) Circuitry with perfect coupling of regulator and effector expression. (b) Design space for the elementary gene circuits. The axes represent the influence of inducer ($g_{13} > 0$) and regulator (g_{12}). Activator-controlled circuits are represented on the right with kinetic orders $g_{12} > 0$; repressor-controlled circuits are represented on the left with $g_{12} < 0$. The kinetic orders can be no larger than the number of binding sites involved in their interactions (the boxed region); typically they have magnitudes less than 2 for dimeric interactions or 4 for tetrameric interactions. Lines that radiate from the common point on the horizontal axis represent circuits with the same capacity for regulation. Performance improves according to quantitative criteria for functional effectiveness for circuits represented by points farthest to the left along any line of equivalent capacity. The dashed line represents the boundary of instability. From W.S. Hlavacek and M.A. Savageau, *J. Mol. Biol.* **248**: 739–755, 1995. With permission.

were all less than or equal to those of the circuit with repressor control and complete uncoupling. Moreover, the differences were all of the form

$$\frac{S(V_3, p_k)^{(p)}}{S(V_3, p_k)^{(u)}} = \frac{h_{11}h_{22}}{h_{11}h_{22} - g_{12}g_{21}} < 1 \tag{2.14}$$

since $g_{12} < 0$ and all other parameters of the model are positive. Conversely, the parameter sensitivities of the circuit with activator control and perfect coupling were all greater than or equal to those of the circuit with activator control and complete uncoupling; in this case the ratio in Equation (2.14) is greater than 1 since now $g_{12} > 0$. Again, these are inherent differences that cannot be eliminated unless $g_{12} = 0$ and the circuits become identical and uncoupled.

These results, together with the systematic differences in response time, yielded the prediction of a simple design principle: perfect coupling in the case of repressor control and complete uncoupling in the case of activator control (Savageau 1974). The results of the analysis also predicted that the regulator of an autonomous system, one without an inducer or corepressor, should be a repressor of its own transcription (Savageau 2009). Becskei and Serrano (2000) provided experimental support for the predicted improvement in robustness with negative autoregulation.

While there was some experimental evidence available at the time to support the predicted forms of coupling, some examples clearly did not fit. However, the anomalies were largely removed when physical constraints on the magnitudes of the regulatory interactions were incorporated into the analysis. The modified design principle made predictions based on two factors, (1) the mode of control (repressor vs. activator) and (2) the capacity for regulation (low vs. high): perfect coupling in the case of repressor control and low capacity or activator control and high capacity; and compete uncoupling in the case of repressor control and high capacity or activator control and low capacity (Hlavacek and Savageau 1995).

2.6.2 Resolution of Anomalies

These analytical results can be seen graphically in the system design space depicted in Figure 2.5b. The predictions of the old design principle were still valid when the capacity for regulation was low. For example, the histidine utilization system of *Salmonella typhimurium*, which has a low capacity (~18-fold), is repressor controlled and exhibits perfect coupling (Smith and Magasanik 1971). However, the predictions of the old principle were invalid when the capacity for regulation was high. With the modified design principle, which accounts for the capacity of a system, anomalous cases were now in agreement. For example, the lactose system of *E. coli* (Jacob and Monod 1961), which has the highest capacity found for a repressor-controlled system (~1000-fold), is repressor controlled and exhibits complete uncoupling. The predictions for activator-controlled systems are just the reverse: complete uncoupling is predicted for low-capacity systems, and perfect coupling is predicted for high-capacity systems.

The qualitative aspect of the predicted improvement in the response time of a repressor-controlled system with perfect coupling can be understood intuitively. For example, such a design allows for very active transcription at the start of induction when repressor levels are low, and as induction proceeds repressor levels increase to eventually rein in the initial high-level transcription. Rosenfeld, Elowitz, and Alon (2002) verified this prediction in an experiment that came close to realizing a mathematically controlled comparison. They compared the response time of a repressor-controlled system with perfect coupling to a mutant that was uncoupled. Although the same state was not maintained experimentally in the alternative systems, Rosenfeld et al. multiplied the lower values of the uncoupled system by a factor that made its final steady state values the same as that of the perfectly coupled system. By this scaling of values they were able to show that the perfectly coupled system had a much faster response time. This would be equivalent to a mathematically controlled result if the systems were linear, but since the two systems were in fact operating over a different range of values the results could be different if the systems were significantly nonlinear. Wong, Bolovan-Fritts, and Weinberger (2009; personal communication) on the major immediate early circuit in human cytomegalovirus provide another example of an experimental study that speaks to the issue of feedback and temporal responsiveness. Wong et al. found a negative feedback within the circuit that speeds response time of the essential viral transactivator IE2-36 in a dose-dependent fashion as feedback strength is increased. Moreover, they also found that the increase in rate of response occurs without significantly amplifying the steady state level of the transactivator (mimicking a mathematically controlled comparison). This mechanism is important because it prevents the cytotoxicity that would occur if the level of transactivator were to increase proportionately, and Wong et al. showed that it confers a selective advantage on the wild-type virus.

2.6.3   Three Forms of Coupling

The intuitive understanding of the faster response time for a repressor-controlled system with perfect coupling does not apply in the case of an activator-controlled system. In this case, the analogous argument would have to go as follows. Such a design allows for very active transcription at the start of induction when activator levels are high, and as induction proceeds activator levels decrease to eventually rein in the initial high-level transcription. Clearly, this scenario cannot be realized by either a completely uncoupled or a perfectly coupled system. More general forms of circuitry must be considered that would allow for the inverse coupling required of activator and effector functions.

In the case of elementary gene circuits involving a single effector and a single regulator the two expression cascades can exhibit one of three logical forms of coupling: completely uncoupled, directly coupled, or inversely coupled (Figure 2.6). A rigorous analytical comparison of these three forms of coupling, again based on a half-dozen quantitative criteria for functional effectiveness, used external equivalence to eliminate many of the extraneous differences between such systems and identify significant irreducible differences regarding local robustness and temporal responsiveness (Hlavacek and Savageau 1996). The result was the prediction of an expanded design principle, as shown in Table 2.1. The results of the analysis also predicted that the regulator should repress its own transcription. Of course in contexts other than these elementary gene circuits, regulators that activate the transcription of their own structural gene (e.g., see Savageau and Fasani 2009; Ray and Igoshin 2010) can be expected.

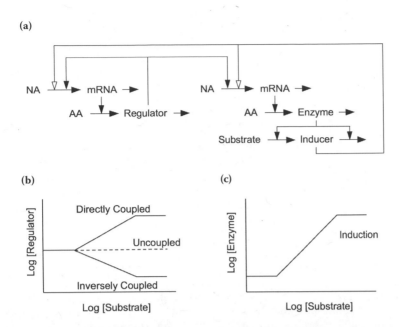

FIGURE 2.6 Model of an elementary gene circuit consisting of two expression cascades, one for the regulator protein and one for the effector protein that catalyzes the processing of the substrate signal. (a) Circuitry showing the connectivity among the cascades. (b) The logical possibilities for regulator expression coupled to (c) effector expression in an inducible system. (From W.S. Hlavacek and M.A. Savageau, *J. Mol. Biol.* **255**: 121–139, 1996. With permission.)

Table 2.1 Design Principle for Coupling of Expression in Elementary Gene Circuits

Mode of Control	Positive			Negative		
Regulatory Capacity	High	Intermediate	Low	High	Intermediate	Low
Predicted Coupling	Direct	Uncoupled	Inverse	Inverse	Uncoupled	Direct

2.6.4 Experimental Evidence

The design principle for coupling of elementary gene circuits could be tested on a large scale when the complete DNA sequence for the *E. coli* genome became available (Blattner et al. 1997). It was quickly shown that most transcription factors in *E. coli*, in agreement with predictions, appear to regulate their own expression as a repressor binding to a site upstream of the promoter for their structural gene (Thieffry et al. 1998). Moreover, by combining transcription factor connectivity and mode of regulation from genomic data with coupling information for effector and regulator expression from data published on a large number of elementary gene circuits (Wall, Hlavacek, and Savageau 2004), support was found for the design principle summarized in Table 2.1. These rules also suggest an experimentally testable hypothesis concerning the evolution of gene regulation based on evidence from comparative genomic approaches (Tan et al. 2001).

The data used to test the predicted design principle for the coupling of expression in elementary gene circuits, shown in Figure 2.7, are in reasonable agreement but raise some additional questions. First, the extent of inverse coupling among activator-controlled systems is rather minimal. Second, there are no examples of inverse coupling

Chapter 2

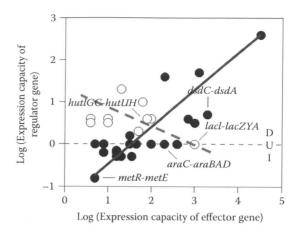

FIGURE 2.7   Coupling of regulator and effector gene expression for elementary gene circuits with various capacities for effector gene regulation. Capacity is measured as the ratio of full expression (induction or repression) to basal expression. (D) Circuits exhibiting direct coupling have full expression greater than basal expression, and thus have capacities greater than 1 (positive log values). (U) Circuits in which regulator expression is invariant while effector expression varies have capacity equal to 1. (I) Circuits exhibiting inverse coupling have full expression less than basal expression and thus have capacities less than 1 (negative log values). Activator-controlled circuits (closed circles); repressor-controlled circuits (open circles). (From W.S. Hlavacek and M.A. Savageau, *J. Mol. Biol.* **255**: 121–139, 1996. With permission.)

among repressor-controlled systems. Third, only the activator-controlled systems show the highest levels of expression.

The minimal extent of inverse coupling among activator-controlled circuits means that other examples might have been missed because the variation may not have risen above the experimental measurement error. In any case, it might also signify that inverse coupling confers only a minor advantage over an uncoupled design, in which case the effective design principle may only involve high and intermediate-low levels of expression.

The absence of repressor-controlled circuits that exhibit inverse coupling may also be related to the issues raised in the preceding paragraph. Another explanation might simply be sampling error. There are ~200 elementary circuits in *E. coli*, but the data in Figure 2.7 represent only ~50 cases. Perhaps when more have been experimentally characterized it will be found that some exhibit inverse coupling. The design space used in the analysis of the extreme forms of coupling (Figure 2.5b) suggests yet a third possibility that might also explain the third observation mentioned at the beginning of this section.

The lines radiating from the common point on the horizontal axis represent the locus of operating points for circuits that have the same capacity for regulation. A low slope corresponds to low capacity, a high slope to high capacity. Since performance of the circuits (based on several of the criteria for functional effectiveness) tends to improve as the operating points move to the left in this space, and since only circuits represented within the boxed area are physically realizable because of limits on the kinetic orders for regulator and inducer, direct coupling is predicted for repressor-controlled circuits and uncoupling for activator-controlled circuits when the capacity for regulation is low.

At the highest capacity that still allows repressor control, the prediction is uncoupling for both repressor-controlled and activator-controlled circuits. Circuits with a still higher capacity for regulation can be realized only with an activator control. Thus, for high-capacity circuits, some fundamental limits may simultaneously explain the presence of activator-controlled circuits and the absence of repressor-controlled circuits.

2.7 Concluding Remarks

Systems seldom present as well-defined or closed; the investigator must make choices. This is an art, not a mechanical process, and the importance of intimate familiarity with the system should not be overlooked. Nevertheless, the choice for delineation of a system must be made according to some criterion (Savageau 2009). For example, let systems be selected to maximize the numerical ratio of internal interactions to external interactions. In this way the resulting system will tend to have a minimal number of interactions with its environment, but at the same time a maximum number of interactions will be included within the system. This choice will tend to preserve the integrity of critical functional groupings, which is important because certain phenomena cannot be adequately understood by simply analyzing the component parts of a system. At certain levels of organization new properties emerge that can be understood only at these levels; such phenomena must be treated as a whole. Interactions between the system chosen in this way and its environment will remain, but the number will have been minimized. Thus, identification of environmental interactions is facilitated, and, when these are disrupted to study the system in isolation, the effects of the normal environment on the system can be most easily approximated by experimental means.

To elucidate these system design principles it is critical that the system is properly delineated. It is not enough to sample a large number of variables, such as one from each of the major classes of molecules in the cell, if these do not constitute an integrated system. Nor is it sufficient to establish exhaustively one type of connected network, such as a transcription factor network, when other connections that are necessary to make a functional system are missing. We see this in the design principles that govern the coupling of elementary gene circuits. Could these principles have been established from a top-down approach? The structural genes could certainly have been found in the DNA sequence. The transcription factor network would have revealed the connections between regulator and effector transcription units. The patterns of coupling between regulator and effector expression would have been found (assuming expression arrays with sufficient range to detect highly abundant effector transcripts and very low abundant regulator transcripts). However, it is unlikely that the coupling rules would have been found empirically, and there would be very little understanding of where the rules come from. Moreover, there would be cases in which expression is all-or-none rather than graded, and this would not have been detected in the typical high-throughput experiments. The critical factor here is the connection between a metabolite signal and the transcription factor. Again, perhaps these critical connections could be discovered by screening transcription factors for their binding of small molecular weight ligands, which is a considerable challenge. There are technical difficulties at all these levels. However, even granting that these can be overcome, this process is hardly going to happen without a hypothesis suggesting what might be important to look for.

Chapter 2

It should be emphasized that mathematically controlled comparisons represent an ideal controlled comparison. It may be possible to closely approximate such a comparison experimentally, but some cases may also not be realized for practical or fundamental reasons. The ultimate test will always be the pragmatic one of whether the predictions of such a comparison are supported by experimental data.

In looking back over the examples presented here, both qualitative and quantitative design principles can be seen. Some of the design principles discussed are qualitative rules that can be stated in words, even though the rule may have been revealed by quantitative mathematical analysis. Other design principles are quantitative rules that are captured in simple mathematical relationships. The latter can be summarized in a verbal description, but the precise nature of the rule is best expressed symbolically. While an empirical approach might lead to qualitative rules, it is unlikely to reveal irreducible quantitative relationships such as

$$(g_{32} - h_{32}) > \frac{h_{11} h_{22} h_{33}}{g_{13} g_{21}} \tag{2.15}$$

which is a determinant of graded versus hysteretic switching of inducible gene circuits (Savageau 2001), because of the confounding effects of accidental differences that typically arise in most studies.

The S-system parameters found in these relationships often provide the simplest expressions of the critical conditions. Moreover, this representation can always be justified for local variations about a nominal steady state. If necessary or desirable, they can be readily related to the parameters of other representations; in the case of the GMA representation, the relationship is simply an averaging of parameters weighted by steady state fluxes; for example, see Equations (2.3) to (2.5).

Finally, a common theme in the irreducible differences is a ratio of the form $A / (A + B) < 1$, in which A and B represent the magnitudes of individual kinetic orders (or constellations of kinetic orders and fluxes) that are always positive. These differences arise from topological considerations involving the connections in the system and the signs of the interactions. In addition, these expressions center on key regulatory signals in the system: one term includes the kinetic order for loss of the signal (biochemical degradation or dilution), often a first-order process; another term includes the kinetic order for the signal's influence on the target (typically an allosteric modification), often a higher-order interaction. Thus, the numerical values for these ratios are in the threefold to sixfold range, which means that the differences should be readily detected in appropriately designed experiments.

References

Alon, U. 2006. *An Introduction to Systems Biology: Design Principles of Biological Circuits.* Boca Raton, FL: Chapman & Hall/CRC.

Alves, R. and M.A. Savageau. 2000. Extending the method of mathematically controlled comparison to include numerical comparisons. *Bioinformatics* **16**: 786–798.

Alves, R. and M.A. Savageau. 2003. Comparative analysis of prototype two-component systems with either bifunctional or monofunctional sensors: differences in molecular structure and physiological function. *Mol. Microbiol.* **48**: 25–51.

Ames, B.N. and P.E. Hartman. 1961. Genes, enzymes, and control mechanisms in histidine biosynthesis. In *Molecular Basis of Neoplasia, Fifteenth Symposium of Fundamental Cancer Research*, 322–345. Austin: University of Texas Press.

Batchelor, E., T.J. Silhavy, and M. Goulian. 2004. Continuous control in bacterial regulatory circuits. *J. Bacteriol.* **186**: 7618–7625.

Becskei, A. and L. Serrano. 2000. Engineering stability in gene networks by autoregulation. *Nature* **405**: 590–593.

Blattner, F.R., G. Plunkett III, C.A. Bloch, N.T. Perna, V. Burland, M. Riley, J. Collado-Vides, J.D. Glasner, C.K. Rode, G.F. Mayhew, J. Gregor, N.W. Davis, H.A. Kirkpatrick, M.A. Goeden, D.J. Rose, B. Mau, and Y. Shao. 1997. The complete genome sequence of *Escherichia coli* K-12. *Science* **277**: 1453–1462.

Bliss, R.D., P.R. Painter, and A.G. Marr. 1982. Role of feedback inhibition in stabilizing the classical operon. *J. Theoret. Biol.* **97**: 177–193.

Çagatay, T., M. Turcotte, M.B. Elowitz, J. Garcia-Ojalvo, and G.M. Süel. 2009. Architecture-dependent noise discriminates functionally analogous differentiation circuits. *Cell* **139**: 512–522.

Cho, D.-Y., K.-H. Cho, and B.-Y. Zhang. 2006. Identification of biochemical networks by S-tree based genetic programming. *Bioinformatics* **22**: 1631–1640.

Chou, I.C. and E.O. Voit. 2009. Recent developments in parameter estimation and structure identification of biochemical and genomic systems. *Math. Biosci.* **219**: 57–83.

Clop, E.M., P.D. Clop, J.M. Sanchez, and M.A. Perillo. 2008. Molecular packing tunes the activity of *Kluyveromyces lactis* β-galactosidase incorporated in Langmuir-Blodgett films. *Langmuir* **24**: 10950–10960.

Coelho, P.M.B.M., A. Salvador, and M.A. Savageau. 2009. Global tolerance of biochemical systems and the design of moiety-transfer cycles. *PLOS Comput. Biol.* **5**(3): e1000319.

Coelho, P.M.B.M., A. Salvador, and M.A. Savageau. 2010. Relating genotype to phenotype via the quantitative behavior of the NADPH redox cycle in human erythrocytes: mutant analysis. *PLOS One.* **5**(9): e13031.

Cotton, R.G.H. and F. Gibson. 1965. The biosynthesis of phenylalanine and tyrosine; enzymes converting chorismic acid into prephenic acid and their relationships to prephenate dehydratase and prephenate dehydrogenase. *Biochim. Biophys. Acta* **100**: 76–88.

Dalton, T.L. and J.R. Scott. 2004. CovS inactivates CovR and is required for growth under conditions of general stress in *Streptococcus pyogenes*. *J. Bacteriol.* **186**: 3928–3937.

Dorf, R.C. and R.H. Bishop. 1995. *Modern Control Systems*, 7th ed. Reading, MA: Addison-Wesley.

Fasani, R.A. and M.A. Savageau. 2010. Automated construction and analysis of the design space for biochemical systems, *Bioinformatics* **26**: 2601–2609.

Gennemark, P. and D. Wedelin. 2009. Benchmarks for identification of ordinary differential equations from time series data. *Bioinformatics* **15**: 780–786.

Gibson, R. and J. Pittard. 1968. Pathways of biosynthesis of aromatic amino acids and vitamins and their control in microorganisms. *Bacteriol. Rev.* **32**: 465–492.

Groban, E.S., E.J. Clarke, H.M. Salis, S.M. Miller, and C.A. Voigt. 2009. Kinetic buffering of cross talk between bacterial two-component sensors. *J. Mol. Biol.* **390**: 380–393.

Hansen, B.S., Vaughan, M.H., and Wang, L.-J. 1972. Reversible inhibition by histidinol of protein synthesis in human cells at the activation of histidine. *J. Biol. Chem.* **247**: 3854–3857.

Hlavacek, W.S. and M.A. Savageau. 1995. Subunit structure of regulator proteins influences the design of gene circuitry: analysis of perfectly coupled and completely uncoupled circuits. *J. Mol. Biol.* **248**: 739–755.

Hlavacek, W.S. and M.A. Savageau. 1996. Rules for coupled expression of regulator and effector genes in inducible circuits. *J. Mol. Biol.* **255**: 121–139.

Igoshin, O.A., R. Alves, and M.A. Savageau. 2008. Hysteretic and graded responses in bacterial two-component signal transduction. *Mol. Microbiol.* **68**: 1196–1215.

Inouye M. and R. Dutta. 2003. *Histidine Kinases in Signal Transduction*. San Diego: Academic Press.

Jacob, F. and J. Monod. 1961. Genetic regulatory mechanisms in the synthesis of proteins. *J. Mol. Biol.* **3**: 318–356.

Jardine, O., J. Gough, C. Chothia, and S.A. Teichmann. 2002. Comparison of the small molecule metabolic enzymes of *Escherichia coli* and *Saccharomyces cerevisiae*. *Genome Res.* **12**: 916–929.

Kato, A. and E.A. Groisman. 2004. Connecting two-component regulatory systems by a protein that protects a response regulator from dephosphorylation by its cognate sensor. *Genes. Dev.* **15**: 2302–2313.

Lewis, D.C. 1991. A qualitative analysis of S-systems: Hopf bifurcations. In *Canonical Nonlinear Modeling. S-System Approach to Understanding Complexity*, ed. E.O. Voit, 304–344. New York: Van Nostrand Reinhold.

Mangan, S. and U. Alon. 2003. Structure and function of the feed-forward loop network motif. *Proc. Natl. Acad. Sci. USA* **100**: 11980–11985.

Martínez-Antonio, A., S.C. Janga, H. Salgado, and J. Collado-Vides. 2006. Internal-sensing machinery directs the activity of the regulatory network in *Escherichia coli*. *Trends. Microbiol.* **14**: 22–27.

Mascher, T., J.D. Helmann, and G. Unden. 2006. Stimulus perception in bacterial signal-transducing histidine kinases. *Microbiol. Mol. Biol. Rev.* **70**: 910–938.

Mizuno, T. 2005. Two-component phosphorelay signal transduction systems in plants: from hormone responses to circadian rhythms. *Biosci. Biotechnol. Biochem.* **69**: 2263–2276.

Nazario, M. 1967. The accumulation of argininosuccinate in *Neurospora crassa*. I. Elevated ornithine carbamoyl transferase with high concentrations of arginine. *Biochim. Biophys. Acta.* **145**: 138–145.

Ovádi, J. 1991. Physiological significance of metabolic channeling. *J. Theoret. Biol.* **152**: 1–22.

Patte, J.-C., P. Truffa-Bachi, and G.N. Cohen. 1966. The threonine-sensitive homoserine dehydrogenase and aspartokinase activities of *Escherichia coli*. I. Evidence that the two activities are carried by a single protein. *Biochim. Biophys. Acta.* **128**: 426–439.

Ray, J.C. and O.A. Igoshin. 2010. Adaptable functionality of transcriptional feedback in bacterial two-component systems. *PLoS Comput. Biol.* **12**: e1000676

Rosenfeld, N., M.B. Elowitz, and U. Alon. 2002. Negative autoregulation speeds the response times of transcription networks. *J. Mol. Biol.* **323**: 785–793.

Savageau, M.A. 1969. Biochemical systems analysis II. The steady state solutions for an n-pool system using a power-law approximation. *J. Theoret. Biol.* **25**: 370–379.

Savageau, M.A. 1972. The behavior of intact biochemical control systems. *Curr. Top. Cell. Reg.* **6**: 63–130.

Savageau, M.A. 1974. Comparison of classical and autogenous systems of regulation in inducible operons. *Nature* **252**: 546–549.

Savageau, M.A. 1975. Optimal design of feedback control by inhibition: dynamic considerations. *J. Mol. Evolution* **5**: 199–222.

Savageau, M.A. 1979. Feed-forward inhibition in biosynthetic pathways: Inhibition of the aminoacyl-tRNA synthetase by the penultimate product. *J. Theoret. Biol.* **77**: 385–404.

Savageau, M.A. 1991. Reconstructionist molecular biology. *New Biologist* **3**: 190–197.

Savageau, M.A. 1995. Michaelis-Menten mechanism reconsidered: Implications of fractal kinetics. *J. Theoret. Biol.* **176**: 115–124.

Savageau, M.A. 1998. Development of fractal kinetic theory for enzyme-catalyzed reactions and implications for the design of biochemical pathways. *BioSystems* **47**: 9–36.

Savageau, M.A. 2001. Design principles for elementary gene circuits: elements, methods, and examples. *Chaos* **11**: 142–159.

Savageau, M.A. 2009. *Biochemical Systems Analysis: A Study of Function and Design in Molecular Biology*, 40th Anniversary Edition [A reprinting of the original edition published by Addison–Wesley, Reading, Mass., 1976].

Savageau, M.A. 2011a. Biomedical engineering strategies in system design space. *Annals Biomed. Engr.* **39**: 1278–1295.

Savageau, M.A. 2011b. Design of the *lac* gene circuit revisited. *Math. Biosci.* **231**: 19–38.

Savageau, M.A. and R.A. Fasani. 2009. Qualitatively distinct phenotypes in the design space of biochemical systems. *FEBS Lett.* **583**: 3914–3922.

Savageau, M.A. and G. Jacknow. 1979. Feed-forward inhibition in biosynthetic pathways: inhibition of the aminoacyl-tRNA synthetase by intermediates of the pathway. *J. Theoret. Biol.* **77**: 405–425.

Savageau, M.A. and E.O. Voit. 1987. Recasting nonlinear differential equations as S-systems: a canonical nonlinear form. *Math. Biosci.* **87**: 83–115.

Savageau, M.A., P.M.B.M. Coelho, R. Fasani, D. Tolla, and A. Salvador. 2009. Phenotypes and tolerances in the design space of biochemical systems. *Proc. Natl. Acad. Sci. USA* **106**: 6435–6440.

Singer, P.A., M. Levinthal, and L.S. Williams. 1984. Synthesis of the isoleucyl- and valyl-tRNA synthetases and the isoleucine-valine biosynthetic enzymes in a threonine deaminase regulatory mutant of *Escherichia coli* K-12. *J. Mol. Biol.* **175**: 39–55.

Siryaporn, A. and M. Goulian. 2008. Cross-talk suppression between the CpxA–CpxR and EnvZ–OmpR two-component systems in *E. coli*. *Mol. Microbiol.* **70**: 494–506.

Smith, G.R. and B. Magasanik. 1971. Nature and self-regulated synthesis of the repressor of the *hut* operons in *Salmonella typhimurium*. *Proc. Nat. Acad. Sci., USA* **68**: 1493–1497.

Stock, A.M., V.L. Robinson, and P.N. Goudreau. 2000. Two-component signal transduction. *Ann. Rev. Biochem.* **69**: 183–215.

Sussenbach, J.S. and P.J. Strijkert. 1969. Arginine metabolism in *Chlamydomonas reinhardi*. on the regulation of the arginine biosynthesis. *Eur. J. Biochem.* **8**: 403–407.

Tan, K., G. Moreno-Hagelsieb, J. Collado-Vides, and G.D. Stormo. 2001. A comparative genomics approach to prediction of new members of regulons. *Genome Res.* **11**: 566–584.

Thieffry, D., H. Salgado, A.M. Huerta, and J. Collado-Vides. 1998. Prediction of transcriptional regulatory sites in the complete genome sequence of *Escherichia coli* K–12. *Bioinformatics* **14**: 391–400.

Tolla, D.A. and M.A. Savageau. 2011. Phenotypic repertoire of the FNR regulatory network in *Escherichia coli*. *Molec. Microbiol.* **79**: 149–165.

Truffa-Bachi, P. and G.N. Cohen. 1966. La β-aspartokinase sensible à la lysine *d'Escherichia coli*; purification et propriétés. *Biochim. Biophys. Acta.* **113**: 531–541.

Voit, E.O., Ed. 1991. *Canonical Nonlinear Modeling: S-System Approach to Understanding Complexity*. New York: Van Nostrand Reinhold.

Wall, M.E. M.J. Dunlop, and W.S. Hlavacek. 2005. Multiple functions of a feed-forward-loop gene circuit. *J. Mol. Biol.* **349**: 501–514

Wall, M.E., W.S. Hlavacek, and M.A. Savageau. 2004. Design of gene circuits: lessons from bacteria. *Nature Reviews Genetics* **5**: 34–42.

Williams, A.L., B.M. Whitfield, and L.S. Williams. 1977. *Abstracts of the Annual Meeting of the American Society for Microbiology*, vol. 213. New Orleans, LA: American Society for Microbiology.

Wong, M.L., C. Bolovan-Fritts, and L.S. Weinberger. 2009. Negative feedback speeds transcriptional response-time in human cytomegalovirus. *Biophys. J.* **96**: 305a.

Yem, D.W. and L.S. Williams. 1971. Inhibition of arginyl-transfer ribonucleic acid synthetase activity of *Escherichia coli* by arginine biosynthetic precursors. *J. Bacteriol.* **107**: 589–591.

Chapter 2

3. Chance and Memory

Theodore J. Perkins

Andrea Y. Weiße

Peter S. Swain

3.1 Introduction

Memory and chance are two properties or, more evocatively, two strategies that cells have evolved. Cellular memory is beneficial for broadly the same reasons that memory is beneficial to humans. How well could we function if we forgot the contents of the refrigerator every time the door was closed or forgot where the children were or who our spouse was? The inability to form new memories, called anterograde amnesia in humans, is tremendously debilitating, as seen in the famous psychiatric patient H. M. [1] and the popular movie *Memento* (Christopher Nolan, 2000).

Quantitative Biology: From Molecular to Cellular Systems edited by Michael E. Wall © 2012 CRC Press / Taylor & Francis Group, LLC. ISBN: 978-1-4398-2722-2

Chapter 3

Cells have limited views of their surroundings. They are able to sense various chemical and physical properties of their immediate environment, but distal information can come only through stochastic processes, such as diffusion of chemicals. Cells lack the ability to sense directly distal parts of their environment, as humans can through vision and hearing. Thus, we might expect that cells have adopted memory of the past as a way of accumulating information about their environment and so gain an advantage in their behavior. In a simple application, memory can be used to track how environmental signals change over time. For example, in the chemotaxis network of *Escherichia coli*, cells use memory of previous measurements of chemical concentrations to estimate the time derivative of the chemical concentration and thus to infer whether the organism is swimming up or down a chemical gradient [2]. Conversely, memory can be useful in enabling persistent behavior despite changing environments. For example, differentiated cells in a multicellular organism must remain differentiated and not redifferentiate despite existing under a variety of conditions. If such cells become damaged or infected, they may undergo apoptosis, but we do not observe neurons becoming liver cells, or liver cells becoming muscle cells, and so on.

Different mechanisms for achieving cellular memory operate at different time scales. For example, the chemotaxis machinery of *E. coli* can change response in seconds. Gene regulatory systems controlling sugar catabolism, such as the *lac* operon in *E. coli* [3] or the galactose network in *Saccharomyces cerevisiae* [4], take minutes or hours to switch state. Terminally differentiated cells can retain their state for years—essentially permanently, although scientists are learning how to reset their memories and return cells to a pluripotent status [5]. Of course, that cells have complex internal chemical and structural states implies a certain amount of inertia. A cell cannot change its behavior instantly, so in a weak sense all cellular systems display some degree of memory. This chapter, however, focuses primarily on forms of cellular memory that operate on longer time scales than the physical limits imposed by changing cellular state. Forms of memory will be described that require active maintenance, as embodied by positive feedback. These memories are more similar to human memories in that their state may represent some specific feature of the cell's environment, such as high lactose.

Although the potential for memory to be advantageous is clear, the benefit of chance behavior is perhaps less obvious. Some human activities provide clear examples, especially in competitive situations. For example, if a tennis player always served to the same spot on the court, the opponent would have a much easier time of returning the ball. Predictable behaviors give competitors an opportunity to adapt and gain an advantage, if they can detect the pattern of the behavior. Thus, intentional randomization of one's own behavior can be beneficial. Unicellular organisms are virtually always in a competitive situation—if not from other species, then from others of their own kind seeking out limited resources such as nutrients or space. One well-known example is the phenomenon of bacterial persistence, where some members of a genetically identical population of bacteria spontaneously and stochastically adopt a phenotype that grows slowly but that is more resistant to environmental insults such as antibiotics [6]. The cells spend some time in the persister state and then spontaneously and stochastically switch back to normal, more rapid growth. This strategy allows most of the population to grow rapidly in ideal conditions but prevents the extinction of the population if an insult should occur. Its stochastic aspect makes it challenging to predict which cells will be persisters at any particular time. Thus, the population as a whole, in its competition with the doctor prescribing the antibiotics, increases its chances of success.

Like memory, chance behavior can have different time scales and specificities. For example, fluctuations in numbers of ribosomes can affect the expression of virtually all genes and persist for hours or longer in microbes. Fluctuations in the number of transcripts of a particular gene may last for only seconds or minutes and may directly influence only the expression of that gene.

The phenomena of memory and chance can be at odds. Memory implies that decision or behaviors are contingent on the past. Chance implies that decisions are randomized, implying a lesser dependence on the past or present. If chance plays a role in the memory mechanism itself, then it tends to degrade the memory, resulting in a neutral or random state that is independent of the past. Yet both memory and chance are important cellular strategies. Intriguingly, they are often found in the same systems. Many of the model systems used as examples of cellular memory, such as gene regulatory networks involved in catabolism, determination of cell type, and chemotaxis, have also been used in studies of cellular stochasticity. This chapter discusses the dual and often related phenomena of cellular memory and chance. First, stochastic effects in gene expression are reviewed, and some of the basic mathematical and computational tools used for modeling and analyzing stochastic biochemical networks are explored. Then, various model cellular systems that have been studied from these perspectives are described.

3.2 Chance, or Stochasticity

Life at the cellular level is inevitably stochastic [7,8]. Biochemistry drives all of life's processes, and reactants come together through diffusion with their motion being driven by rapid and frequent collisions with other molecules. Once together, these same collisions alter the reactants' internal energies and thus their propensities to react. Both effects cause individual reaction events to occur randomly and the overall process to be stochastic. Such stochasticity need not always be significant. Intuitively, we expect the random timings of individual reactions to matter only when typical numbers of molecules are low, because then each individual event, which at most changes the numbers of molecules by one or two, generates a substantial change. Low numbers are not uncommon intracellularly: gene copy numbers are usually one or two, and transcription factors, at least for microbes, frequently number in the tens [9,10].

3.2.1 Intrinsic and Extrinsic Stochasticity

Two types of stochasticity affect any biochemical system [11,12]. Intuitively, intrinsic stochasticity is stochasticity that arises inherently in the dynamics of the system from the occurrence of biochemical reactions. Extrinsic stochasticity is additional variation that is generated because the system interacts with other stochastic systems in the cell or in the extracellular environment (Figure 3.1a).

In principle, intrinsic and extrinsic stochasticity can be measured by creating a copy of the system of interest in the same cellular environment as the original network (Figure 3.1a) [12]. Intrinsic and extrinsic variables can be defined with fluctuations in these variables together generating intrinsic and extrinsic stochasticity [11]. The intrinsic

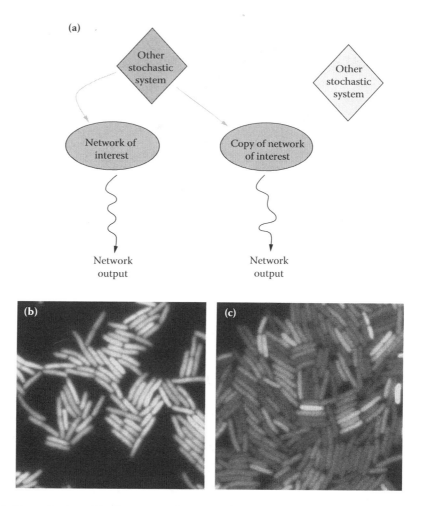

FIGURE 3.1 (See color insert.)   Intrinsic and extrinsic fluctuations. (a) Extrinsic fluctuations are generated by interactions of the system of interest with other stochastic systems. They affect the system and a copy of the system identically. Intrinsic fluctuations are inherent to each copy of the system and cause the behavior of each copy to be different. (b) Strains of *Escherichia coli* expressing two distinguishable alleles of Green Fluorescent Protein controlled by identical promoters. Intrinsic noise is given by the variation in color (differential expression of each allele) over the population and extrinsic noise by the correlation between colors. (c) The same strains under conditions of low expression. Intrinsic noise increases. (From Elowitz et al. [12]. With permission.)

variables will typically specify the copy numbers of the molecular components of the system. For the expression of a particular gene, the level of occupancy of the promoter by transcription factors, the numbers of mRNA molecules, and the number of proteins are all intrinsic variables. Imagining a second copy of the system—an identical gene and promoter elsewhere in the genome—then the instantaneous values of the intrinsic variables of this copy of the system will frequently differ from those of the original system. At any point, for example, the number of mRNAs transcribed from the first copy of the gene will usually be different from the number of mRNAs transcribed from the second copy. Extrinsic variables, however, describe processes that equally affect each copy. Their values are therefore the same for each copy. For example, the number of cytosolic

RNA polymerases is an extrinsic variable because the rate of gene expression from both copies of the gene will increase if the number of cytosolic RNA polymerases increases and decrease if the number of cytosolic RNA polymerases decreases.

Stochasticity is quantified by measuring an intrinsic variable for both copies of the system. For our example of gene expression, the number of proteins is typically measured by using fluorescent proteins as markers [12–15] (other techniques are described in Chapter 10). Imaging a population of cells then allows estimation of the distribution of protein levels, typically at steady state. Fluctuations of the intrinsic variable will have both intrinsic and extrinsic sources in vivo. The term *noise* means an empirical measure of stochasticity—for example, the coefficient of variation (the standard deviation divided by the mean) of a stochastic variable.

Intrinsic noise is defined as a measure of the difference between the value of an intrinsic variable for one copy of the system and its counterpart in the second copy. For gene expression, the intrinsic noise is typically the mean absolute difference (suitably normalized) between the number of proteins expressed from one copy of the gene, which we will denote I_1, and the number of proteins expressed from the other copy [12,16], denoted I_2:

$$\eta_{\text{int}}^2 = \frac{\langle (I_1 - I_2)^2 \rangle}{2\langle I \rangle^2} \tag{3.1}$$

where we use angled brackets to denote an expectation over all stochastic variables, and $\langle I_1 \rangle = \langle I_2 \rangle = \langle I \rangle$. Such a definition supports the intuition that intrinsic fluctuations cause variation in one copy of the system to be uncorrelated with variation in the other copy. Extrinsic noise is defined as the correlation coefficient of the intrinsic variable between the two copies:

$$\eta_{\text{ext}}^2 = \frac{\langle I_1 I_2 \rangle - \langle I \rangle^2}{\langle I \rangle^2}. \tag{3.2}$$

Extrinsic fluctuations equally affect both copies of the system and consequently cause correlations between variation in one copy and variation in the other. Intuitively, the intrinsic and extrinsic noise should be related to the coefficient of variation of the intrinsic variable of the original system of interest. This so-called total noise is given by the square root of the sum of the squares of the intrinsic and the extrinsic noise [11]:

$$\eta_{\text{tot}}^2 = \frac{\langle I^2 \rangle - \langle I \rangle^2}{\langle I \rangle^2} = \eta_{\text{int}}^2 + \eta_{\text{ext}}^2 \tag{3.3}$$

which can be verified from adding Equations (3.1) and (3.2).

Such two-color measurements of stochasticity have been applied to bacteria and yeast where gene expression has been characterized by using two copies of a promoter integrated in the genome with each copy driving a distinguishable allele of Green Fluorescent Protein [12,15] (Figures 3.1b and 3.1c). Both intrinsic and extrinsic noise can be substantial giving, for example, a total noise of around 0.4: the standard deviation of protein numbers is 40% of the mean. Extrinsic noise is usually higher than intrinsic noise. There are some experimental caveats: both copies of the system should be placed equally in the

Chapter 3

genome so that the probabilities of transcription and replication are equal. This equality is perhaps best met by placing the two genes adjacent to each other [12]. Although conceptually there are no difficulties, practical problems arise with feedback. If the protein synthesized in one system can influence its own expression, the same protein will also influence expression in a second copy of the system. The two copies of the system have then lost the (conditional) independence they require to be two simultaneous measurements of the same stochastic process.

3.2.2   Understanding Stochasticity

Although measurements of fluorescence in populations of cells allow us to quantify the magnitude of noise, time-series measurements, where protein or mRNA levels are followed in single cells over time, best determine the mechanisms that generate noise. With the adoption of microfluidics by biologists [17,18], such experiments are becoming more common, but, at least for bacteria, simpler approaches have also been insightful.

3.2.2.1   Translational Bursts

Following over time the expression of a repressed gene in *E. coli*, where occasionally and stochastically repression was spontaneously lifted, revealed that translation of mRNA can occur in bursts [19]. Using a *lac* promoter and a fusion of Yellow Fluorescent Protein to a membrane protein so that slow diffusion in the membrane allowed single-molecule sensitivity, Yu et al. [19] showed that an individual mRNA is translated in a series of exponentially distributed bursts (Figure 3.2a). Such bursting was expected because of the competition between ribosomes and the RNA degradation machinery for the 5' end of the mRNA [20]. If a ribosome binds first to the mRNA, then translation occurs, and the mRNA is protected from degradation because the ribosome physically obstructs binding of the degradation machinery; if the degradation machinery binds first, then further ribosomes can no longer bind and the mRNA is degraded from its 5' end. Let v_1 be the probability of translation per unit time of a particular mRNA and d_0 be its probability of degradation (Figure 3.2b). Then, the probability that a ribosome binds rather than the degradation machinery is $\dfrac{v_1}{v_1 + d_0}$ or $\dfrac{b}{1+b}$, where $b = v_1/d_0$ is the typical number of proteins translated from one mRNA during the mRNA's lifetime—the so-called burst size [21]. The probability of r proteins being translated from a particular mRNA, P_r, requires that ribosomes win the competition with the mRNA degradation machinery r times and lose once [20]:

$$P_r = \left(\frac{b}{1+b} \right)^r \left(1 - \frac{b}{1+b} \right). \tag{3.4}$$

The number of proteins translated from one mRNA molecule thus has a geometric distribution, which if $b < 1$ can be approximated by an exponential distribution with parameter b. Such exponential distributions of bursts of translation (Figure 3.2c) were measured by Yu et al. [19].

If the mRNA lifetime is substantially shorter than the protein lifetime, then all the proteins translated from any particular mRNA appear at once if viewed on the timescale associated with typical changes in the number of proteins: translational bursting

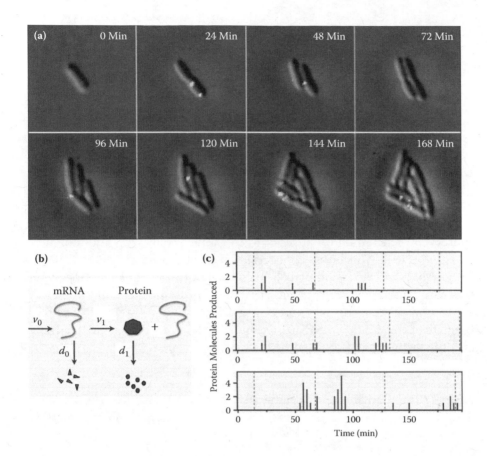

FIGURE 3.2 (See color insert.) Stochasticity is generated by bursts of translation. (a) Time series showing occasional expression from a repressed gene in *E. coli*, which results in the synthesis of a single mRNA but several proteins. Proteins are membrane proteins tagged with Yellow Fluorescent Protein (yellow dots). (b) The two-stage model of constitutive gene expression. Transcription and translation are included. (c) Quantization of the data from (a) showing exponentially distributed bursts of translated protein. (From Yu et al. [19]. With permission.)

occurs [22]. This separation of timescales is common, being present, for example, in around 50% of genes in budding yeast [22]. It allows for an approximate formula for the distribution of proteins [23], which if transcription occurs with a constant probability per unit time is a negative binomial distribution (Figure 3.2b): Where a is the number of ribosomes synthesized per protein lifetime (v_0/d_1).

$$P_n = \frac{\Gamma(a+n)}{\Gamma(n+1)\Gamma(a)}\left(\frac{b}{1+b}\right)^n \left(1-\frac{b}{1+b}\right)^a. \tag{3.5}$$

This distribution is generated if translation from each mRNA is independent and if translation occurs in geometrically distributed translational bursts [22]. Such bursting allows the master equation for the two-dimensional system of mRNA and protein to be reduced to a one-dimensional equation just describing proteins. This equation is, however, more complex than a typical master equation for a one-step process because it contains a term that is effectively a convolution between the geometric distribution of bursts and the distribution for the number of proteins.

Chapter 3

3.2.2.2 Transcriptional Bursts

Transcription can also occur in bursts. By using the MS2 RNA-binding protein tagged with a fluorescent protein and by inserting tens of binding sites for MS2 in the mRNA of a gene of interest, Golding et al. [24] were able to follow individual mRNAs in *E. coli* over time. They observed that these tagged mRNAs are not synthesized with a constant probability per unit time even for a constitutive promoter but that synthesis is consistent with a random telegraph process where the promoter stochastically transitions between an on state, capable of transcription, and an off state, incapable of transcription (Figure 3.3a). These on and off states of the promoter could reflect binding of RNA polymerase or a transcription factor, changes in the structure of chromatin in higher organisms, or even transcriptional traffic jams caused by multiple RNA polymerases transcribing simultaneously [24,25]. Whatever its cause, transcriptional bursting has been observed widely in budding yeast [14,15], slime molds [26], and human cells [27] as well as in bacteria. This broad applicability has led to what might be called the standard model of gene expression, which includes stochastic transitions of the promoter and translation bursting if the mRNA lifetime is substantially smaller than the protein lifetime (Figure 3.3b). An exact expression for the distribution of mRNAs [27] and an approximate expression for the distribution of proteins are known [22], although both include only intrinsic but not extrinsic fluctuations.

3.2.2.3 Simulation Methods

We may gain intuition about the behavior of more complex networks using stochastic simulation. For intrinsic fluctuations or for networks where all the potential chemical

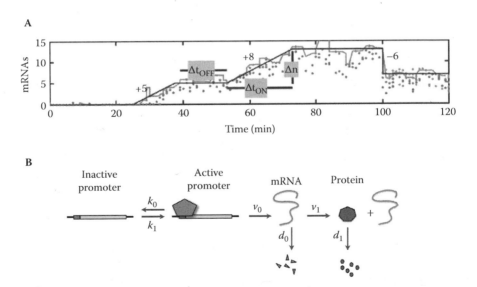

FIGURE 3.3 Stochasticity arises from transcriptional bursting. (a) Time-lapse data of the number of mRNAs from a phage lambda promoter in single cells of *E. coli*. The dots are the number of mRNAs in one cell; the jagged line is a smoothed version of the dots; the dots appearing after the 100 min are the number of mRNAs in either the original cell or the sister cell after division; and the black line is a piecewise linear fit. (b) The three-stage model of gene expression with activation of the promoter, transcription, and translation and the potential for both transcriptional and translational bursting. The rate k_0 is inversely related to the mean Δt_{ON} in (a) and the rate k_1 is inversely related to the mean Δt_{OFF}. (From Golding et al. [24]. With permission.)

reactions are known, Gillespie's [28] Stochastic Simulation Algorithm (SSA) is most commonly used (see also Chapters 10 and 12). At its simplest, this algorithm involves rolling the equivalent of two dice for each reaction that occurs. One die determines which reaction out of the potential reactions happens, with the propensity for each reaction being determined by the probability per unit time of occurrence of a single reaction times the number of ways that reaction can occur given the numbers of molecules of reactants present. The second die determines the time when this reaction occurs by generating a sample from an exponential distribution. Usually extrinsic fluctuations enter into simulations because kinetic rate constants are often implicitly functions of the concentration of a particular type of protein. For example, degradation is commonly described as a first-order process with each molecule having a constant probability of degradation per time. Yet the rate of degradation is actually a function of the concentration of proteasomes, and if the number of proteasomes fluctuates the rate of degradation fluctuates. Such fluctuations are extrinsic and can also occur, for example, in translation if the number of ribosomes or tRNAs fluctuates or in transcription if the number of RNA polymerases or nucleotides fluctuates. Extrinsic fluctuations can be simulated by including, for example, a model for the availability of proteasomes, but such models can be complicated containing unknown parameters and, in principle, other extrinsic fluctuations. It therefore can be preferable to have an effective model of extrinsic fluctuations by letting the rate of a reaction, or potentially many rates, fluctuate directly in the Gillespie algorithm. These fluctuations can be simulated with a minor modification of the algorithm to include piecewise linear changes in reaction rates and then by simulating the extrinsic fluctuations of a particular reaction rate ahead of time and feeding these fluctuations, or more correctly a piecewise linear approximation of the fluctuations, into the Gillespie algorithm as it runs [16]. Such an approach also has the advantage of potentially generating extrinsic fluctuations with any desired properties. Where measured, extrinsic fluctuations have a lifetime comparable to the cell-cycle time [29]—they are *colored*—and it is important to include this long lifetime because such fluctuations are not typically averaged away by the system and can generate counterintuitive phenomena [16].

3.3 Bistability and Stochasticity

Given that life is fundamentally biochemistry, how do biochemical systems generate memory? One mechanism is positive feedback that is sufficiently strong to create bistability [31]. At long times, a bistable system tends toward one of two stable steady states. If the system starts with an initial condition in one particular collection of initial conditions, it will tend to one steady state; if it starts with any other initial condition, it will tend to the other steady state. The space of initial conditions—the space of all possible concentrations of the molecular components of the system—has two so-called domains of attraction, one for each stable steady state (Figure 3.4a). Sufficiently strong positive feedback can create such bistable behavior. If the output of a system can feedback on the system and encourage more output, then intuitively we can see that an output sufficiently low only weakly activates such positive feedback, and the system remains at a low level of output. If, however, the output becomes high, then the positive feedback is strongly activated and the output becomes higher still generating more feedback and more output. This runaway

Chapter 3

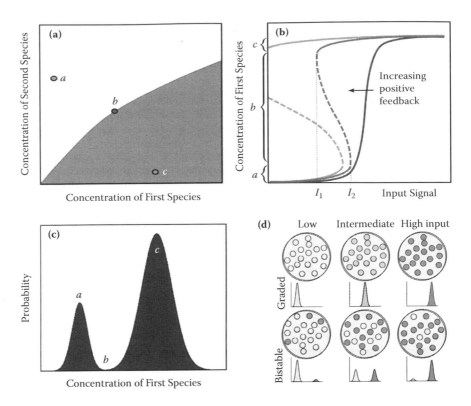

FIGURE 3.4 Bistability. (a) A steady state has its own domain of attraction. The unshaded region is the domain of attraction of stable steady state a; the shaded region is the domain of attraction of stable steady state c. The steady state b is unstable. (b) Bifurcation diagram as a function of the level of input signal. With positive feedback an otherwise monostable system (dark line) can become bistable (medium shaded line). Two saddle-node bifurcations occur when the input is at I_1 and I_2. The system is bistable for inputs between I_1 and I_2 and will tend to steady state at either a or c depending on initial concentrations. It is monostable otherwise. For sufficiently strong positive feedback, the system becomes irreversible (light line); it becomes locked at steady state c once it jumps from a to c. (c) Stochasticity undermines bistability. Although deterministically the system is either at steady state a or c, stochastic effects generate nonzero probabilities to be at either steady state in the bistable regime. (d) Single-cell measurements can reveal differences between graded and bistable responses that are lost in population-averaged measurements [30].

feedback causes the system to tend to a high level of output. No intermediate output level is possible: either output is high enough to start runaway positive feedback or it is not.

3.3.1 Bistable Systems Can Have Memory

Biochemical systems that are bistable have the potential for history-dependent behavior and thus for memory. Imagine a typical dose-response curve: as the input to the system increases, so does the output. A linear change in input often gives a nonlinear change in output and thus a sigmoidal input–output relationship. For bistable systems, the output increases with respect to the input for low levels of input (Figure 3.4b). Such levels are not strong enough to generate output that can initiate runaway positive feedback, and all initial conditions tend to one *inactivated* steady state. As the input increases, however, the system can undergo what is called a saddle-node bifurcation that creates a second steady state (at I_1 in Figure 3.4b): the input has become sufficiently high to generate positive feedback strong enough to create bistability. This second, *activated* steady state is not apparent at this level of input though, and only on further increases is there a dramatic, discontinuous change in output at a critical threshold of input when the system jumps from one steady state to the other (at I_2 in Figure 3.4b). This jump occurs at another saddle-node bifurcation where the inactivated steady state is lost because the positive feedback is so strong that all systems have sufficient output to generate runaway feedback. Regardless of the initial condition, there is just one steady state, and the system is no longer bistable. At this new steady state, the output again increases monotonically with the input. If the input is then lowered, the strength of the feedback falls, a saddle-node bifurcation occurs (at I_2), and the low, inactivated steady state is again created at the same threshold of input where it was lost. This bifurcation does not typically perturb the dynamics of the system until the input is lowered sufficiently that a second saddle-node bifurcation removes the activated state and the bistability of the system (at I_1). At this threshold of input, the steady state behavior of the system changes discontinuously to the inactivated steady state, and there its output decreases continuously with decreasing input. The system has memory, and the past affects whether the system jumps. For example, if the system has already experienced levels of input higher than I_2, there will be a jump at I_1 as the input decreases; if the system has not previously experienced input higher than I_2, there will be no jump at I_1. Such history-dependent behavior is called hysteresis. It allows information to be stored despite potentially rapid turnover of the system's molecules [32].

Permanent memory can be created by sufficiently strong feedback. If the feedback is so strong that the system is bistable even with no input then once the input is enough to drive the system from the inactivated to the activated state then no transition back to the inactivated state is possible. Although biochemical bistability has been proposed to describe neural memories [33,34], permanent memory is perhaps more appropriate for differentiation where cells should remain differentiated once committed to differentiation [35].

3.3.2 Stochasticity Undermines Memory

Biochemical systems are stochastic, and stochasticity undermines bistability by generating fluctuations that drive the system from one stable steady state to the other. Typically, we can think of two sets of timescales in the system: (1) fast timescales that control the return of the system to steady state in response to small, frequent perturbations and determine the lifetime of typical fluctuations around a steady state; and (2) slow timescales generated by rare fluctuations between steady states that determine the time spent by the system at each steady state. Stochasticity in a bistable system can cause the distribution for the numbers of molecules to become bimodal (Figure 3.4c). Frequent (fast) fluctuations determine the width of this distribution close to each steady state; rare fluctuations determine the relative height of each peak. If both fluctuations are common, they can drive a bimodal distribution to be unimodal. With measurements of populations of cells, however, averaging can obscure bistable behavior [30]: the average response of a population of bistable cells, particularly if stochastic effects cause each cell to respond at a different threshold of input concentration, and the average response of cells with a monostable, graded behavior can be indistinguishable (Figure 3.4d).

Nevertheless, a bimodal distribution of, for example, the numbers of protein does not mean that the system's underlying dynamics are necessarily bistable. Even expression from a single gene can generate a bimodal distribution if the promoter transitions slowly between an active and an inactive state [22,23,36–38]. This bimodality is lost if fluctuations at the promoter are sufficiently fast because they are then averaged by other processes of gene expression [39].

3.3.3 Estimating the Time Spent at Each Steady State

A challenge for theory is to calculate the fraction of time spent near each steady state in a bistable system. Such a calculation requires a description of the rare fluctuations that drive the system from one steady state to the other. It therefore cannot be tackled with Van Kampen's linear noise approximation [40], the workhorse of stochastic systems biology, because this approximation requires that fluctuations generate only small changes in numbers of molecules compared with the deterministic behavior and therefore can be applied only near one of the system's steady states. Formally, the master equation for a vector of molecules $\mathbf{n}$ obeys

$$\frac{\partial}{\partial t} P(\mathbf{n},t) = A P(\mathbf{n},t) \tag{3.6}$$

where the operator A is determined by the chemical reactions (see also Chapter 10). This equation has a solution in terms of an expansion of the eigenfunctions of A:

$$P(\mathbf{n},t) = a_0 \psi_0(\mathbf{n}) + \sum_{i=1} a_i e^{-\lambda_i t} \psi_i(\mathbf{n}) \tag{3.7}$$

where $\psi_i(t)$ is the eigenfunction of A with eigenvalue λ_i. The steady state distribution $P^{(s)}(\mathbf{n})$ is determined by the zero-th eigenfunction, $P^{(s)}(\mathbf{n}) = a_0\psi_0(\mathbf{n})$, which has a zero eigenvalue. If the system is bistable, this distribution will typically be bimodal with maxima at $\mathbf{n} = \mathbf{a}$ and $\mathbf{n} = \mathbf{c}$, say, and a minimum at $\mathbf{n} = \mathbf{b}$, corresponding deterministically to stable steady states at a and c and an unstable steady state at b (Figure 3.4c). The next smallest eigenvalue λ_1 determines the slow timescales. As the system approaches steady state, peaks at a and c will develop, and we expect metastable behavior with molecules mostly being near the steady state at either a or c but undergoing occasional transitions between a and c. The probability distribution at this long-lived metastable stage will be approximately

$$P(\mathbf{n},t) \simeq P^{(s)}(\mathbf{n}) + a_1 e^{-\lambda_1 t} \psi_1(\mathbf{n}). \tag{3.8}$$

Defining Π_a as the probability of the system being near the steady state at $\mathbf{a}$ and Π_c as the probability of the system being near the steady state at $\mathbf{c}$:

$$\Pi_a = \int_0^b d\mathbf{n} P(\mathbf{n}, t) \quad ; \quad \Pi_c = \int_b^\infty d\mathbf{n} P(\mathbf{n}, t) \tag{3.9}$$

then we expect

$$\frac{\partial \Pi_a}{\partial t} = -\frac{\partial \Pi_c}{\partial t} = \kappa_{c \to a} \Pi_c - \kappa_{a \to c} \Pi_a \tag{3.10}$$

because $\Pi_a + \Pi_c = 1$, and if $\kappa_{a \to c}$ is the rate of fluctuating from the neighborhood of **a** to the neighborhood of **c** and $\kappa_{c \to a}$ is the rate of fluctuating back. At steady state, we have $\Pi_a^{(s)} \kappa_{a \to c} = \Pi_c^{(s)} \kappa_{c \to a}$. If we integrate Equation (3.8) with respect to **n** following Equation (3.9), insert the resulting expressions into Equation (3.10), and simplify with the steady state equation, we find that $\kappa_{a \to c} + \kappa_{c \to a} = \lambda_1$ [41]: the rate of fluctuating from one steady state to another is, indeed, determined by the slow timescale, $1/\lambda_1$.

For a one-dimensional system where chemical reactions, which may occur with nonlinear rates, increase or decrease the number of molecules by only one molecule (bursting, e.g., cannot occur), an analytical expression for the average time to fluctuate from one steady state to another can be calculated [40]. The exact initial and final numbers of molecules are not important because the time to jump between steady states is determined by the slow timescales and the additional time, determined by the fast timescales, to move the molecules between different initial conditions or between different final conditions is usually negligible. If n molecules are present, the probability of creating a molecule is $\Omega w_+(n)$ for a system of volume Ω and the probability of degrading a molecule is $\Omega w_-(n)$, then the steady state probability of the concentration $x = n/\Omega$ is approximately [42]

$$P^{(s)}(x) \sim e^{-\Omega \phi(x)} \tag{3.11}$$

for large Ω and where $\phi(x)$ is an effective potential:

$$\phi(x) = -\int_0^x dx' \log \frac{w_+(x')}{w_-(x')}. \tag{3.12}$$

To calculate the average time to fluctuate between steady states, typically the average time, $t_{a \to b}$, to fluctuate from $x = a$ to b, where b is the deterministically unstable steady state, is calculated because this time is much longer than the time to fluctuate from b to c, which is determined by the fast timescales. Then $\kappa_{a \to c} = \frac{1}{2} t_{a \to b}^{-1}$ because at $x = b$ a fluctuation to either a or c is equally likely. For large Ω, the rates are [42]

$$\kappa_{a \to c} \simeq \frac{1}{2\pi} w_+(a) \sqrt{\phi''(a) |\phi''(b)|} \, e^{-\Omega[\phi(b) - \phi(a)]} \tag{3.13}$$

Chapter 3

and

$$\kappa_{c \to a} \simeq \frac{1}{2\pi} w_+(c)\sqrt{\phi''(c)|\phi''(b)|}\, e^{-\Omega[\phi(b)-\phi(c)]} \tag{3.14}$$

and depend exponentially on the difference in the effective potential at the unstable and a stable steady state. They are exponentially small in the size of the system, Ω, and therefore vanish as expected in the deterministic limit. In this deterministic limit, the dynamics of x obey

$$\dot{x} = w_+(x) - w_-(x) \tag{3.15}$$

and the potential of motion, $-\int dx'[w_+(x') - w_-(x')]$, is not the same as the effective potential of Equation (3.12) [42]. The rates are exponentially sensitive to changes in the parameters of the system because the parameters determine the effective potential. Such exponential sensitivity implies a large variation in fluctuation-driven switching rates between cells, perhaps suggesting that other, more robust switching mechanisms also exist [43].

To estimate transition rates between steady states for systems that cannot be reduced to one variable, numerical methods are most often used, although some analytical progress is possible [44]. Perhaps the most efficient numerical algorithm developed to estimate transition rates is forward flux sampling [45]. The two steady states, a and c, are defined through a parameter λ, which may be a function of all the chemical species in the system. The system is at steady state a if $\lambda \leq \lambda_a$ for some λ_a and at steady state c if $\lambda \geq \lambda_c$ for some λ_c. A series of interfaces is then defined between λ_a and λ_c. If a trajectory starts from steady state a and reaches one of the interfaces, then we can consider the probability that this trajectory will reach the next interface closest to λ_c on its first crossing of the current interface. The transition rate $\kappa_{a \to c}$ is approximated by a suitably normalized product of these probabilities for all interfaces between λ_a and λ_c [45]. The probabilities can be estimated sequentially using the Gillespie algorithm. Initially simulations are started from a to estimate the first probability and generate a collection of initial conditions for the estimation of the second probability. These initial conditions are given by the state of the system each time the simulation trajectory comes from a and crosses the first interface. Further simulation from these initial conditions then allows estimation of the second probability and generation of the initial conditions for estimation of the next probability, and so on. The efficiency of the method is determined by the choice of the parameter λ and the accuracy by the number of interfaces between λ_a and λ_c and the number of initial conditions used to estimate each probability. In general, forward flux sampling can be highly efficient [45].

3.4 Examples

3.4.1 The *lac* operon in *E. coli*

The *lac* operon of *E. coli* was first studied by Jacob and Monod (see [31] for an early summary of some of their and others' work) and indeed was the first system for which the basic mechanism of transcriptional regulation was understood (see Wilson et al. [46]

and references therein). The operon includes genes important in the intake and metabolism of lactose: *lacZ* and *lacY* (Figure 3.5a). In particular, LacZ is an enzyme that breaks lactose down into glucose and galactose, and LacY encodes a permease that transports lactose into the cell. Expression of the *lac* operon is repressed by the transcription factor lacI, appropriately called the lac repressor. LacI binds to the promoter of the *lac* operon, where it blocks transcription. If intracellular allolactose, a metabolite of lactose, binds to lacI, however, then LacI loses its ability to bind to the promoter, and the *lac* operon becomes de-repressed. These qualitative influences are summarized in Figure 3.5a, with some abridgement in intracellular lactose metabolism [46].

The system has the potential for strong positive feedback. In a low lactose environment, intracellular lactose, and thus allolactose, is negligible. LacI remains unbound by allolactose and acts to repress expression of the *lac* operon. Conversely, in a high lactose environment, intracellular lactose, and thus allolactose, becomes substantial, and binds up all the free molecules of LacI. Without LacI binding the promoter, the *lac* operon

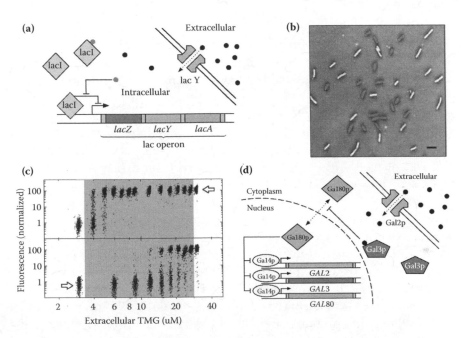

FIGURE 3.5 Sugar utilization networks in *E. coli* and *S. cerevisiae*. (a) The *lac* operon of *E. coli*. The *lac* operon comprises three genes, one of which (*lacY*) encodes a lactose permease. The permease allows lactose (black circles) to enter the cell. Intracellular allolactose (gray circles), a metabolite of lactose, binds the transcription factor LacI, preventing it from repressing the operon [46]. (b) Bistable responses observed in uninduced *E. coli* cells upon exposure to 18 μM of TMG a non-metabolizable analog of lactose. Levels of induction in each cell is monitored by a Green Fluorescent Protein reporter controlled by a chromosomally integrated *lac* promoter. (From Ozbudak et al. [47]. With permission.) (c) A hysteresis curve is an unequivocal proof of bistability. Scatterplots of the GFP reporter and another estimating catabolite repression are shown. In the top panel, cells are initially fully induced; in the lower panels, cells are initially uninduced. Bistable behavior occurs in the gray regime, and there are two separate thresholds of TMG concentration for the on–off and off–on threshold transitions. (From Ozbudak et al. [47]. With permission.) (d) The galactose utilization network in *S. cerevisiae*. Galactose enters the cell via Gal2p and binds the sensor Gal3p, which then inhibits transcriptional repression of the GAL genes by the repressor Gal80p. This inhibition of repression increases synthesis of both Gal2p and Gal3p giving positive feedback and of Gal80p giving negative feedback. Gal4p is here a constitutively expressed activator.

becomes expressed and leads to the synthesis of more LacY permeases. These permeases increase the cell's potential to import lactose and therefore the potential for greater expression of the *lac* operon: the syste m has positive feedback through the double negative feedback of lactose on LacI and LacI on expression.

Nevertheless, bistable behavior has been observed only with thio-methylgalactoside (TMG), a nonmetabolizable analog of lactose [47,48]. TMG also binds LacI and derepresses the *lac* operon. When the amount of TMG in the extracellular environment is at an intermediate level and if there is only a small number of LacY permeases in the cell membrane, perhaps because the *lac* operon has not been expressed for some time, then only a small amount of TMG will enter the cell—not enough to de-repress the *lac* operon. On the other hand, if the membrane has many LacY permeases, perhaps because of recent *lac* operon expression, then enough TMG can enter the cell to enhance expression of the operon. The exact shape of this bistable behavior has been mapped in some detail, along with the influence of other factors (Figures 3.5c and 3.5d) [47,49,50].

How does stochasticity influence the function of the *lac* regulatory machinery? The early work of Novick and Weiner [51] was remarkably prescient. They were the first to recognize the bimodal expression of the *lac* operon at intermediate extracellular TMG concentrations as well as the hysteretic nature of its regulation. Moreover, by detailed study of the kinetics of the induction process, they concluded that a single, random chemical event leads to a runaway positive feedback that fully induces the operon—a hypothesis that is essentially correct. They hypothesized that this single event was a single lactose permease entering the cell membrane, whereupon the TMG floodgate was opened. This explanation is incorrect, though it took decades of biochemical research as well as advances in biotechnology to establish the true mechanism [52]. A typical uninduced *E. coli* harbors a handful of lactose permeases, perhaps as many as 10, in its membrane. These are the result of small bursts of expression from the operon. Induction at an intermediate level of extracellular TMG occurs when a random much larger burst of expression produces on the order of hundreds of permeases, which then sustain expression. The mechanism supporting these large and small bursts is intimately connected with the DNA binding behavior of LacI. There are binding sites on either side of the operon, and, through DNA looping a single LacI molecule is able to bind on both sides. Short-lived dissociations from one side or the other allow small bursts of expression. Complete dissociation from the DNA, a much rarer event, leads to a much longer period of de-repression and a large burst of expression. Thus, as Novick and Weiner hypothesized, a single, rare, random chemical—complete dissociation of LacI from the DNA—flips the switch from uninduced to induced [51].

There are, of course, other stochastic aspects to the system. LacI levels fluctuate, as do permease levels. These fluctuations are much smaller than the difference between off and on states [47]. Nevertheless, Robert et al. [53] showed that various measurable properties of individual cells can be used to predict whether the cells will become induced at intermediate levels of inducer. Moreover, they show that these proclivities can be inherited epigenetically. Thus, the stochastic flipping of the *lac* operon switch is driven by both intrinsic and extrinsic fluctuations, with the latter just beginning to be identified.

3.4.2   The GAL System in Budding Yeast

The theme of positive feedback via permeases resulting in bistability and hysteresis also occurs in the arabinose utilization system in *E. coli* [54], but a better-known example is the galactose utilization network of *S. cerevisiae*, which we treat here briefly. In this system, the master activator is the transcription factor Gal4p, which turns on a number of regulatory and metabolism-related genes, including the galactose permease *GAL2* and the galactose sensor *GAL3* [4]. In the absence of intracellular galactose, the repressor Gal80p binds to Gal4p and inhibits its activation of expression. When galactose binds to Gal3p, however, Gal3p becomes activated and can bind to Gal80p and so remove its ability to repress Gal4p (Figure 3.5b). The system has two positive feedbacks through expression of both the permease, *GAL2*, and the sensor, *GAL3*, once Gal4p is no longer repressed by Gal80p, and one negative feedback through expression of *GAL80*.

As with the *lac* operon, yeast cells in an intermediate extracellular galactose concentration may be either stably on or off. However, that stability is relative to the stochasticity of the biochemistry, so that at sufficiently long timescales any cell may turn on or turn off. Indeed, Acar et al. [55] and Song et al. [56] observed bimodal and history-dependent distributions for the galactose response of *S. cerevisiae* (analogous to those of Figure 3.5d). Acar et al., by sorting a bimodal population to select only the off or on cells and subsequently assaying the population distribution over time, furthermore estimated the duration of memory in the system could be at least on the order of tens if not hundreds of hours, depending on experimental parameters. Perhaps surprisingly, the *GAL2* gene is not necessary for bistability. Upon its deletion, bistability is still observed, albeit with an on state with significantly less expression than in the wild-type. Galactose is still able to enter the cell via other means, and the positive feedback loop via *GAL3* mediates bistability. Removing the feedbacks to *GAL3* or *GAL80*, by putting the genes under the control of different promoters, eliminates bistability. When *GAL3* is not activated by Gal4p, but instead by an alternate activator, cell populations show a unimodal distribution of induction, and one that loses any history dependence. Interestingly, when the gene encoding the repressor Gal80p is not activated by Gal4p, cell populations can show a bimodal induction distribution—some cells off and some on—but one with little history dependence. That is, individual cells switch comparatively rapidly between the two states, though there is still some tendency to be in one state or the other and not in between. Acar et al. dubbed this phenomenon destabilized memory.

3.4.3   Bistability in Higher Organisms

Perhaps the most understood example of bistable behavior in higher organisms is oocyte maturation in the frog *Xenopus laevis*. After S-phase, immature oocytes carry out several early steps of meiotic prophase but proceed only with the first meiotic division in response to the hormone progesterone. The commitment to meiosis is all-or-none and irreversible. It is generated biochemically by two mechanisms: a mitogen-activated protein (MAP) kinase cascade of three kinases and a positive feedback loop from the last kinase, MAP kinase, to the first kinase, MAP kinase kinase kinase (Figure 3.6a). The MAP kinase cascade converts the input signal of progesterone into an ultrasensitive or switch-like response with each level of the cascade increasing the degree of ultrasensitivity [59]. This ultrasensitivity is necessary, but not sufficient, to generate bistable behavior

Chapter 3

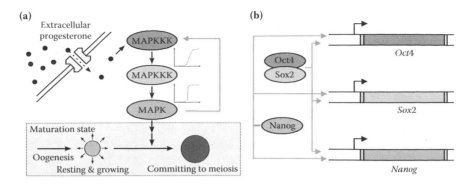

FIGURE 3.6 Positive feedback in higher organisms. (a) Progesterone-induced oocyte maturation in *X. laevis*. Through both the activation of a MAP kinase cascade that converts a graded signal into a switch-like response and positive feedback from MAPK to MAPKKK, progesterone induces all-or-none, irreversible maturation of oocytes. Inset: the degree of ultrasensitivity in the response increases down the cascade [57]. (b) The core genetic network driving pluripotency in human embryonic stem cell. Arrows represent positive transcriptional regulation [58].

and the additional positive feedback is required [30]. When this feedback is blocked, the cells exhibit intermediate responses that are not irreversible (Figure 3.4b) [35].

Such bistability with strong positive feedback might be expected in differentiated cells in a multicellular organism. Differentiation represents the locking-in of an extremely stable state—one that lasts until the death of the cell, potentially many years. How then do stem cells fit into this framework?

Accumulating evidence suggests that stem cells are poised for differentiation and must be actively maintained in their stem state [60]. Stem cells in culture must be treated with various chemical factors to prevent differentiation (e.g., [61,62]). In this sense, the stem state is not naturally stable. In vivo, a variety of signals affect stem cell function. For example, many stem cell systems become less active and thus less able to repair or maintain tissue as an animal ages [63–65]. However, this loss of activity does not appear to be from changes in the stem cells themselves. Subjecting the cells to signals present in younger animals reenlivens the aged cells [65,66]. Given that the right signals are provided, are stem cells at a kind of conditional steady state? Stem cell identity is usually defined by one or a few master regulator genes. For example, human embryonic stem cells require the coordinated expression of three genes—Sox2, Oct4, and Nanog—which are connected in positive feedback loops (Figure 3.6b) [58]. Several works have found bimodal expression distributions for these genes in populations of cultured cells, suggesting some form of stochastic bistability [67,68], and various models have been proposed [69-71]. Although thinking of the pluripotency properties of stem cells as driven by stochastic, multistable genetic networks is perhaps the best approach [72], there are many challenges to link these ideas to mechanisms and quantitative predictions.

3.5 Conclusions

Bistability is one mechanism that cells can use to retain memory. If the positive feedback driving the bistability is sufficiently strong, then once a biochemical network switches it can never switch back (Figure 3.4b). Such permanent memories were once thought to be maintained as long as the cell has sufficient resources to generate the positive feedback. We now recognize that stochasticity is pervasive both intra- and extracellularly and that fluctuations can undermine memory with rare events causing networks that are permanently on to switch off.

Stochasticity also challenges our study of bistable systems. We should study single cells because extrinsic fluctuations can wash out bistable responses when averaged over measurements of populations of cells [30]. We should look for hysteresis because bimodal distributions can be generated by slow reactions in biochemical networks that have no positive feedback [22,23,36–38]. Such short-term memory, however, may be sufficient for some systems, and indeed positive feedback need not generate bistability but simply long-lived transient behavior [73]. Single-cell studies, and particularly the time-lapse studies useful for investigating cellular memory, are challenging. Nevertheless, developments in microfluidic technology are easing the generation of dynamic changes in the cellular environment and the quantification and tracking of individual cellular responses over many generations [17,18]. We expect that such experiments will reveal new insights into cellular memory and its role in cellular decision making [74].

By having switches of greater complexity, cells can limit the effects of stochastic fluctuations but can also evolve biochemistry that exploits stochasticity. Bet-hedging by populations of microbes is one example [75], where some cells in the population temporarily and stochastically switch to a behavior that is inefficient in the current environment but potentially efficient if a sudden change in the environment occurs. Perhaps the most intriguing example is, however, in our own cells. Mammalian embryonic stem cells appear to use stochasticity to become temporarily primed for different cell fates and thus to become susceptible over time to the different differentiation signals for all possible choices of cell fate [60].

References

[1] Scoville WB, Milner B (2000) Loss of recent memory after bilateral hippocampal lesions. *J Neuropsychiatry Clin Neurosci* 12:103–13.

[2] Berg HC, Brown DA (1972) Chemotaxis in *Escherichia coli* analysed by three-dimensional tracking. *Nature* 239:500–4.

[3] Muller-Hill B (1996) *The lac operon: a short history of a genetic paradigm* (Walter de Gruyter, Berlin).

[4] Lohr D, Venkov P, Zlatanova J (1995) Transcriptional regulation in the yeast *GAL* gene family: a complex genetic network. *FASEB J* 9:777–87.

[5] Takahashi K, Yamanaka S (2006) Induction of pluripotent stem cells from mouse embryonic and adult fibroblast cultures by defined factors. *Cell* 126:663–76.

[6] Lewis K (2007) Persister cells, dormancy and infectious disease. *Nat Rev Micro* 5:48–56.

[7] Shahrezaei V, Swain PS (2008) The stochastic nature of biochemical networks. *Curr Opin Biotechnol* 19:369–74.

[8] Raj A, van Oudenaarden A (2008) Nature, nurture, or chance: stochastic gene expression and its consequences. *Cell* 135:216–26.

[9] Ghaemmaghami S, et al. (2003) Global analysis of protein expression in yeast. *Nature* 425:737–41.

[10] Taniguchi Y, et al. (2010) Quantifying *E. coli* proteome and transcriptome with single-molecule sensitivity in single cells. *Science* 329:533–8.

[11] Swain PS, Elowitz MB, Siggia ED (2002) Intrinsic and extrinsic contributions to stochasticity in gene expression. *Proc Natl Acad Sci USA* 99:12795–800.

[12] Elowitz MB, Levine AJ, Siggia ED, Swain PS (2002) Stochastic gene expression in a single cell. *Science* 297:1183–6.

Chapter 3

[13] Ozbudak EM, Thattai M, Kurtser I, Grossman AD, van Oudenaarden A (2002) Regulation of noise in the expression of a single gene. *Nat Genet* 31:69–73.

[14] Blake WJ, Kaern M, Cantor CR, Collins JJ (2003) Noise in eukaryotic gene expression. *Nature* 422:633–7.

[15] Raser JM, O'Shea EK (2004) Control of stochasticity in eukaryotic gene expression. *Science* 304:1811–4.

[16] Shahrezaei V, Ollivier JF, Swain PS (2008) Colored extrinsic fluctuations and stochastic gene expression. *Mol Syst Biol* 4:196.

[17] Bennett MR, Hasty J (2009) Microfluidic devices for measuring gene network dynamics in single cells. *Nat Rev Genet* 10:628–38.

[18] Wang CJ, Levchenko A (2009) Microfluidics technology for systems biology research. *Methods Mol Biol* 500:203–19.

[19] Yu J, Xiao J, Ren X, Lao K, Xie XS (2006) Probing gene expression in live cells, one protein molecule at a time. *Science* 311:1600–3.

[20] McAdams HH, Arkin A (1997) Stochastic mechanisms in gene expression. *Proc Natl Acad Sci USA* 94:814–9.

[21] Thattai M, van Oudenaarden A (2001) Intrinsic noise in gene regulatory networks. *Proc Natl Acad Sci USA* 98:8614–9.

[22] Shahrezaei V, Swain PS (2008) Analytical distributions for stochastic gene expression. *Proc Natl Acad Sci USA* 105:17256–61.

[23] Friedman N, Cai L, Xie XS (2006) Linking stochastic dynamics to population distribution: an analytical framework of gene expression. *Phys Rev Lett* 97:168302.

[24] Golding I, Paulsson J, Zawilski SM, Cox EC (2005) Real-time kinetics of gene activity in individual bacteria. *Cell* 123:1025–36.

[25] Voliotis M, Cohen N, Molina-París C, Liverpool TB (2008) Fluctuations, pauses, and backtracking in DNA transcription. *Biophys J* 94:334–48.

[26] Chubb JR, Trcek T, Shenoy SM, Singer RH (2006) Transcriptional pulsing of a developmental gene. *Curr Biol* 16:1018–25.

[27] Raj A, Peskin CS, Tranchina D, Vargas DY, Tyagi S (2006) Stochastic MRNA synthesis in mammalian cells. *Plos Biol* 4:e309.

[28] Gillespie DT (1977) Exact stochastic simulation of coupled chemical reactions. *J Phys Chem* 81:2340–2361.

[29] Rosenfeld N, Young JW, Alon U, Swain PS, Elowitz MB (2005) Gene regulation at the single-cell level. *Science* 307:1962–5.

[30] Ferrell JE, Machleder EM (1998) The biochemical basis of an all-or-none cell fate switch in xenopus oocytes. *Science* 280:895–8.

[31] Monod J, Jacob F (1961) General conclusions: Teleonomic mechanisms in cellular metabolism, growth, and differentiation. *Cold Spring Harb Symp Quant Biol* 26:389.

[32] Lisman JE (1985) A mechanism for memory storage insensitive to molecular turnover: a bistable auto phosphorylating kinase. *Proc Natl Acad Sci USA* 82:3055–7.

[33] Lisman JE, Goldring MA (1988) Feasibility of long-term storage of graded information by the Ca^{2+}/calmodulin-dependent protein kinase molecules of the postsynaptic density. *Proc Natl Acad Sci USA* 85:5320–4.

[34] Bhalla US, Iyengar R (1999) Emergent properties of networks of biological signaling pathways. *Science* 283:381–7.

[35] Xiong W, Ferrell JE (2003) A positive-feedback-based bistable "memory module" that governs a cell fate decision. *Nature* 426:460–5.

[36] Pirone JR, Elston TC (2004) Fluctuations in transcription factor binding can explain the graded and binary responses observed in inducible gene expression. *J Theor Biol* 226:111–21.

[37] Karmakar R, Bose I (2004) Graded and binary responses in stochastic gene expression. *Physical Biology* 1:197–204.

[38] Hornos JEM, et al. (2005) Self-regulating gene: an exact solution. *Phys Rev E* 72:051907.

[39] Qian H, Shi PZ, Xing J (2009) Stochastic bifurcation, slow fluctuations, and bistability as an origin of biochemical complexity. *Phys Chem Chem Phys* 11:4861–70.

[40] Van Kampen NG (1981) *Stochastic processes in physics and chemistry* (North-Holland, Amsterdam, The Netherlands).

[41] Procaccia I, Ross J (1977) Stability and relative stability in reactive systems far from equilibrium. ii. kinetic analysis of relative stability of multiple stationary states. *J Chem Phys* 67:5565–5571.

[42] Hanggi P, Grabert H, Talkner P, Thomas H (1984) Bistable systems: master equation versus Fokker-Planck modeling. *Phys Rev, A* 29:371.

[43] Mehta P, Mukhopadhyay R, Wingreen NS (2008) Exponential sensitivity of noise-driven switching in genetic networks. *Phys Biol* 5:26005.

[44] Roma DM, O'Flanagan RA, Ruckenstein AE, Sengupta AM, Mukhopadhyay R (2005) Optimal path to epigenetic switching. *Phys Rev E* 71:011902.

[45] Allen RJ, Warren PB, ten Wolde PR (2005) Sampling rare switching events in biochemical networks. *Phys Rev Lett* 94:018104.

[46] Wilson CJ, Zhan H, Swint-Kruse L, Matthews KS (2007) The lactose repressor system: paradigms for regulation, allosteric behavior and protein folding. *Cell Mol Life Sci* 64:3–16.

[47] Ozbudak EM, Thattai M, Lim HN, Shraiman BI, van Oudenaarden A (2004) Multistability in the lactose utilization network of *Escherichia coli*. *Nature* 427:737–40.

[48] Santillán M, Mackey MC, Zeron ES (2007) Origin of bistability in the *lac* operon. *Biophys J* 92:3830–42.

[49] Setty Y, Mayo AE, Surette MG, Alon U (2003) Detailed map of a cis-regulatory input function. *Proc Natl Acad Sci USA* 100:7702–7.

[50] Kuhlman T, Zhang Z, Saier MH, Hwa T (2007) Combinatorial transcriptional control of the lactose operon of *Escherichia coli*. *Proc Natl Acad Sci USA* 104:6043–8.

[51] Novick A, Weiner M (1957) Enzyme induction as an all-or-none phenomenon. *Proc Natl Acad Sci USA* 43:553–66.

[52] Choi PJ, Cai L, Frieda K, Xie XS (2008) A stochastic single-molecule event triggers phenotype switching of a bacterial cell. *Science* 322:442–6.

[53] Robert L, et al. (2010) Pre-dispositions and epigenetic inheritance in the *Escherichia coli* lactose operon bistable switch. *Mol Syst Biol* 6:357.

[54] Megerle JA, Fritz G, Gerland U, Jung K, Rädler JO (2008) Timing and dynamics of single cell gene expression in the arabinose utilization system. *Biophys J* 95:2103–15.

[55] Acar M, Becskei A, van Oudenaarden A (2005) Enhancement of cellular memory by reducing stochastic transitions. *Nature* 435:228–32.

[56] Song C, et al. (2010) Estimating the stochastic bifurcation structure of cellular networks. *PLoS Comput Biol* 6:e1000699.

[57] Huang CY, Ferrell JE (1996) Ultrasensitivity in the mitogen-activated protein kinase cascade. *Proc Natl Acad Sci USA* 93:10078–83.

[58] Boyer LA, et al. (2005) Core transcriptional regulatory circuitry in human embryonic stem cells. *Cell* 122:947–56.

[59] Ferrell JE (1996) Tripping the switch fantastic: how a protein kinase cascade can convert graded inputs into switch-like outputs. *Trends Biochem Sci* 21:460–6.

[60] Silva J, Smith A (2008) Capturing pluripotency. *Cell* 132:532–6.

[61] Smith AG, et al. (1988) Inhibition of pluripotential embryonic stem cell differentiation by purified polypeptides. *Nature* 336:688–90.

[62] Ying QL, Nichols J, Chambers I, Smith A (2003) BMP induction of Id proteins suppresses differentiation and sustains embryonic stem cell self-renewal in collaboration with Stat3. *Cell* 115:281–92.

[63] Kuhn HG, Dickinson-Anson H, Gage FH (1996) Neurogenesis in the dentate gyrus of the adult rat: age-related decrease of neuronal progenitor proliferation. *J Neurosci* 16:2027–33.

[64] Morrison SJ, Wandycz AM, Akashi K, Globerson A, Weissman IL (1996) The aging of hematopoietic stem cells. *Nat Med* 2:1011–6.

[65] Conboy IM, Conboy MJ, Smythe GM, Rando TA (2003) Notch-mediated restoration of regenerative potential to aged muscle. *Science* 302:1575–7.

[66] Conboy IM, et al. (2005) Rejuvenation of aged progenitor cells by exposure to a young systemic environment. *Nature* 433:760–4.

[67] Chambers I, et al. (2007) Nanog safeguards pluripotency and mediates germline development. *Nature* 450:1230–4.

[68] Davey RE, Onishi K, Mahdavi A, Zandstra PW (2007) LIF-mediated control of embryonic stem cell self-renewal emerges due to an autoregulatory loop. *FASEB J* 21:2020–32.

[69] Chickarmane V, Troein C, Nuber UA, Sauro HM, Peterson C (2006) Transcriptional dynamics of the embryonic stem cell switch. *PLoS Comput Biol* 2:e123.

[70] Kalmar T, et al. (2009) Regulated fluctuations in Nanog expression mediate cell fate decisions in embryonic stem cells. *PLoS Biol* 7:e1000149.

Chapter 3

[71] Glauche I, Herberg M, Roeder I (2010) Nanog variability and pluripotency regulation of embryonic stem cells – insights from a mathematical model analysis. *PLoS ONE* 5:e11238.

[72] Huang S (2009) Reprogramming cell fates: reconciling rarity with robustness. *Bioessays* 31:546–60.

[73] Weinberger LS, Dar RD, Simpson ML (2008) Transient-mediated fate determination in a transcriptional circuit of *HIV. Nat Genet* 40:466–70.

[74] Perkins TJ, Swain PS (2009) Strategies for cellular decision-making. *Mol Syst Biol* 5:326.

[75] Veening JW, Smits WK, Kuipers OP (2008) Bistability, epigenetics, and bet-hedging in bacteria. *Annu Rev Microbiol* 62:193–210.

4. Information Theory and Adaptation

Ilya Nemenman

4.1 Life Is Information Processing

All living systems have evolved to perform certain tasks in specific contexts. There are a lot fewer tasks than there are different biological solutions that nature has created. Some of these problems are universal, while the solutions may be organism specific. Thus, a lot can be understood about the structure of biological systems by focusing on understanding of *what* they do and *why* they do it, in addition to *how* they do it on molecular or cellular scales. In particular, this way we can uncover phenomena that generalize across different organisms, thus increasing the value of experiments and building a coherent understanding of the underlying physiological processes.

This chapter takes this point of view while analyzing what it takes to do one of the most common, universal functions performed by organisms at all levels of organization: signal or information processing and shaping of a response, variously known in different contexts as learning from, for example, observations, signal transduction, regulation, sensing, and adaptation. Studying these types of phenomena poses a series of well-defined, physical questions:

Quantitative Biology: From Molecular to Cellular Systems edited by Michael E. Wall © 2012 CRC Press / Taylor & Francis Group, LLC. ISBN: 978-1-4398-2722-2

Chapter 4

How can organisms deal with noise, whether extrinsic or generated by intrinsic stochastic fluctuations within molecular components of information processing devices?
How long should the world be observed before a certain inference about it can be made?
How is the internal representation of the world made and stored over time?
How can organisms ensure that the information is processed fast enough for the formed response to be relevant in the ever-changing world?
How should the information processing strategies change when the properties of the environment surrounding the organism change?

In fact, such *information processing* questions have been featured prominently in studies on all scales of biological complexity, from learning phenomena in animal behavior [1–7] to analysis of neural computation in small and large animals [8-16] to molecular information processing circuits [17–26], to name just a few.

In what follows, we will not try to embrace the unembraceable but will instead focus on just two questions, fundamental to the study of signal processing in biology:

What is the right way to measure the quality of information processing in a biological system?
What can real-life organisms do to improve their performance in these tasks?

The field of study of biological information processing has undergone a dramatic growth in recent years, and it is expanding at an ever-growing rate. As evidence of this growth, biological information processing has been a major theme of recent conferences (perhaps the best example is *The Annual q-bio Conference*, http://q-bio.org). Hence, in this short chapter, we have neither an ability nor a desire to provide an exhaustive literature review. Instead the reader should keep in mind that the selection of references cited here is a biased sample of important results in the literature; much deserving work has been omitted in this overview.

4.2   Quantifying Biological Information Processing

In the most general context, a biological system can be modeled as an input–output device (Figure 4.1) that observes a time-dependent state of the world $s(t)$ (where s may

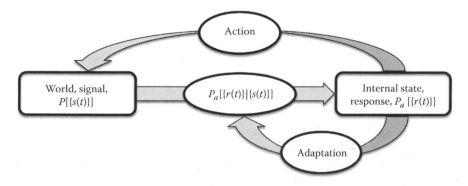

FIGURE 4.1   Biological information processing and interactions with the world. In this review we leave aside the feedback action between the organism internal state and the state of the world and focus on the signal processing and the adaptation arrows.

be intrinsically multidimensional, or even formally infinite-dimensional), processes the information, and initiates a response $r(t)$ (which can also be very large-dimensional). In some cases, in its turn, the response changes the state of the world and hence influences the future values of $s(t)$, making the whole analysis so much harder [27]. In view of this, analyzing the information processing means quantifying certain aspects of the mapping $s(t) \to r(t)$. This section discusses the properties that this quantification should possess and introduces the quantities that satisfy them.

4.2.1 What Is Needed?

Typically, an attempt is made to model molecular or other physiological *mechanisms* of the response generation. For example, in well-mixed biochemical kinetics approaches, where $s(t)$ may be a ligand concentration and $r(t)$ may be an expression level of a certain protein

$$\frac{dr(t)}{dt} = F_a(r,s,h) - G_a(r,s,h) + \eta_a(r,s,h,t) \tag{4.1}$$

where the nonnegative functions F_a and G_a stand for the production and degradation, respectively, of the response influenced by the level of the signal s. The term η_a is a random forcing due to the intrinsic stochasticity of chemical kinetics at small molecular copy numbers [28,29]. The subscript a stands for the values of adjustable parameters that define the response (e.g., various kinetic rates, concentrations of intermediate enzymes). These can change themselves, but on timescales much slower than the dynamics of s and r. Finally, h stands for the activity of other, hidden cellular state variables that change according to their own dynamics, similar to Equation (4.1). These dynamics can be written for many diverse biological information processing systems, including the neural dynamics, where r would stand for the firing rate of a neuron induced by the stimulus s [30].

Importantly, because of the intrinsic stochasticity in Equation (4.1) and because of the effective randomness due to the hidden variables, the mapping between the stimulus and the response is nondeterministic (see also Chapters 3 and 10). It is summarized in the following probability distribution:

$$P[\{r(t)\} \mid \{s(t)\}, \{h(t)\}, a] \text{ or, marginalizing over } h, \tag{4.2}$$

$$P[\{r(t)\} \mid \{s(t)\}, a] \equiv P_a[\{r(t)\} \mid \{s(t)\}] \tag{4.3}$$

In addition, $s(t)$ is not deterministic either: other agents, chaotic dynamics, statistical physics effects, and, at a more microscopic level, even quantum mechanics conspire to ensure that $s(t)$ can be specified only probabilistically. Therefore, a simple mapping $s \to r$ is replaced by a joint probability distribution (note that we will drop the index a in the future where it doesn't cause ambiguities)

$$P[\{r(t)\} \mid \{s(t)\}, a] P[\{s(t)\}] = P[\{r(t)\}, \{s(t)\} \mid a] \equiv P_a[\{r(t)\}, \{s(t)\}] \tag{4.4}$$

Hence, the measure of the quality of biological information processing must be a *functional* of this joint distribution. Biological information processing is almost always probabilistic.

Now consider, for example, a classical system studied in cellular information processing: *Escherichia coli* chemotaxis (see Chapters 8 and 13) [31]. This bacterium is capable of swimming up gradients of various nutrients. In this case, the signal $s(t)$ is the concentration of such extracellular nutrients. The response of the system is the activity levels of various internal proteins (e.g., *cheY, cheA, cheB, cheR*), which combine to modulate the cellular motion through the environment. It is possible to write the chemical kinetics equations that relate the stimulus to the response accurately enough and eventually produce the sought after conditional probability distribution $P_a[\{r(t)\}|\{s(t)\}]$. However, are the ligand concentrations the variables that the cell cares about? In this system, it is reasonable to assume that all protein expression states that result in the same intake of the catabolite are functionally equivalent. That is, the goal of the information processing performed by the cell likely is not to serve as a passive transducer of the signal into the response but to make an active computation that extracts only the part of the signal relevant to making behavioral decisions. The design of the chemotaxis system appears to reflect this [24,25]. We will denote such *relevant* aspects of the world as $e(t)$. For example, for the chemotactic bacterium, $e(t)$ can be the maximum nutrient intake realizable for a particular spatiotemporal pattern of the nutrient concentration.

In general, e is not a subset of s, or vice versa, and instead the relation between s and e is also probabilistic, $P[\{e(t)\}|\{s(t)\}]$, and hence the relevant variable, the signal, and the response form a Markov chain:

$$P\big[\{e(t)\},\{s(t)\},\{r(t)\}\big]=P\big[\{e(t)\}\big]P\big[\{s(t)\}|\{e(t)\}\big]P\big[\{r(t)\}|\{s(t)\}\big] \tag{4.5}$$

The measure chosen to characterize the quality of biological information processing must respect this aspect of the problem. That is, its value must depend explicitly on the choice of the relevance variable: a computation resulting in the same response will be either *good* or *bad* depending on what this response is used for. In other words, it is necessary to know what the problem is before determining whether a solution is good or bad. It is impossible to quantify the information processing without specifying the purpose of the device—the relevant quantity that it is supposed to compute.

4.2.2   Introducing the Right Quantities

The question of how much can be inferred about a state of a variable X from measuring a variable Y was answered by Claude Shannon over 60 years ago [32]. Starting with basic, uncontroversial axioms that a measure of information must obey, he derived an equation for the uncertainty S in a state of a variable,

$$S[X]=-\sum_x P(x)\log P(x)=-\langle\log P(x)\rangle_{P(x)} \tag{4.6}$$

which we know now as the *Boltzmann-Shannon entropy*. Here $\langle\cdots\rangle_{P(x)}$ denotes averaging over the variable x using the probability distribution $P(x)$. When the logarithm in Equation (4.6) is binary (which we always assume in this chapter), then the unit of entropy is a *bit*: one bit of uncertainty about a variable is equivalent to the latter being in one of two states with equal probabilities.

Observing the variable Y (i.e., *conditioning* on it) changes the probability distribution of X, $P(x) \to P(x|y)$, and the difference between the entropy of X prior to the measurement and the average conditional entropy yields I, which tells how informative Y is about X:

$$I[X;Y] = S[X] - \langle S[X|Y] \rangle_{P(y)} \tag{4.7}$$

$$= -\langle \log P(x) \rangle_{P(x)} + \left\langle \langle \log P(y|x) \rangle_{P(x|y)} \right\rangle_{P(y)} \tag{4.8}$$

$$= -\left\langle \log \frac{P(x,y)}{P(x)P(y)} \right\rangle_{P(x,y)} \tag{4.9}$$

The quantity $I[X;Y]$ in Equation (4.9) is known as *mutual information*. As entropy, it is measured in bits. Mutual information of one bit means that specifying the variable Y provides us with the knowledge to answer one yes–no question about X.

Entropy and information are additive quantities. That is, when considering entropic quantities for time series data defined by $P[\{x(t)\}]$, for $0 \le t \le T$, the entropy of the entire series will diverge linearly with T. Therefore, it makes sense to define entropy and information rates [32]

$$\mathcal{S}[X] = \lim_{T \to \infty} \frac{S[x(0 \le t < T)]}{T} \tag{4.10}$$

$$\mathcal{I}[X;Y] = \lim_{T \to \infty} \frac{I[\{x(0 \le t < T)\};\{y(0 \le t < T)\}]}{T} \tag{4.11}$$

which measure the amount of uncertainty in the signal per unit time, and the reduction of this rate of uncertainty when given the response, respectively.

Entropy and mutual information possess some simple, important properties [33]:

1. Both quantities are nonnegative, $0 \le S[X]$ and $0 \le I[X;Y] \le \min(S[X], S[Y])$.
2. Entropy is zero if and only if (*iff*) the studied variable is not random. Further, mutual information is zero *iff* $P(x,y) = P(x)P(y)$, that is, there are no any kind of statistical dependences between the variables.
3. Mutual information is symmetric, $I[X;Y] = I[Y;X]$.
4. Mutual information is well defined for continuous variables; one needs only to replace sums by integrals in Equation (4.9). On the contrary, entropy formally diverges for continuous variables (any truly continuous variable requires infinitely many bits to be specified to an arbitrary accuracy). However, many properties of entropy are also exhibited by the *differential entropy*:

$$S[X] = -\int_x dx\, P(x) \log P(x) \tag{4.12}$$

Chapter 4

which measures the entropy of a continuous variable relative to the uniformly distributed one. In this chapter, $S[X]$ will always mean the differential entropy if x is continuous.

5. For a Gaussian distribution with a variance of σ^2,

$$S = 1/2 \log \sigma^2 + \text{const.} \tag{4.13}$$

For a bivariate Gaussian with a correlation coefficient ρ,

$$I[X;Y] = -1/2 \log(1-\rho^2) \tag{4.14}$$

Thus entropy and mutual information can be viewed as generalizations of more familiar notions of variance and covariance.

6. Unlike entropy, mutual information is invariant under reparameterization of variables:

$$I[X;Y] = I[X';Y'] \text{ for all invertible } x' = x'(x),\ y' = y'(y) \tag{4.15}$$

That is, I provides a measure of statistical dependence between X and Y that is independent of our subjective choice of the measurement device.*

4.2.3 When the Relevant Variable Is Unknown: The Value of Information about the World

One of the most fascinating properties of mutual information is the data processing inequality [33].

Suppose three variables X, Y, and Z form a Markov chain, $P(x,y,z) = P(x)P(y|x)P(z|y)$. In other words, Z is a probabilistic transformation of Y, which, in turn is a probabilistic transformation of X. Then it can be proven that

$$I[X;Z] \leq \min(I[X;Y], I[Y;Z]) \tag{4.16}$$

That is, new information cannot be obtained about the original variable by further transforming the measured data; any such transformation cannot increase the information.

Together with the fact that mutual information is zero *iff* the variables are completely statistically independent, the data processing inequality suggests that if the variable of interest that the organism cares about is unknown to the experimenter, then analyzing the mutual information between the entire input stimulus (sans noise) and the response may serve as a good proxy. Indeed, due to the data processing inequality, if $I[S;R]$ is small, then $I[E;R]$ is also small for any mapping $S \to E$ of the signal into the relevant variable, whether deterministic, $e = e(s)$, or probabilistic, $P(e|s)$. In many cases, such as [16,21,35], this allows us to stop guessing which calculation the organism is trying to

---

* From these properties, it is clear that mutual information is in some sense a nicer, more fundamental quantity than entropy. Indeed, even the famous Gibbs paradox in statistical physics [34] is related to the fact that entropy of continuous variables is ill-defined. Therefore, it is a pity that standard theoretical developments make entropy a primary quantity and derive mutual information from it. We believe that it should be possible to develop an alternative formulation of information theory with mutual information as the primary concept, without introducing entropy at all.

perform and to put an upper bound on the efficiency of the information transduction, whatever an organism cares about. However, as was recently shown in the case of chemotaxis in *E. coli*, when e and s are substantially different (resource consumption rate vs. instantaneous surrounding resource concentration), maximizing $I[S;R]$ is not necessarily what organisms do [25].

Another reason to study the information about the outside world comes from the old argument that relates information and game theories [36]. Namely, consider a zero-sum probabilistic betting game. Think of a roulette without the zeros, where the red and the black are two outcomes and betting on the right outcome doubles one's investment whereas betting on the wrong one leads to a loss of the bet. Then the logarithmic growth rate of one's capital is limited from above by the mutual information between the outcome of the game and the betting strategy. This was recently recast in the context of population dynamics in fluctuating environments [37–40]. Suppose the environment surrounding a population of genetically identical organisms fluctuates randomly with no temporal correlations among multiple states with probabilities $P(s)$. Each organism, independently of the rest, may choose among a variety of phenotypical decisions d, and the log-growth rate depends on the pairing of s and d. Evolution is supposed to maximize this rate, averaged over long times. However, the current state of the environment is not directly known, and the organisms might need to respond probabilistically. While the short-term gain would suggest choosing the response that has the highest growth rate for the most probable environment, the longer term strategy would require bet hedging [41], with different individuals making different decisions. Such bet hedging has been experimentally evolved [42].

Suppose an individual now observes the environment and gets an imperfect internal representation of it, r, with the conditional probability of $P(r|s)$. What is the value of this information? Under very general conditions, this information about the environment can improve the log-growth rate by as much as $I[S;R]$ [37]. In more general scenarios, the maximum log-growth advantage over uninformed peers needs to be discounted by the cost of obtaining the information, by the delay in getting it [38], and, more trivially, by the ability of the organism to use it. Therefore, while these brief arguments are far from painting a complete picture of relation between information and natural selection, it is clear that maximization of the information between the surrounding world and the internal response to it is not an academic exercise but is directly related to fitness and will be evolutionary advantageous.* Information about the outside world puts an upper bound on the fitness advantage of an individual over uninformed peers.

It is now well-known that probabilistic bet hedging is the strategy taken by bacteria for survival in the presence of antibiotics [44,45] and for genetic recombination [46–48]. In both cases, cell division (and hence population growth) must be stopped either to avoid DNA damage by antibiotics or to incorporate newly acquired DNA into the chromosome. Still, a small fraction of the cells choose not to divide even in the absence of antibiotics to reap the much larger benefits if the environment turns sour

---

* Interestingly, it was recently argued [43] that natural selection, indeed, serves to maximize the information that a population has about its environment, providing yet another evidence for the importance of information-theoretic considerations in biology.

(these are called the *persistent* and the DNA uptake *competent* bacteria for the two cases, respectively). However, it remains to be seen in an experiment if real bacteria can reach the maximum growth advantage allowed by the information-theoretic considerations. Another interesting possibility is that cancer stem cells and mature cancer cells also are two probabilistic states that hedge bets against interventions of immune systems, drugs, and other surveillance mechanisms [49].

4.2.4 Time–Dependent Signals: Information and Prediction

In many situations, such as persistence in the face of the aforementioned antibiotics treatment, an organism settles into a certain response much faster than the time scale of changes in environmental stimuli. In these cases, it is sufficient to consider the same-time mutual information between the signals and the responses, as in [21], $I[s(t);r(t)] = I[S;R]$, which is what we've been doing up to now.

More generally, the response time may be significant compared with the timescale of changes in the stimuli. What are the relevant quantities to characterize biological information processing in such situations? Traditionally, one either considers delayed informations [11,21,26,50],

$$I_\tau[S; R] = I[s(t); r(t + \tau)] \tag{4.17}$$

where τ may be chosen as $\tau = \mathrm{argmax}_{t'} I[s(t); r(t+t')]$, or information rates, as in Equation (4.11). The first choice measures the information between the stimulus and the response most constrained by it. Typically, this would be the response formed a certain characteristic signal transduction time after the stimulus occurrence. The second approach looks at correlations between all possible pairs of stimuli and responses as a function of the delay.

While there are plenty of examples of biological systems where one or the other of these quantities is worthy of analysis, both of the approaches are insufficient. I_τ does not represent all of the gathered information since bits at different moments of time are not independent of each other. Further, it does not take into the account that temporal correlations in the stimulus allow to predict it, and hence the response may be formed even before the stimulus occurs. On the other hand, the information rate does not distinguish among predicting the signal, knowing it soon after it happens, or having to wait, at an extreme, for $T \to \infty$ to be able to estimate it from the response.

To avoid these pitfalls, one can consider the information available to an organism that is relevant for specifying not all of the stimulus but only its future. Namely, we can define the *predictive information* about the stimulus available from observation of a response to it of a duration T,

$$I_{\mathrm{pred}}[R(T); S] = I[\{r(-T \le t \le 0)\}; \{s(t > 0)\}] \tag{4.18}$$

This definition is a generalization of the one used in [51], which had $r(t) = s(t)$ and hence calculated the upper bound on I_{pred} over all possible sensory channels $P[\{r(t)\}|\{s(t)\}]$. All of the I_{pred} bits are available to be used instantaneously, and there is no specific delay τ chosen a priori and semiarbitrarily. The predictive information is nonzero only to the extent that the signal is temporally correlated, and hence the response to its past values can say something about its future. Thus, focusing on predictability may resolve a traditional criticism of information theory that bits do not have an intrinsic meaning and value, and some are more useful than the others: since any action takes time, only those bits have value that can be used to predict the stimulus at the time of action, that is, in the future [51,52].

The notion of predictive information is conceptually appealing, and there is clear experimental and computational evidence that behavior of biological systems, from bacteria to mammals, is consistent with attempting to make better predictions (see, e.g., [16,53–58] for just a few of the results). However, even over 10 years after I_{pred} was first introduced, it still remains to be seen experimentally whether optimizing predictive information is one of the objectives of biological systems (Chapter 2 discusses several other important performance criteria), and whether population growth rates in temporally correlated environments can be related to the amount of information available to predict them. One of the reasons for the relative lack of progress might be a practical consideration that estimation of informations among nonlinearly related multidimensional variables [16,59,60] or extracting the predictive aspects of the information from empirical data is hard [61]. Further, for simple Gaussian signals and responses with finite correlation times, it is hard to distinguish between optimization of predictive information and the much simpler matching of Wiener extrapolation filters [62].

4.3 Improving Information-Processing Performance

Understanding the importance of information about the outside world and knowing which quantities can be used to measure it, we are faced with the next question: how can the available information be increased in view of the limitations imposed by the physics of the signal and of the processing device, such as stochasticity of molecular numbers and arrival times or energy constraints?

4.3.1 Strategies for Improving the Performance

We start with three main theorems of information theory due to Shannon [32]. In the *source-coding theorem*, he proved that, to record a signal without losses, one needs only $\mathcal{S}$, the signal entropy rate, bits per unit time. In the *channel-coding theorem*, he showed that the maximum rate of errorless transmission of information through a channel specified by $P[\{r(t)\}|\{s(t)\}]$ is given by $C = \max_{P(\{s(t)\})} \mathcal{I}[R;S]$, which is called the channel capacity. Finally, the *rate distortion theorem* calculates the minimum size of the message that must be sent error-free to recover the signal with an appropriate mean level of some prespecified distortion measure. None of these theorems considers the time delay before the message can be decoded, and typically one would need to wait for very long times and accumulate long message sequences to reach the bounds predicted by the theorems.

Leaving aside the complication of dynamics, which may hopefully be solved someday using the predictive information ideas, these theorems tell us exactly what an organism can do to optimize the amount of information it has about the outside world. First, the measured signal needs to be compressed, removing as many redundancies as possible. There is evidence that this happens in a variety of signaling systems, starting with the classical [53,63–65]. Second, the signal needs to be encoded in a way that allows the transmitted information to approach the channel capacity limit by remapping different values of signal into an intermediate signaling variable whose states are easier to transmit without an error. Again, there are indications that this happens in living systems [22,66–70]. Finally, only important aspects of the signal may be focused on, such as communicating changes in the signal, or thresholding its value [16,31,71,72].

If the references in the previous paragraph look somewhat thin, it is by choice since none of these approaches are unique to biology, and, in fact, most artificial communication systems use them: for example, a cell phone filters out audio frequencies that are irrelevant to human speech, compresses the data, and encodes it for sending with the smallest possible errors. A lot of engineering literature discusses these questions [33], and we will not touch them here anymore. What makes the problem unique in biology is the ability of biological systems to improve the information transmission by modifying their own properties in the course of their life. This adjusts the a in $P_a[\{r(t)\}|\{s(t)\}]$ and hence modifies the conditional probability distribution itself. This would be equivalent to a cell phone being able to change its physical characteristics on the fly. Human engineered systems are no match to biology in this regard: they are largely incapable of adjusting their own design if the original turns out to be flawed. Unlike most artificial systems, living organisms can change their own properties to optimize their information processing.

The property of changing one's own characteristics in response to the observed properties of the world is called *adaptation*, and the remainder of this section will be devoted to its overview. In principle, we make no distinction whether this adaptation is achieved by natural selection or by physiological processes that act on much faster timescales (comparable to the typical signal dynamics), and sometimes the latter may be as powerful as the former [21,73]. Further, adaptation of the response probability distribution and formation of the response are, in principle, a single process of formation of the response on multiple timescales. Our ability to separate it into a fast response and a slow adaptation (and hence much of the following discussion) depends on the existence of two well-separated timescales in the signal and in the mechanism of the response formation. While such clear separation is possible in some cases, it is harder in others, and especially when the timescales of the signal and the fast response may be changing themselves. Cases without a clear separation of scales raise a variety of interesting questions [74], but we will leave them aside for this discussion.

4.3.2   Three Kinds of Adaptation in Information Processing

In this section we largely follow the exposition of [75].

We often can linearize the dynamics, Equation (4.1), to get the following equation describing formation of small responses:

$$\frac{dr}{dt} = f[s(t)] - kr + \eta(t,r,s) \tag{4.19}$$

Here r may be, for example, an expression of an mRNA following activation by a transcription factor s, or the firing rate of a neuron following stimulation. In the previous expression, f is the response activation function, which depends on the current value of the signal, k is the rate of the first-order relaxation or degradation, and η is some stochastic process representing the intrinsic noisiness of the system. In this case, $r(t)$ depends on the entire history of $\{s(t')\}$, $t' < t$, and hence carries some information about it as well.

For quasi-stationary signals (i.e., the correlation time of the signal, $\tau_s \gg 1/k$), we can write the steady state dose response (or firing rate, or ...) curve

$$r_{ss} = f[\, s(t)\,]/k \tag{4.20}$$

and this will be smeared by the noise η. A typical monotonic sigmoidal f is characterized by only a few large-scale parameters: the range, f_{min} and f_{max}; the argument $s_{1/2}$ at the midpoint value $(f_{min} + f_{max})/2$; and the width of the transition region, Δs (see Figure 4.2). If the mean signal $\mu \equiv \langle s(t) \rangle_t \gg s_{1/2}$, then, for most signals, $r_{ss} \approx f_{max}/k$ and responses to two typical different signals s_1 and s_2 are indistinguishable as long as

$$\left. \frac{dr_{ss}(s)}{ds} \right|_{s=s_1} (s_2 - s_1) < \sigma_\eta / k \tag{4.21}$$

where σ_η/k is the precision of the response resolution expressed through the standard deviation of the noise. A similar situation happens when $\mu \ll s_{1/2}$ and $r_{ss} \approx f_{min}/k$. Thus, to reliably communicate information about the signal, f should be tuned such that $s_{1/2} \approx \mu$.

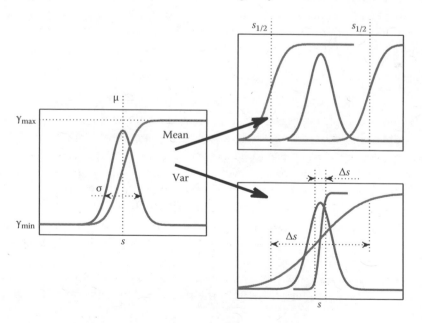

FIGURE 4.2 (See color insert.) Parameters characterizing response to a signal. Left panel: the probability distribution of the signal, $P(s)$ (blue), and the best-matched steady state dose-response curve r_{ss} (green). Top right: mismatched response midpoint. Bottom right: mismatched response gain. (From [75]. With permission.)

A real biological system that can perform this adjustment is known as *adaption to the mean* of the signal, *desensetization*, or *adaptation of the first kind*. If $s_{1/2}(\mu) = \mu$, then the adaptation is *perfect*. This kind of adaptation has been observed experimentally and predicted computationally in many more systems than we can list here, including phototransduction, neural and molecular sensing, multistate receptor systems, immune response, and so on, with active work persisting to date (see, e.g., [31,70,76–84] for a very incomplete list of references on the subject). For example, the best-studied adaptive circuit in molecular biology, the control of chemotaxis of *E. coli* (see Chapter 13), largely produces adaptation of the first kind [85,86]. Further, various applications in synthetic biology require matching the typical protein concentration of the input signal to the response function that maps this concentration into the rate of mRNA transcription or protein translation (cf. [87] and Chapter 9). Thus, an active community of researchers is now working on endowing these circuits with proper adaptive matching abilities of the first kind.

Consider now the quasi-stationary signal taken from the distribution with $\sigma \equiv (\langle s(t)^2 \rangle_t - \mu^2)^{1/2} \gg \Delta s$. Then the response to most of the signals is indistinguishable from the extremes, and it will be near the midpoint $\sim (r_{max}+r_{min})/2$ if $\sigma \ll \Delta s$. Thus, to use the full dynamic range of the response, a biological system must tune the width of the sigmoidal dose-response curve to $\Delta s \approx \sigma$. This is called *gain control, variance adaptation*, or *adaptation of the second kind*. Experiments and simulations show that a variety of systems exhibit this adaptive behavior as well [88], especially in the context of neurobiology [11,74] and maybe even of evolution [89].

These matching strategies are well-known in signal processing literature under the name of histogram equalization. Surprisingly, they are nothing but a special case of optimizing the mutual information $I[S;R]$, as has been shown first in the context of information processing in fly photoreceptors [8]. Indeed, for quasi-steady state responses, when noises are small compared with the range of the response, the arrangement that optimizes $I[S;R]$ is the one that produces $P(r) \propto 1/\sigma_{r|s}$. In particular, when σ_η is independent of r and s, this means that each r must be used equiprobably, that is, $f^*(s) = \int_{-\infty}^{s} P(s')ds'$.

Adaptation of the first and the second kind follows from these considerations immediately. In more complex cases, when the noise variance is not small or not constant, derivation of the optimal response activation function cannot be done analytically, but numerical approaches can be used instead. In particular, in transcriptional regulation of the early *Drosophila* embryonic development, the matching between the response function and the signal probability distribution has been observed for nonconstant $\sigma_{r|s}$ [22]. However, even though adaptation can potentially have this intimate connection to information maximization, and it is essentially omnipresent, the number of systems where the adaptive strategy has been analyzed quantitatively to show that it results in optimal information processing is not that large.

We now relax the requirement of quasi-stationarity and return to dynamically changing stimuli. We rewrite Equation (4.19) in the frequency domain (a technique also used in Chapter 8),

$$r_\omega = \frac{[f(s)]_\omega + \eta_\omega}{k + i\omega} \tag{4.22}$$

which shows that the simple first-order (or linearized) kinetics performs low-pass filtering of the nonlinearly transformed signal [18,19]. As discussed long ago by Wiener [62], for given temporal correlations of the stimulus and the noise (which we summarize here for simplicity by correlation times τ_s and τ_η), there is an optimal cutoff frequency k that allows to filter out as much noise as possible without filtering out much of the signal. Change of the parameter k to match the temporal structure of the problem is called the *timescale adaptation* or *adaptation of the third kind*. Just like the first two kinds, timescale adaptation also can be related to maximization of the stimulus-response mutual information by means of a simple observation that minimization of the quadratic prediction error of the Wiener filter is equivalent to maximizing information about the signal, at least when the signals are GAUSSIAN, and the system has no long-term memory, cf. Equation (4.14).

This adaptation strategy is difficult to study experimentally since (1) detection of variation of the integration cutoff frequency k potentially requires observing the adaptation dynamics on very long timescales, and (2) prediction of optimal cutoff frequency requires knowing the temporal correlation properties of signals, which are hard to measure (see, e.g., [90] for a review of literature on analysis of statistical properties of natural signals). Nonetheless, experimental systems as diverse as turtle cones [91], rats in matching foraging experiments [3], mice retinal ganglion cells [92], and barn owls adjusting auditory and visual maps [93] show adaptation of the filtering cutoff frequency in response to changes in the relative timescales or the variances of the signal and the noise. In a few rare cases, including fly self-motion estimation [13] and *E. coli* chemotaxis [94] (numerical experiment), it was shown that the timescale matching not only improves but also optimizes the information transmission. (See also Chapter 8 for a discussion of the relationship between stimulus dynamics and system response times in *E. coli* chemotaxis.)

Typically, adaptation as a phenomenon is considered different from redundancy reduction, and this view is accepted here. However, there is a clear relation between the two mechanisms. For example, adaptation of the first kind can be viewed as subtracting out the mean of the signal, stopping its repeated, redundant transmission, and allowing for a focus on the nonredundant, changing components of the signal. As any redundancy reduction procedure, this may introduce ambiguities: a perfectly adapting system will respond in the same fashion to different stimuli, preventing unambiguous identification of the stimulus based on the instantaneous response. Knowing statistics of responses on the scale of adaptation itself may be required to resolve the problem. This interesting complication has been explored in a few model systems [13,92].

4.3.3   Mechanisms of Different Adaptations

The three kinds of adaptation considered here can all be derived from the same principle of optimizing the stimulus–response mutual information. They are all evolutionary accessible. However, the mechanisms behind these adaptations on physiological, nonevolutionary timescales and their mathematical descriptions can be substantially different.

Adaptation of the first kind has been studied extensively. On physiological scales, it is implemented typically using negative feedback loops or incoherent feedforward loops, as illustrated in Figure 4.3. In all of these cases, the fast activation of the response by the signal is then followed by a delayed suppression mediated by a memory node. This

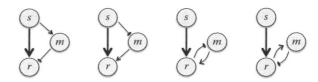

FIGURE 4.3   Different network topologies able to perform adaptation to the mean. *s*, signal. *r*, response. *m*, memory. Sharp arrows indicate activation/excitation and blunt ones stand for deactivation/suppression. The thickness of arrows denotes the speed of action (faster action for thicker arrows).

allows the system to transmit changes in the signal yet to desensetize and return close (and sometimes perfectly close) to the original state if the same excitation persist. This response to *changes* in the signal earns adaptation of the first kind the name of *differentiating filter*. In particular, the feedback loop in *E. coli* chemotaxis [31,85] (see Chapter 13) or yeast signaling [95] can be represented as the right-most topology in the figure, and different models of *Dictyostelium* adaptation include both feed-forward and feedback designs [76,96].

The different network topologies have different sensitivities to changes in the internal parameters, different trade-offs between the sensitivity to the stimulus change and the quality of adaptation, and so on. However, fundamentally they are similar. This can be seen by noting that since the goal of these adaptive systems is to keep the signal within the *small* transition region between the minimum and the maximum activation of the response, it makes sense to linearize the dynamics of the networks near the mean values of the signal and the corresponding response. Defining $\xi = s - \bar{s}$, $\zeta = r - \bar{r}$, and $\chi = m - \bar{m}$, one can write, for example, for the feedback topologies in Figure 4.3

$$\frac{d\zeta}{dt} = -k_{\zeta\zeta}\zeta + k_{\zeta\xi}\xi - k_{\zeta\chi}\chi + \eta_\zeta \tag{4.23}$$

$$\frac{d\chi}{dt} = k_{\chi\zeta}\zeta - k_{\chi\chi}\chi + \eta_\chi \tag{4.24}$$

where η are noises, and the coefficients k_{**} change sign depending on the network topology. In our notation $k_{**} > 0$ for the fourth topology. Doing the usual Fourier transform of these equations (see [79] for a very clear, pedagogical treatment) and expressing ζ in terms of ξ, η_ζ, and η_χ, it is evident that only the product of $k_{\chi\zeta}k_{\zeta\chi}$ matters for the properties of the filter with feedback topologies (Equations (4.23) and (4.24)). Hence, both the feedback systems in Figure 4.3 are essentially equivalent in this regime. Furthermore, as argued in [76,97], a simple linear transformation of ζ and χ enables the incoherent feed-forward loops (the two first topologies in Figure 4.3) to be recast into a feedback design, again arguing that, at least in the linear regime, the differences among all of these organizations are rather small from the mathematical point of view.*

---

* Sontag has recently considered the case where the linear term in the feed-forward or feedback interactions is zero, and the leading coupling term in the dynamics of ζ is bilinear; there the distinctions among the topologies are somewhat more tangible [97].

The reason so much progress can be made in analyzing adaptation to the mean is that the mean is a linear function of the signal, and hence it can be accounted for in a linear approximation. Different network topologies differ in their actuation components (i.e., how the measured mean is then fed back into changing the response generation), but averaging a linear function of the signal over a certain timescale is the common description of the sensing component of essentially all adaptive mechanisms of the first kind. Adaptation to the mean can be analyzed linearly, and many different designs become similar in this regime.

Variance and timescale adaptations are fundamentally different from adaptation to the mean. While the actuation part for them is not any more difficult than for adaptation to the mean, adapting to the variance requires averaging the square or another nonlinear function of the signal to sense its current variance. Further, estimation of the timescale of the signal requires estimation of the spectrum or of the correlation function (both are bilinear averages). Therefore, maybe it is not surprising that the literature on mathematical modeling of mechanisms of these types of biological adaptation is rather scarce. While functional models corresponding to a bank of filters or estimators of environmental parameters operating at different timescales can account for most of the experimentally observed data about changes in the gain and in the scale of temporal integration [3,11,13,92,98], to our knowledge these models largely have not been related to nonevolutionary, mechanistic processes at molecular and cellular scales that underlie them.

The largest inroads in this direction have been achieved when integration of a nonlinear function of a signal results in an adaptation response that depends not just on the mean but also on higher-order cumulants of the signal, effectively mixing different kinds of adaptation together. One example is the recently analyzed fold-change-detection mechanism [99]. This mixing may be desirable in the cases of photoreception [79] and chemosensing [88,100], where the signal mean is tightly connected to the signal or the noise variances. For example, the standard deviation of brightness of a visual scene scales linearly with the background illumination, while the noise in the molecular concentration is proportional to the square root of the latter. Similarly, mixing means and variances allows the budding yeast to respond to *fractional* rather than additive changes of a pheromone concentration [101]. In other situations, like adaptation by a receptor with state-dependent inactivation properties, the purpose of mixing the mean signal with its temporal correlation properties to form an adaptive response is not yet understood [81].

In a similar manner, integration of a strongly nonlinear function of a signal may allow a system to respond to signals in a gain-insensitive fashion, effectively adapting to the variance without a true adaptation. Specifically, the stimulus can be thresholded around its mean value and then integrated to count how long it has remained positive. For any temporally correlated stimulus, the time since the last mean-value crossing is correlated to the instantaneous stimulus value (it takes a long time to reach high stimulus values), and this correlation is independent of the gain. It has been argued that adaptation to the variance in fly-motion estimation can be explained at least in part by this nonadaptive process [74]. Similar mechanisms can be implemented in molecular signaling systems [75].

Chapter 4

4.4 What's Next?

It is clear that information theory has an important role in biology. It is a mathematically correct construction for analysis of signal processing systems. It provides a general framework to recast adaptive processes on scales from evolutionary to physiological in terms of a (constrained) optimization problem. Sometimes it even makes (correct!) predictions about responses of living systems following exposure to various signals. So, what's next for information theory in the study of signal processing in living systems?

The first and the most important problem that still remains to be solved is that many of the aforementioned stories are incomplete. Since we never know for sure which specific aspect of the world, $e(t)$, an organism cares about and the statistics of signals are hard to measure in the real world, an adaptation that seems to optimize $I[S;R]$ may be an artifact of our choice of S and of assumptions about $P(s)$ but not a consequence of the quest for optimality by an organism. For example, the timescale of filtering in *E. coli* chemotaxis [94] may be driven by the information optimization, or it may be a function of very different pressures. Similarly, a few standard deviations mismatch between the cumulative distribution of light intensities and a photoreceptor response curve in fly vision [8] can be a sign of an imperfect experiment, or it can mean that we simply got (almost) lucky and the two curves nearly matched by chance. It is difficult to make conclusions based on one data point.

Therefore, to complete these and similar stories, the information arguments must be used to make predictions about adaptations in novel environments, and such adaptations must be observed experimentally. This has been done in some contexts in neuroscience [2,11,13,102], but molecular sensing lags behind. This is largely because evolutionary adaptation, too slow to observe, is expected to play a major role here and because careful control of dynamic environments or characterization of statistical properties of naturally occurring environments [90] needed for such experiments is not easy. New experimental techniques such as microfluidics [103] and artificially sped-up evolution [104] are about to solve these problems, opening the proverbial doors wide open for a new class of experiments.

The second important research direction, which will require combined progress in experimental techniques and mathematical foundations, is likely going to be the return of dynamics. This has had a revolutionary effect in neuroscience [10], revealing responses unimaginable for quasi-steady state stimuli. Dynamical stimulation is starting to take off in molecular systems as well [70,105]. How good are living systems in filtering out those aspects of their time-dependent signals that are not predictive and are therefore of no use? What is the evolutionary growth bound when signals change in a continuous, somewhat predictive fashion? None of these questions have been touched yet, whether theoretically or experimentally.

Finally, we need to start building mechanistic models of adaption in living systems that are more complex than a simple subtraction of the mean. How are the amazing adaptive behaviors of the second and the third kind achieved in practice on physiological scales? Does it even make sense to distinguish the three different adaptations, or can some molecular or neural circuits achieve them all? How many and which parameters of the signal do neural and molecular circuits estimate and how? Some of these questions may be answered if one is capable of probing the subjects with high-frequency, controlled signals [106], and the recent technological advances will be a game changer as well.

Overall, studying biological information processing over the next 10 years will be an exciting pastime.

Acknowledgments

I am grateful to Sorin Tanase Nicola and H. G. E. Hentschel for insightful comments about the manuscript.

References

[1] S Kakade and P Dayan. Acquisition in autoshaping. *Advances in Neural Information Processing Systems,* 12:24, 2000.

[2] CR Gallistel and J Gibbon. Time, rate, and conditioning. *Psychol Rev,* 107(2):289–344, 2000.

[3] CR Gallistel, T Mark, A King, and P Latham. The rat approximates an ideal detector of changes in rates of reward: implications for the law of effect. *J Exp Psychol: Anim Behav Process,* 27(4):354–72, 2001.

[4] CR Gallistel, S Fairhurst, and P Balsam. The learning curve: implications of a quantitative analysis. *Proc Natl Acad Sci USA,* 101(36):13124–31, 2004.

[5] L Sugrue, G Corrado, and WT Newsome. Matching behavior and the representation of value in the parietal cortex. *Science,* 304(5678):1782–7, 2004.

[6] P Balsam and CR Gallistel. Temporal maps and informativeness in associative learning. *Trends Neurosci,* 32(2):73–8, 2009.

[7] P Balsam, M Drew, and CR Gallistel. Time and associative learning. *Compar Cognition Behavior Rev,* 5:1–22, 2010.

[8] S Laughlin. A simple coding procedure enhances a neuron's information capacity. *Z Naturforsch, C, Biosci,* 36:910–2, 1981.

[9] S Laughlin, R de Ruyter van Steveninck, and Anderson J. The metabolic cost of neural information. *Nat Neurosci,* 1:36–41, 1998.

[10] F Rieke, D Warland, R de Ruyter van Steveninck, and W Bialek. *Spikes: Exploring the Neural Code.* MIT Press, 1999.

[11] N Brenner, W Bialek, and R de Ruyter van Steveninck. Adaptive rescaling optimizes information transmission. *Neuron,* 26:695, 2000.

[12] P Reinagel and R Reid. Temporal coding of visual information in the thalamus. *J Neurosci,* 20:5392–400, 2000.

[13] A Fairhall, G Lewen, W Bialek, and R de Ruyter van Steveninck. Efficiency and ambiguity in an adaptive neural code. *Nature,* 412:787, 2001.

[14] R Liu, S Tzonev, S Rebrik, and K Miller. Variability and information in a neural code of the cat lateral geniculate nucleus. *J Neurophysiol,* 86:2789–806, 2001.

[15] J Victor. Binless strategies for estimation of information from neural data. *Phys Rev E,* 66:51903, 2002.

[16] I Nemenman, GD Lewen, W Bialek, and R de Ruyter van Steveninck. Neural coding of natural stimuli: information at sub-millisecond resolution. *PLoS Comput Biol,* 4(3):e1000025, 2008.

[17] S Forrest and S Hofmeyr. Immunology as information processing. In L Segel and I Cohen, editors, *Design Principles for the Immune System and Other Distributed Autonomous Systems.* New York: Oxford University Press, 2000.

[18] A Arkin. Signal processing by biochemical reaction networks. In J Walleczek, editor, *Self-Organized Biological Dynamics and Nonlinear Control: Toward Understanding Complexity, Chaos and Emergent Function in Living Systems.* Cambridge, UK: Cambridge University Press, 2000.

[19] M Samoilov, A Arkin, and J Ross. Signal processing by simple chemical systems. *J Phys Chem A,* 106:10205–21, 2002.

[20] B Andrews and P Iglesias. An information-theoretic characterization of the optimal gradient sensing response of cells. *PLoS Comput Biol,* 3(8):e153, 2007.

[21] E Ziv, I Nemenman, and CH Wiggins. Optimal signal processing in small stochastic biochemical networks. *PLoS One,* 2(10):e1077, 2007.

Chapter 4

[22] G Tkacik, C Callan, and W Bialek. Information flow and optimization in transcriptional regulation. *Proc Natl Acad Sci USA*, 105:12265–70, 2008.

[23] F Tostevin and PR ten Wolde. Mutual information between input and output trajectories of biochemical networks. *Phys Rev Lett*, 102:218101, 2009.

[24] A Celani and M Vergassola. Bacterial strategies for chemotaxis response. *Proc Natl Acad Sci USA*, 107:1391–6, 2010.

[25] M Skoge, Y Meir, and N Wingreen. Dynamics of cooperativity in chemical sensing among cell-surface receptors. *Phys Rev Lett*, 107:178101, 2011.

[26] R Cheong, A Rhee, CJ Wang, I Nemenman, and A Levchenko. Information transduction capacity of noisy biochemical signaling networks. *Science*, 334:354–358, 2011.

[27] S Still. Information-theoretic approach to interactive learning. *Europhys Lett (EPL)*, 85:28005, 2009.

[28] J Paulsson. Summing up the noise in gene networks. *Nature*, 427:415–8, 2004.

[29] N van Kampen. *Stochastic Processes in Physics and Chemistry*, 3d ed. Amsterdam: North-Holland, 2007.

[30] P Dayan and L Abbott. *Theoretical Neuroscience: Computational and Mathematical Modeling of Neural Systems*. Cambridge, MA: MIT Press, 2005.

[31] H Berg. *E. coli in motion*. New York: Springer, 2004.

[32] CE Shannon and W Weaver. *A mathematical theory of communication*. Urbana: University of Illinois Press, 1949.

[33] T Cover and J Thomas. *Elements of Information Theory*, 2d ed. New York: Wiley-Interscience, 2006.

[34] JW Gibbs. *The Scientific Papers of J. Willard Gibbs*, vol. 1. Woodbridge, CT: Ox Bow Press, 1993.

[35] S Strong, R Koberle, R de Ruyter van Steveninck, and W Bialek. Entropy and information in neural spike trains. *Phys Rev Lett*, 80(1):197–200, 1998.

[36] J Kelly. A new interpretation of information rate. *IRE Transactions on Information Theory*, 2:185–9, 1956.

[37] C Bergstrom and M Lachmann. Shannon information and biological fitness. In *Information Theory Workshop, 2004*, pages 50–4. IEEE, 2005.

[38] E Kussell and S Leibler. Phenotypic diversity, population growth, and information in fluctuating environments. *Science*, 309:2075–8, 2005.

[39] M Donaldson-Matasci, M Lachmann, and C Bergstrom. Phenotypic diversity as an adaptation to environmental uncertainty. *Evo Ecology Res*, 10:493–515, 2008.

[40] O Rivoire and S Leibler. The value of information for populations in varying environments. *J Stat Phys*, 142:1124–1166, 2011.

[41] J Seger and H Brockmann. *What Is Bet Hedging?*, vol. 4. New York: Oxford University Press, 1987.

[42] H Beaumont, J Gallie, C Kost, G Ferguson, and P Rainey. Experimental evolution of bet hedging. *Nature*, 462:90–93, 2009.

[43] S Frank. Natural selection maximizes Fisher information. *J Evol Biol*, 22:231–44, 2009.

[44] N Balaban, J Merrin, R Chait, L Kowalik, and S Leibler. Bacterial persistence as a phenotypic switch. *Science*, 305:1622–5, 2004.

[45] E Kussell, R Kishony, N Balaban, and S Leibler. Bacterial persistence: a model of survival in changing environments. *Genetics*, 169:1807–14, 2005.

[46] H Maamar, A Raj, and D Dubnau. Noise in gene expression determines cell fate in *Bacillus subtilis*. *Science*, 317:526–9, 2007.

[47] T Cağatay, M Turcotte, M Elowitz, J Garcia-Ojalvo, and G Süel. Architecture-dependent noise discriminates functionally analogous differentiation circuits. *Cell*, 139:512–22, 2009.

[48] CS Wylie, A Trout, D Kessler, and H Levine. Optimal strategy for competence differentiation in bacteria. *PLoS Genet*, 6:e1001108, 2010.

[49] N Bowen, L Walker, L Matyunina, S Logani, K Totten, B Benigno, and J McDonald. Gene expression profiling supports the hypothesis that human ovarian surface epithelia are multipotent and capable of serving as ovarian cancer initiating cells. *BMC Med Genomics*, 2:71, 2009.

[50] A Arkin and J Ross. Statistical construction of chemical reaction mechanisms from measured time-series. *J Phys Chem*, 99:9709, 1995.

[51] W Bialek, I Nemenman, and N Tishby. Predictability, complexity, and learning. *Neural Comput*, 13:2409–63, 2001.

[52] F Creutzig, A Globerson, and N Tishby. Past-future information bottleneck in dynamical systems. *Phys Rev E*, 79:041925, 2009.

[53] M Srinivasan, S Laughlin, and A Dubs. Predictive coding: a fresh view of inhibiton in retina. *Proc Roy Soc B*, 216:427–59, 1982.

[54] T Hosoya, S Baccus, and M Meister. Dynamic predictive coding by the retina. *Nature*, 436:71–77, 2005.

[55] M Vergassola, E Villermaux, and B Shraiman. "Infotaxis" as a strategy for searching without gradients. *Nature*, 445:406–9, 2007.

[56] G Schwartz, R Harris, D Shrom, and MJ Berry. Detection and prediction of periodic patterns by the retina. *Nature Neurosci*, 10:552–4, 2007.

[57] I Tagkopoulos, Y-C Liu, and S Tavazoie. Predictive behavior within microbial genetic networks. *Science*, 320:1313–7, 2008.

[58] A Mitchell, G Romano, B Groisman, A Yona, E Dekel, M Kupiec, O Dahan, and Y Pilpel. Adaptive prediction of environmental changes by microorganisms. *Nature*, 460(7252):220–4, 2009.

[59] I Nemenman, F Shafee, and W Bialek. Entropy and inference, revisited. In TG Dietterich, S Becker, and Z Gharamani, editors, *Advances in Neural Information Processing Systems*, vol. 14, 2002.

[60] L Paninski. Estimation of entropy and mutual information. *Neural Comput*, 15:1191–253, 2003.

[61] N Tishby, F Pereira, and W Bialek. The information bottleneck method. In B Hajek and RS Sreenivas, editors, *Proc 37th Annual Allerton Conference on Communication, Control and Computing*, pages 368–77. Urbana: University of Illinois, 1999.

[62] N Wiener. *Extrapolation, Interpolation, and Smoothing of Stationary Time Series*. Cambridge, MA: MIT Press, 1964.

[63] H Barlow. Sensory mechanisms, the reduction of redundancy, and intelligence. In D Blake and A Utlley, editors, *Proc Symp Mechanization of Thought Processes*, vol. 2, page 53774. London: HM Stationery Office, 1959.

[64] H Barlow. Possible principles underlying the transformation of sensory messages. In W Rosenblith, editor, *Sensory Communication*, page 21734. Cambridge, MA: MIT Press, 1961.

[65] J Atick and A Redlich. Toward a theory of early visual processing. *Neural Comput*, 2:30820, 1990.

[66] C Marshall. Specificity of receptor tyrosine kinase signaling: transient versus sustained extracellular signal-regulated kinase activation. *Cell*, 80:17985, 1995.

[67] W Sabbagh Jr, L Flatauer, A Bardwell, and L Bardwell. Specificity of MAP kinase signaling in yeast differentiation involves transient versus sustained MAPK activation. *Mol Cell*, 8:68391, 2001.

[68] G Lahav, N Rosenfeld, A Sigal, N Geva-Zatorsky, A Levine, M Elowitz, and U Alon. Dynamics of the p53-Mdm2 feedback loop in individual cells. *Nat Genet*, 36:147–50, 2004.

[69] L Ma, J Wagner, J Rice, W Hu, A Levine, and G Stolovitzky. A plausible model for the digital response of p53 to DNA damage. *Proc Natl Acad Sci USA*, 102:14266–71, 2005.

[70] P Hersen, M McClean, L Mahadevan, and S Ramanathan. Signal processing by the HOG MAP kinase pathway. *Proc Natl Acad Sci USA*, 105:7165–70, 2008.

[71] C-Y Huang and J Ferrel. Ultrasensitivity in the mitogen-activated protein kinase cascade. *Proc Natl Acad Sci USA*, 93:10078, 1996.

[72] N Markevich, J Hock, and B Kholodenko. Signaling switches and bistability arising from multisite phosphorylation in protein kinase cascades. *J Cell Biol*, 164:354, 2004.

[73] A Mugler, E Ziv, I Nemenman, and C Wiggins. Quantifying evolvability in small biological networks. *IET Syst Biol*, 3:379–87, 2009.

[74] A Borst, V Flanagin, and H Sompolinsky. Adaptation without parameter change: dynamic gain control in motion detection. *Proc Natl Acad Sci USA*, 102:6172, 2005.

[75] I Nemenman. Gain control in molecular information processing: lessons from neuroscience. *Phys Biol*, Volume 9, Article No. 026003, 2012.

[76] P Iglesias. Feedback control in intracellular signaling pathways: regulating chemotaxis in *Dictyostelium discoideum*. *European J Control*, 9:227–36, 2003.

[77] R Normann and I Perlman. The effects of background illumination on the photoresponses of red and green cells. *J Physiol*, 286:491, 1979.

[78] D MacGlashan, S Lavens-Phillips, and K Miura. IgE-mediated desensitization in human basophils and mast cells. *Front Biosci*, 3:d746–56, 1998.

[79] P Detwiler, S Ramanathan, A Sengupta, and B Shraiman. Engineering aspects of enzymatic signal transduction: photoreceptors in the retina. *Biophys J*, 79:2801, 2000.

[80] C Rao, J Kirby, and A Arkin. Design and diversity in bacterial chemotaxis: a comparative study in *Escherichia coli* and *Bacillus subtilis*. *PLoS Biol*, 2:E49, 2004.

[81] T Friedlander and N Brenner. Adaptive response by state-dependent inactivation. *Proc Natl Acad Sci USA*, 106:22558–63, 2009.

[82] D Muzzey, C Gomez-Uribe, J Mettetal, and van Oudenaardem A. A systems-level analysis of perfect adaptation in yeast osmoregulation. *Cell*, 138:160–71, 2009.

Chapter 4

[83] A Anishkin and S Sukharev. State-stabilizing interactions in bacterial mechanosensitive channel gating and adaptation. *J Biol Chem*, 284:19153–7, 2009.

[84] V Belyy, K Kamaraju, B Akitake, A Anishkin, and S Sukharev. Adaptive behavior of bacterial mechanosensitive channels is coupled to membrane mechanics. *J Gen Physiol*, 135:641–52, 2010.

[85] U Alon, M Surette, N Barkai, and S Leibler. Robustness in bacterial chemotaxis. *Nature*, 397:168–71, 1999.

[86] C Hansen, R Enders, and N Wingreen. Chemotaxis in *Escherichia coli*: a molecular model for robust precise adaptation. *PLoS Comput Biol*, 4:e1, 2008.

[87] H Salis, E Mirsky, and C Voigt. Automated design of synthetic ribosome binding sites to control protein expression. *Nat Biotech*, 27:946–50, 2009.

[88] R Endres, O Oleksiuk, C Hansen, Y Meir, V Sourjik, and N Wingreen. Variable sizes of *Escherichia coli* chemoreceptor signaling teams. *Mol Syst Biol*, 4:211, 2008.

[89] N Kashtan and U Alon. Spontaneous evolution of modularity and network motifs. *Proc Natl Acad Sci USA*, 102:13773–8, 2005.

[90] P Reinagel and S Laughlin. Editorial: Natural stimulus statistics. *Network: Comput Neura Syst*, 12:237–40, 2001.

[91] D Baylor and A Hodgkin. Changes in time scale and sensitivity in turtle photoreceptors. *J Physiol*, 242:729–58, 1974.

[92] B Wark, A Fairhall, and F Rieke. Timescales of inference in visual adaptation. *Neuron*, 61:750–61, 2009.

[93] E Knudsen. Instructed learning in the auditory localization pathway of the barn owl. *Nature*, 417:322–8, 2002.

[94] B Andrews, T-M Yi, and P Iglesias. Optimal noise filtering in the chemotactic response of *Escherichia coli*. *PLoS Comput Biol*, 2(11):e154, Nov 2006.

[95] RC Yu, CG Pesce, A ColmanLerner, L Lok, D Pincus, E Serra, M Holl, K Benjamin, A Gordon, and R Brent. Negative feedback that improves information transmission in yeast signalling. *Nature*, 456:75561, 2008.

[96] L Yang and P Iglesias. Positive feedback may cause the biphasic response observed in the chemoattractant-induced response of *Dictyostelium cells*. *Syst Control Lett*, 55:329–37, 2006.

[97] E Sontag. Remarks on feedforward circuits, adaptation, and pulse memory. *IET Syst Biol*, 4:39–51, 2010.

[98] M DeWeese and A Zador. Asymmetric dynamics in optimal variance adaptation. *Neural Comp*, 10:1179–202, 1998.

[99] O Shoval, L Goentoro, Y Hart, A Mayo, and E Sontag a U Alon. Fold-change detection and scalar symmetry of sensory input fields. *Proc Natl Acad Sci USA*, 107:15995–16000, 2010.

[100] M Lazova, T Ahmed, D Bellomo, R Stocker, and T Shimizu. Response rescaling in bacterial chemotaxis. *Proc Natl Acad Sci USA*, 108:13870–13875, 2011.

[101] S Paliwal, P Iglesias, K Campbell, Z Hilioti, A Groisman, and A Levchenko. MAPK-mediated bimodal gene expression and adaptive gradient sensing in yeast. *Nature*, 446(7131):46–51, 200vv7.

[102] I Witten, E Knudsen, and H Sompolinsky. A hebbian learning rule mediates asymmetric plasticity in aligning sensory representations. *J Neurophysiol*, 100:1067–79, 2008.

[103] J Melin and S Quake. Microfluidic large-scale integration: the evolution of design rules for biological automation. *Annu Rev Biophys Biomol Struct*, 36:213–31, 2007.

[104] F Poelwijk, D Kiviet, D Weinreich, and S Tans. Empirical fitness landscapes reveal accessible evolutionary paths. *Nature*, 445:383–6, 2007.

[105] J Mettetal, D Muzzey, C Gomez-Uribe, and A van Oudenaarden. The frequency dependence of osmo-adaptation in *Saccharomyces cerevisiae*. *Science*, 319:482–4, 2008.

[106] I Nemenman. Fluctuation-dissipation theorem and models of learning. *Neural Comput*, 17:2006–33, 2005.

5. Quantitative In Vitro Biochemistry, One Molecule at a Time

Jeffrey A. Hanson

Haw Yang

5.1   Introduction

A first-principle understanding of the molecular basis for enzyme function has long been a major goal of the field of biochemistry. Due to the complexity of the problem, achieving this goal will require collaborative efforts of biologists, chemists, physicists, and mathematicians. The focus of this chapter discusses contributions of high-resolution in vitro single-molecule studies toward a quantitative understanding the functional role of large-amplitude conformational dynamics to enzyme biochemistry.

The conventional paradigm in understanding enzyme function is the *structure–function relationship*, which dictates that a thorough understanding of molecular structure is sufficient for understanding enzyme function. The revolution in structural biology techniques, primarily X-ray crystallography, nuclear magnetic resonance (NMR), and electron microscopy—all capable of providing snapshots of protein molecules at atomic

Chapter 5

Quantitative Biology: From Molecular to Cellular Systems edited by Michael E. Wall © 2012 CRC Press / Taylor & Francis Group, LLC. ISBN: 978-1-4398-2722-2

resolution—has enabled important discoveries about the molecular details of many enzymes. The molecular picture availed by these methods is averaged over an ensemble of molecules. Such an ensemble-averaged picture is incomplete since each averaged structure represents a single relatively easily populated, low-energy conformation of the protein. The existence and functional roles of higher energy conformational substates and transition states that are not significantly populated under experimental conditions cannot be directly studied by such methods. A number of general questions remain regarding the room-temperature solution structure of enzymes, including the following: Are other additional conformational states populated? What are the relative energetics and rates of transitions between states? Do conformational substates play important roles for enzyme function?

In contrast to the static models provided by crystallographic methods, the solution structure of proteins is highly dynamic. Thermal fluctuations and collisions with solvent molecules drive spontaneous structural transitions which span a wide range of time and length scales (Figure 5.1). Given the ubiquitous nature of conformational fluctuations in enzymes, the hypothesis that protein structures have evolved to harness productive motions that promote catalysis or provide other benefits is intuitively attractive. The large array of time and length scales for molecular motions in proteins makes their complete characterization difficult using ensemble methods because it is nearly impossible to synchronize the motion of a large number of molecules at longer timescales. To fully understand the biological ramifications of molecular motions, it could be argued that the most interesting will be the motions that occur on a timescale similar to those in biological processes. These relatively slow motions in the μs-min time regime are also predicted to coincide with relatively large-amplitude structural changes in proteins (Figure 5.1). Until all-atom molecular dynamics simulations have reached a point where they are capable of spanning the entire range of relevant timescales, our understanding of motions in enzymes must be cobbled together using complementary experimental techniques. The fastest ranges of motion are easily accessible via many time-resolved methods such as neutron scattering [1,2], time-resolved crystallography [3], and optical spectroscopes [4–6]. NMR has to date been the most informative

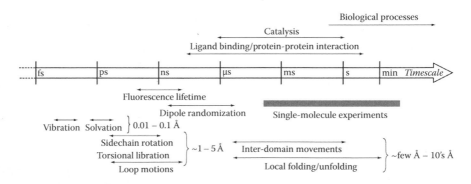

FIGURE 5.1   Timescale and length scale of molecular motions in proteins. The types of molecular motion in proteins and corresponding length scales are summarized below the time axis. Tasks that a protein may perform are indicated above the time axis. Also included are the processes that are relevant to single-molecule FRET measurements: fluorescence lifetime, dipole randomization, and single-molecule measurement time. (From Yang [37]. With permission.)

experimental technique for studying longer timescale functionally important large-amplitude protein dynamics since motions in the micro-to-millisecond regime can be observed with the help of spectral analysis and modeling [7].

Optical single-molecule methods offer great promise toward resolving large-amplitude conformational dynamics in enzymes [8,9]. By studying one molecule at a time, the issues of ensemble averaging can be avoided altogether. In combination with fluorescence resonance energy transfer (FRET), they are readily able to resolve nanometer changes in distance on a timescale of milliseconds to minutes. Thus, single-molecule experiments are complementary to existing methods and offer a means to improve our understanding of the molecular constraints that govern large-amplitude protein conformational motions at biologically relevant timescales. The advantage of studying one molecule at a time also creates the greatest challenge for the experimentalist. Fluctuations that would otherwise be averaged out in ensemble experiments, for instance, the stochastic photo-physics of photon absorption, emission and detection in a fluorescence experiment, and the random movements of molecular conformations, are present in nontrivial levels when a single molecule is observed. Thus, one of the most challenging aspects of single-molecule experiments is reliable analysis of intrinsically noisy experimental data. A quantitative approach to the analysis of single-molecule data requires pushing the technique to its limits to obtain the highest possible resolution data about the molecular fluctuations of enzyme molecules: a goal requiring careful experimental practices as well as an advanced statistical treatment to obtain accurate results with the highest possible resolution [10–12].

The range of problems to which single-molecule experiments have been applied is broad enough to fill multiple volumes; thus, this chapter is restricted to the application of optical single-molecule experiments using fluorescence detection by confocal microscopy. For a more complete picture of the applications of single-molecule FRET to biology, the reader is referred to many excellent reviews in the related fields of protein folding [13], molecular motors [14–16], and RNA folding and ribozymes [17].

5.2   Single-Molecule Fluorescence Spectroscopy

Single-molecule fluorescence experiments are rapidly becoming one of the most popular techniques in the biophysical toolbox for the characterization of biological molecules. They may be readily be performed by a nonspecialist since they require relatively inexpensive instrumentation and only a basic understanding of spectroscopy and optics [18,19]. In addition, fluorescence experiments are highly configurable allowing a wide array of experimental strategies that are highly adaptable in terms of experimental design and setup. Indeed, there are so many permutations to single-molecule fluorescence experiments that a comprehensive review of all available techniques is beyond the scope of this chapter. Instead, this section will focus on describing the basics of single-molecule fluorescence experiments as well as introducing the common challenges that the experimentalist must contend regardless of the specifics of experimental design. Specific emphasis will be placed on high-resolution single-molecule FRET experimental and data analysis methods, which are required for the quantitative analysis of large-amplitude conformational transitions in enzymes.

Though many types of optical spectroscopes are in principle amenable to single-molecule detection [20,21], florescence is perhaps the most ideal due to its intrinsically

Chapter 5

high signal-to-noise ratio. This allows single fluorescent molecules to be optically detected on a standard laboratory microscope with the appropriate detectors and optics despite the fact that a typical protein is two orders of magnitude smaller than the diffraction limit of light. In addition, fluorescent probes are sensitive reporters of their local environment [22], making them ideal for detecting subtle changes in protein structure due to conformational dynamics. Although many aspects of molecular fluorescence, such as emission intensity, fluorescence lifetime, and fluorescence polarization [23], have been used in single-molecule studies, this chapter will focus primarily on the application of single molecule FRET experiments to the study of large-amplitude conformational dynamics in enzymes.

5.2.1   Fluorescent Probes

Many fluorescent chromophores have been employed in single-molecule experiments, with the choice of fluorescent probe generally dictated by experimental requirements (also see Chapter 7). Probes can be broadly divided in to two categories: (1) intrinsically fluorescent probes, such as fluorescent cofactors [24,25] or proteins [26]; and (2) extrinsic probes, generally synthetic dyes that must be attached to the enzyme after expression and purification. Though intrinsically fluorescent probes are indispensible for in vivo fluorescence experiments [27] and cofactors have the advantage of not significantly perturbing the enzyme structure, extrinsic probes offer far better optical properties for the purposes of high-resolution single-molecule spectroscopy [28]. Synthetic dyes are significantly brighter than naturally occurring chromophores, since they have a higher absorbance cross section and emission quantum yield, leading to more detected photons and higher-resolution experimental data. In addition, synthetic dyes can be engineered to absorb longer wavelength light than the naturally occurring chromophores, significantly increasing the experimental signal-to-noise and reducing the risk of photo damage to the sample.

Synthetic dyes require a method to specifically attach them to the protein of interest to facilitate single-molecule experiments. The most common approaches employ amino acid specific chemistry, for instance to the primary amine in lysine or the thiol of cysteine [29]. However, many other novel approaches to this problem have been introduced including catalytic addition of dyes [30], unnatural amino acid incorporation [31], intien mediated protein [32], or noncovalent specific chelation [33]. Attachment of synthetic dyes to cysteine residues is advantageous since many thiol-reactive dyes are commercially available and since proteins typically contain very few surface-exposed cysteine residues. Once the desired probe location has been identified, cysteine residues can readily be genetically added to the protein sequence through site directed mutagenesis. After derivitizing the enzyme with any fluorescent probe, it is critically important to properly design control experiments to ensure that neither the activity of the enzyme nor the photo-physical characteristics of the chromophore have been perturbed.

Two important photo-physical effects must also be considered in single-molecule fluorescence experiments: photobleaching and triplet-state blinking. Photobleaching occurs from the irreversible destruction of the dye, typically caused by reaction of the excited state of the dye with another molecule. Photobleaching of the chromophores is inevitable and can significantly limit the duration of single-molecule experiments. Triplet-state blinking occurs when the singlet electronic excited state of the chromophore undergoes intersystem crossing to a long-lived triplet state. Since the relaxation of the triplet excited state to the singlet electronic ground state is quantum mechanically forbidden. In stark contrast to the nanosecond lifetime singlet excited state, the triplet excited state can have a lifetime of seconds in the absence of triple quenchers such as triplet oxygen, resulting in the blinking phenomenon of the chromophore. This blinking effect can easily be confused with more interesting events and may results in erroneous interpretation of experimental results. Both of these effects can be mitigated with the addition of deoxygenating agents, which prevent reaction of the chromaphore excited-state with molecular oxygen, and triplet-state sensitizers, which can help the dye relax to its singlet ground state [34,35].

5.2.2   Förster–Type Resonance Energy Transfer

FRET is the optimal method for studying large-amplitude fluctuations in protein structure at the single-molecule level (see also Chapter 7). FRET is a distance-dependent, non-radiative energy transfer process between two chromophores, a donor and an acceptor, which has been commonly used as a molecular ruler [36]. Importantly, single-molecule FRET is useful for measuring distances in the nanometer range with typical timescales ranging from milliseconds to minutes, limited, respectively, by the rate at which photons can be acquired and the rate of dye photobleaching, thus placing the technique in a range that is useful for monitoring functionally relevant protein motions since these are the timescales and length scales expected for these types of fluctuations.

Accuracy of FRET distance measurements is of particular interest in high-resolution single-molecule experiments. One of the major assumptions underpinning the equations governing FRET is that the orientations of the donor and acceptor dipoles are randomized on fast timescales. Though complete randomization will never be achieved for the case of dyes attached to proteins, it has been shown that measurement errors resulting from incomplete randomization should be less than ~10% given a relatively unrestricted motion of the dipole moment on the single-molecule experimental timescale of milliseconds [37]. In addition, the proper choice of labeling positions and FRET probes is also required for quantitative distance measurements. Due to the characteristic R^{-6} distance dependence for dipole–dipole energy transfer, measurements made at distances close to R_0 (a FRET efficiency of 0.5) where the distance versus efficiency curve is the steepest will be more accurate than measurements of distances much shorter or greater than R_0 ($E = 0$ or 1) where the distance versus efficiency curve is nearly flat. Thus, the experimentalist should carefully choose locations for FRET probes so that the measured distances will fall in the most accurate region of the FRET regime, approximately $0.7 < R/R_0 < 1.5$ [38].

5.2.3   Single–Molecule Microscopy

Single-molecule experiments have a seemingly limitless number of permutations since it is possible to imagine a wide variety of physical observables depending on experimental design and the questions asked. Commonly used parameters include fluorescence intensity, fluorescence lifetime, emission spectra, and dipole orientation [23]. Many good references are available describing the details of experimental configurations

Chapter 5

[18,19]; however, this section briefly discusses general single-molecule instrumentation as it pertains to high-resolution studies of protein conformational dynamics.

Single-molecule fluorescence experiments can be generally broken into two categories depending on the microscope configuration: confocal and wide field. Confocal microscopy uses a laser point source to illuminate a single diffraction-limited volume of sample, which is then imaged onto a detector, commonly a single-photon counting avalanche photo diode (APD). APDs have a nominal time resolution of hundreds of picoseconds, thus enabling detection of single photon arrival events with a high temporal resolution, including measurements of single-molecule fluorescence lifetimes given a pulsed laser source. Collection of individual photon arrival times preserves the information contained in each event and enables advanced statistical treatment of the raw data, which is critical for achieving the highest time and special resolution in single-molecule measurements [38–40]. Confocal microscopy is primarily limited by the long data acquisition time needed for accumulating sufficient statistics. Since data can be obtained only on one molecule at a time, collection of a complete data set can take days to weeks because many of the collected trajectories are discarded during analysis. Rejected single-molecule trajectories typically arise from trivial experimental issues such as multiple molecules in the illumination volume or proteins that do not contain an active FRET donor and acceptor pair. In addition, trajectories containing significant blinking or other uncorrelated changes in intensity are frequently removed. The other common type of single-molecule experiment involves wide-field detection with total internal reflection fluorescence (TIRF) collected using a high-speed charge coupled device (CCD). TIRF illumination uses light at a higher incidence angle than the critical angle reflected off of the microscope slide to create an evanescent field that penetrates ~100 nm into the sample [41]. The time resolution is limited by the frame rate of the CCD, commonly ~1–5 ms, which generally limits effective experimental time resolution to even longer values. Wide field-based approaches offer a great advantage since data can be collected on many molecules simultaneously. However, the limitations imposed by CCD frame rate as well as the vendor-dependent noise characteristics of the cameras prevent the use of wide-field methods for high-resolution FRET studies of enzyme conformational dynamics since the motions are expected to be on the millisecond timescale.

Sample immobilization is another concern in single-molecule FRET experiments. Regardless of experimental configuration, high-resolution experiments of large-amplitude conformational dynamics demand immobilized samples since the average residence time of a freely diffusing molecule through the confocal illumination volume would be ~1 ms (see Chapter 8 for a discussion of protein diffusion). Extreme care must be used in designing an immobilization strategy, however, since proteins commonly interact with the quartz surfaces used in single-molecule experiments. The most common technique is passivation of the quartz with an inert molecule such as polyethylene glycol (PEG) [42,43], to which proteins can be bound using specific chemistry such as covalent bonds, Ni-NTA, or biotin/streptavidin. Though many different immobilization strategies have been employed in single-molecule experiments, the general requirements are that they be specific and inert. Regardless of the method used, it is of critical importance to demonstrate that proximity to the quartz cover slip and other components of the immobilization scheme have not perturbed the conformational dynamics or reactivity of the enzyme being studied.

5.2.4   Single-Molecule Data Analysis

Data analysis is the most challenging part of single-molecule fluorescence experiments. Single-molecule experiments have an intrinsically low signal-to-noise ratio due to the Poisson statistics arising from each element of the experiment including the excitation source, the chromophore, and the detection system. If not properly accounted for in data processing, this noise can be mistaken for actual events in the single-molecule time trace, leading to erroneous conclusions. For studies of conformational dynamics with single-molecule FRET data, processing can be subdivided into three basic steps: (1) transformation of fluorescence emission into distance; (2) determination of the number of states; and (3) determination of the rates of interconversion. This section briefly discusses the most common method for analyzing single-molecule data and reviews advanced techniques that can push FRET measurements to their highest possible resolution.

Interpretation of noisy single-molecule FRET data relies on temporal averaging or filtering to facilitate interpretation of the signal. The simplest form of temporal averaging is to divide the single-molecule trajectory into equal time segments and to calculate the FRET efficiency for each time point using a ratiometric calculation such as

$$E = \frac{I_A}{\gamma I_D + I_A} \tag{5.1}$$

where I_A and I_D are acceptor and acceptor intensities, respectively, and γ is an experimentally determined correction factor accounting for differences in collection efficiency and quantum yield between the two chromophores. This technique of constant-time binning has been used extensively in single-molecule experiments with great success; however, it is most appropriate for studying slow dynamics (> 50 ms) corresponding to large distance changes (> 1 nm) due to uncertainties in the analysis that prevent it from truly being quantitative. The choice of time-averaging window size is an arbitrary choice and can significantly affect the results of the analysis: If the time window is too small, the distance trajectory will be dominated by noise; however, if the time window is too large, the noise can be mitigated but interesting dynamics of the molecule may also be missed. Constant-time analysis also lacks a way to accurately access the error associated with distance measurements, which is a prerequisite for quantitative experimental methods. Due to the R^{-6} dependence between energy transfer efficiency and distance the error associated with FRET measurements is not linearly related to measured distance and in practice depends on both the number of photons collected as well as the FRET efficiency being measured [38]. Thus, single-molecule measurements based on constant-time binning have an unknown amount of error that is different for each distance determination in the time series. The uncertainty in distance measurements further complicates accurate determination of both the number of states accessible to the molecule and the determination of rates since the results may be highly dependent on the initial choice of time-averaging window.

To achieve quantitative analysis of single-molecule data, particularly when the timescales of molecular motion are similar to the intrinsic experimental time resolution, several groups have developed more advanced data analysis methods based on

Chapter 5

application of information theoretical treatment of the raw photon arrival time serves to accurately account for the complexities of single-molecule photon-arrival data [38,39,44]. The methods proposed by Watkins et al. [40] are best suited to extracting sub-millisecond distance-time trajectories from single-molecule intensity time traces and will be discussed here. This method uses a maximum likelihood estimator to calculate the distance from a series of photon arrival events while explicitly accounting for the Poisson statistics of photon counting at the single molecule level, background, and cross-talk between the donor and acceptor channels:

$$\hat{E} = \frac{I_d^\beta n_{a-} I_a^\beta n_d \beta_a^{-1}}{I_d^\beta n_a (1 - \beta_d^{-1}) + I_d^\beta n_a (1 - \beta_a^{-1})} \tag{5.2}$$

where I_d^β is the full intensity of the donor in the presence of background, n_a is the number of acceptor photons, and β_a is the acceptor signal-to-noise ratio. Information theory allows the calculation of the variance associated which each distance measurement by applying the Cramér-Rao-Fréchet inequality, which states that the variance of an unbiased estimator is greater than or equal to its Fisher information [45]. An algorithm is then used to determine the optimum number of photons for each measurement so that each distance determination in the time series has the same variance. Since the number of photons and the distance both determine the error, each measurement in this analysis method has a different time window rather than enforcing a constant time interval for averaging. In other words, this method enforces a temporal averaging scheme where each measurement has a constant amount of information but is integrated over a variable time window.

The number of states present in single-molecule experiments is commonly assessed using distance distributions created by constructing histograms of distances measured in an experiment. These histograms are usually quite broad due to the presence of a large amount of noise in the underlying data, and it is often impossible to determine the number of conformational states accessible to the system without significant assumptions of the number of underlying conformational states and curve fitting. The maximum information method presents unique a solution to this problem: since all of the distance measurements have the same amount of relative error, it is possible to use maximum entropy deconvolution to remove this well-characterized error and recover the true underlying distribution [40]. This method assumes that the measured distribution is a convolution of the true distribution with the well-characterized experimental error. The maximum entropy deconvolution procedure then searches for the simplest solution to the problem by maximizing the information entropy, which corresponds to the distribution with the smoothest curve and fewest numbers of states. Recovery of the denoised true probability density function is performed in an unbiased manner and can often visually report the number of conformational states in the system in a way that no longer requires a priori knowledge of the system or curve fitting.

Once the number of states accessible to the system has been identified, the experimentalist is interested in their rates of interconversion. Conventionally this is done by choosing a threshold and assigning a state to each distance measurement in the time series. This is not appropriate in the case of experiments where the transition rate is comparable to the experimental timescale since the results will be highly dependent on user-determined parameters such as the bin size or the position of the threshold. Instead, several groups have proposed methods where the rates of interconversion can be determined by modeling of shape the time-dependent probability distance distributions [46,47]. This process is similar to motional narrowing in NMR and optical spectroscopies [48,49] and has been solved explicitly for the case of single-molecule experiments [50,51]. In the method proposed by Hanson et al. [52], maximum entropy deconvoluted distance distributions are constructed over a range of timescales by changing the variance in the maximum information analysis over a range of 7–15% relative error. Since a less accurate measurement in general requires fewer photons than a more accurate determination this method allows for the timescale of the analysis to be changed over a small range, whereas the additional experimental error introduced by less accurate measurements should be eliminated by the maximum entropy deconvolution without introducing serious artifacts. This series of time-dependent distributions can then be fit to a two-state motional narrowing model to extract the mean positions of two states as well as their rates of interconversion [47,52].

5.3 Enzyme Conformational Dynamics

The three-dimensional solution structure of enzyme molecules at room temperature is highly dynamic. The current physical picture of protein structure is an ensemble of rapidly interconnecting low-energy conformations rather than a single discrete structure. Thus, the conformational coordinate of proteins can be considered a rough energy landscape with peaks and valleys of a wide range of sizes corresponding to the rates of interconversion between different states (see Chapter 1 for a review of energy landscapes, as well as [53]).

The basic question regarding enzyme conformational dynamics is understanding the relationship between these motions and enzyme function. Have enzymes evolved their structure to harness spontaneous structural fluctuations in a productive way, or is a single enzyme conformation active while other structures are nonfunctional? Due to the complexity of proteins, it seems inevitable that both of these scenarios are partially correct. However, despite the relative simplicity of the hypothesis, unambiguous experimental determination of these issues is complicated by the difficulties of measuring enzyme conformational dynamics, particularly on the relevant microsecond-minute timescales.

5.3.1 Single-Molecule Enzyme Turnover

Early single-molecule experiments used direct observation of catalytic turnover to relate enzyme function to conformational dynamics. Though not direct observations of protein conformational dynamics, these studies shed fascinating light on the possible role of slow conformational fluctuations in relation to enzyme function (for recent reviews see [23,54]). By observing single turnover events, the behavior of individual molecules can be compared to that expected from the conventional Michaelis-Menten kinetic treatment. Deviation from ideal behavior is expected to arise from coupling of the catalysis to other events in the system: notably slow conformational transitions between substates.

In these experiments, a fluorescent reporter—such as a cofactor [24,55], labeled enzyme [56], or fluorescent product [57,58]—is used to monitor the time dependent turnover of a single enzyme molecule and characterize the statistics of individual events.

The first of such experiments was performed on cholesterol oxidase, whose fluorescent FAD cofactor is reduced to nonfluorescent $FADH_2$ during enzymatic turnover [24]. The mechanism for this reaction is

$$E\text{-FAD} + S \rightleftharpoons E\text{-FAD-S} \rightarrow E\text{-FADH}_2 + P$$

$$\text{on} \qquad\qquad \text{on} \qquad\qquad \text{off} \qquad\qquad\qquad\qquad (5.3)$$

$$E\text{-FADH}_2 + O_2 \rightleftharpoons E\text{-FADH}_2\text{-}O_2 \rightarrow E\text{-FAD} + H_2O_2$$

$$\text{off} \qquad\qquad \text{off} \qquad\qquad \text{on}$$

By trapping the enzyme molecules in an agarose gel through which the substrate and product could freely diffuse, Lu et al. [24] were able to monitor the fluorescent transitions of the FAD cofactor between on and off states as the enzyme underwent catalysis (Figure 5.2a). According to the standard Michaelis-Menten interpretation, each catalytic event should be independent, which would result in an exponential decay of the intensity autocorrelation function. However, when such an analysis was performed for the emission from a single enzyme, it was found that a single exponential did not give a satisfactory fit (Figure 5.2b). Analysis of joint probability distributions between sequential catalytic events revealed a molecular memory effect in which short catalytic events are likely to be followed by short events and vice versa for long events. This correlation is revealed by a diagonal element in a 2-D joint probability distribution monitoring the length of neighboring turnover events (left panel, Figure 5.2c). While this holds for subsequent turnover events, if one looks at the correlation between turnovers separated by 10 events the diagonal element disappears (right panel, Figure 5.2c). Interestingly, this single-molecule experiment was able to identify and characterize both static heterogeneity (molecule-to-molecule variation) and dynamic heterogeneity (variation within an individual molecule as a function of time). These two phenomena are nearly impossible to differentiate in conventional ensemble-averaged experiments. The dynamic disorder observed in individual molecules is due to a time-dependent catalytic rate suggested to result from conformational transitions in the enzyme itself on a timescale slower that the average turnover rate. It was found that a two-state model, with two distinct catalytic rates as well as interconversion between the conformational states, was adequate to explain observed intensity autocorrelation functions from single molecules [24]. Further evidence supporting the existence of conformational changes in cholesterol oxidase was obtained by observing the spectral fluctuations of emission from the FAD cofactor in the absence of substrate. An autocorrelation function of the spectral mean could be fit to a single exponential with a lifetime comparable to the interconversion rates used to fit the two-state model for catalysis.

Subsequent experiments measuring the catalytic turnover of other enzymes have found dynamic heterogeneity in many different systems [56,57,59,60]. Though these experiments do not directly prove the presence of conformational changes in proteins,

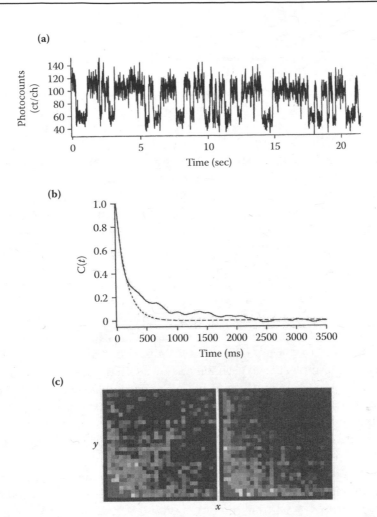

FIGURE 5.2 (See color insert.) Single-molecule study of FAD fluorescence in cholesterol oxidase (COx). (a) Sample intensity versus time trajectory showing characteristic on and off blinking behavior as the FAD cofactor is oxidized and reduced during its catalytic cycle. (b) Intensity autocorrelation function for trajectory in (a) (black line). The data have been fit to a single exponential model (dashed line). (c) 2-D conditional probability distribution for the length of subsequent on times showing a diagonal feature (left) and with a lag of 10 on times showing no correlation in lengths (right). (From Lu et al. [24]. With permission.)

they certainly provide intriguing evidence. A key question is therefore whether such slow conformational fluctuations exist in proteins.

5.3.2 Single-Molecule Studies of Protein Structure

To illustrate that enzyme conformational dynamics are playing a direct role in the proteins function, we need an experimental measure of the protein's structure as a function of time. As previously discussed single-molecule FRET is ideally suited for this task, though the timescales expected for enzyme conformational dynamics require high-resolution single-molecule approaches described in Section 5.2.

Chapter 5

5.3.2.1 A Direct Role for Conformational Dynamics in Catalysis

Adenylate kinase (AK) is a model system that has played an important role in our understanding conformational dynamics in enzymes and has been extensively studied as a model system [47,61–64]. This ubiquitous enzyme catalyzes the reversible reaction ATP + AMP ↔ 2·ADP and is responsible for maintaining the energy balance in cells [65]. Crystal structures of the *E. coli* enzyme demonstrate the presence of two lid domains: (1) the ATP_{lid} domain covering the ATP binding site; and (2) the NMP_{bind} domain covering the AMP binding site (Figure 5.3). From crystallographic evidence it was proposed that the function of the lid domain is to protect the active site from water during catalysis [66].

To directly observe the conformational motions of the ATP_{lid}, high-resolution single-molecule FRET experiments were used to monitor a protein coordinate between the lid domain and the core of the enzyme as the enzyme fluctuated in solution [47]. A sample single-molecule intensity versus time trajectory and subsequent maximum likelihood analysis [38] for ligand-free AK are displayed in Figures 5.4a–b. The observed distances fluctuate over a range consistent with the open and closed states of the ATP_{lid} domain observed in crystal structures [64]. Maximum entropy deconvoluted probability density functions (PDFs) [40] created from over 200 individual single-molecule trajectories clearly demonstrated the existence of two distinct states in both the ligand-free protein and when bound to the nonhydrolyzable substrate analogs AMP-PNP and AMP (Figure 5.4d). Single-molecule experiments showed that the closed conformation is favored for both ligand-free and ligand-bound enzymes, the effect of ligand-binding being a redistribution of the conformational subpopulations to favor the closed state as well as a compaction of the closed state to shorter distances.

To demonstrate the functional importance of the AK lid dynamics it is necessary to show that the two states are interconverting and that this rate is consistent with the overall catalytic rate of the enzyme. By changing the time resolution at which the deconvoluted PDFs are constructed over a range of ~2–10 ms, the two conformational modes observed at the fastest time resolution gradually coalesce into a single distribution (Figure 5.5), indicating exchange between the open and closed state on a timescale similar to our experimental time resolution. Global fitting of the time-dependent

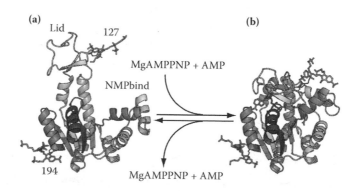

FIGURE 5.3 (See color insert.) Adenylate kinase (AK) crystal structures. (a) Substrate-free structure with lids in an open state. Dyes have been modeled in at the labeling positions used in single-molecule experiments (A127C and A194C). (b) AMP-PNP and AMP bound AK crystal structure with both lids in the closed state. (From Hanson et al. [47]. With permission.)

PDFs to a two-state motional narrowing model were used to determine the opening and closing rates for the AK lid. This analysis demonstrated that AK is indeed fluctuating between its open and closed state on a millisecond timescale, even in the presence of ligand. This analysis also demonstrated that the binding of nucleotide analogs reweights the AK conformational populations by changing the relative closing rates ($k_{close,apo}$ = 220 ± 110 s^{-1}, $k_{close,AMPPNP}$ = 440 ± 220 s^{-1}) while the opening rates remain relatively unchanged.

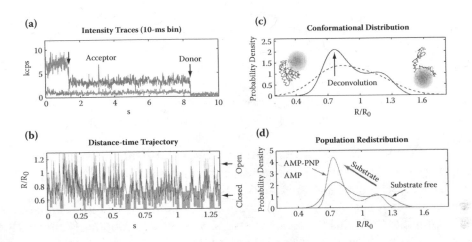

FIGURE 5.4   AK single-molecule data. (a) Intensity versus time trajectory for a single substrate-free AK molecule, arrows indicate the time at which each fluorophore is irreversibly photobleached. (b) Distance versus time trajectory from (a) constructed using the maximum likelihood method. Gray boxes represent the uncertainty in distance and time for each measurement. (From Watkins and Yang [38]. With permission.) (c) Probability density function constructed from > 400 substrate-free AK molecules. The dotted line shows the raw probability density function, whereas the solid has had photon counting error removed by entropy-regularized deconvolution. (From Watkins et al. [40]. With permission.) (c) Comparison of deconvoluted probability density functions for substrate-free AK (blue) and AK in the presence of substrate analogs AMPPNP and AMP (red). All distances are normalized, R/R_0, where R_0 is the Forster radius or 51 Å in the present work. (From Hanson et al. [47]. With permission.)

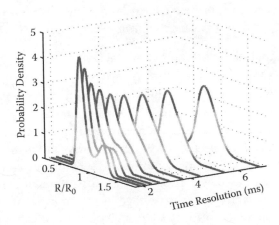

FIGURE 5.5 (See color insert.)   Time-dependent deconvoluted probability density functions for AK with AMPNP and AMP showing motional narrowing. (From Hanson et al. [47]. With permission.)

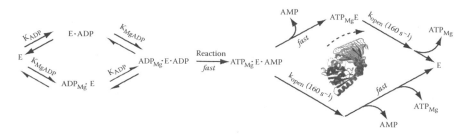

FIGURE 5.6 Mechanistic model for AK's reverse reaction integrating lid opening as the rate-limiting step of the reaction. (From Hanson et al. [47]. With permission.)

Since it was found that the opening rate in the presence of ligand was similar to the catalytic rate for the AK reverse reaction (2·ADP → ATP + AMP), a reasonable hypothesis is that the opening rate in the presence of nucleotide is controlling the overall catalytic rate of the enzyme. A kinetic model was constructed in which phosphate transfer is fast in the ternary, closed-lid enzyme-substrate complex with the overall rate of the reaction determined by the lid opening required for product release (Figure 5.6). This model was found to be consistent with ensemble, initial-rate kinetic data using the opening rate derived from single-molecule experiments as the rate-limiting step [47].

This experiment demonstrates an application of high-resolution single-molecule FRET techniques to the field of biochemistry, since it was quantitatively shown that the proteins' conformational dynamics are likely to control AK's catalytic rate. In addition, single-molecule experiments enable a detailed view of the energy landscape governing the AK solution structure. Though the induced fit picture from AK crystal structures predicts that the enzyme should favor the open state in the absence of ligand and the closed state once ligands have bound, single-molecule experiments reveal that the solution structure of ligand-free AK actually favors the a relatively closed state. Though it is seemingly counterintuitive that the substrate-free enzyme would favor conformation in which there is no access for substrates to the active site, this observation likely reflects the evolutionary adaptation for the conflicting pressures of rapid turnover and active site protection from solvent. Indeed, Nussinov et al. [67,68] argued that all enzyme conformations must be part of the protein energy landscape before and after ligand binding, as has been demonstrated for the case of *E. coli* AK.

5.3.2.2 Novel Adaptation of Conformational Dynamics

The protein tyrosine phosphatase B (PtpB) from *Mycobacterium tuberculosis* represents another demonstration of the unique ability high-resolution single-molecule methods to provide a novel perspective on the functional role of large-amplitude conformational dynamics in enzymes [52]. *M. tb* PtpB is an intriguing system about which relatively little is known where single-molecule studies can greatly aid the molecular picture of the enzyme's function. The *M. tb* genome contains two tyrosine phosphatases yet no protein tyrosine kinases [69] and evidence exists that PtpB is secreted from the cells [70] and required for long-term viability in macrophage infection [71]. Interestingly, the protein structure has a novel sequence forming a two-helix lid covering the protein active site in available the crystal structures (Figure 5.7) [72,73]. Though the overall protein fold is similar to eukaryotic protein tyrosine phosphatases, only proteins from the *Mycobacterium*

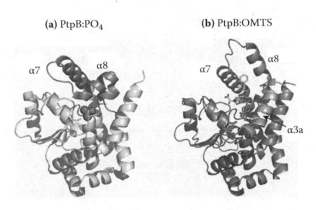

FIGURE 5.7 Ribbon diagrams of (a) the PtpB:product complex, PtpB:PO$_4$ (1YWF[72]), (b) the PtpB:inhibitor complex PtpB:OMTS (2OZ5[73]). Lid helices α7 and α8 are labeled as well as helix α3a, which is only observed in the PtpB:inhibitor complex structure. (From Flynn et al. [52]. With permission.)

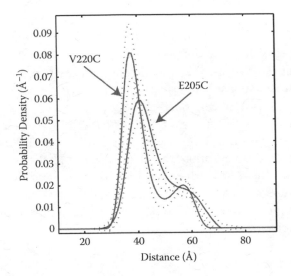

FIGURE 5.8 Distance probability density functions comparing the two lid helices, α7 (E205C) and α8 (V220C). The sharper distribution of α8 reflects ~twofold slower opening and closing rates for helix α8 compared with helix α7. (From Flynn et al. [52]. With permission.)

species have this unique lid insertion. While the protein is active in solution, no activity is detected in crystals and the structure reveals that the active site is completely buried, with no obvious routes for substrate access [72]. This implies that there must be a conformational reconfiguration required in the protein's lid for it to be active.

To determine whether the PtpB lid can spontaneously open in solution, two FRET constructs were designed to monitor fluctuations between each of the two lid helice—α7 and α8 (Figure 5.7)—and the core of the enzyme [52]. As predicted, both lid helixes were capable of sampling an open and a closed position (Figure 5.8). Although the closed position was consistent with observed crystallographic evidence, these data represented the first experimental demonstration that the PtpB lid is capable of opening in solution. Interestingly, though the measured PDFs for both the α7 and α8 lid helices are similar, measurement of the rates by fitting time-dependent distributions to a two-state

Chapter 5

motional narrowing model indicated that the opening and closing rates of the α7 lid helix are roughly twice that of the α8 helix. Though the flexible linker between the two helices can be seen in the inhibitor bound crystal structure (Figure 5.7b) [73], this discovery raises the possibility that the slower rates of the α8 lid helix are due to interactions with the α3a helix, which is only seen in the inhibitor bound crystal structure when the α8 lid helix is in an extended conformation (Figure 7) [72,73].

Though single-molecule experiments revealed the existence of an open conformation of the PtpB lid, an even more fundamental issue is to understand the functional significance of the lid. Since PtpB may be secreted into host macrophage cells [70,71], it was suggested that the role of the lid may be a novel adaptation that helps protect the enzyme from the host's chemical defenses [72]. Eukaryotes commonly control the activity of protein tyrosine phosphatases through a reversible oxidation of the active site cysteine with reactive oxygen species, such as H_2O_2 [74,75]. Thus the role of the PtpB lid may be to protect the protein active site from this oxidative inactivation by the host. Consistent with this hypothesis PtpB was found to have a 5- to 10-fold slower rate of oxidative inactivation compared with other protein lidless tyrosine phosphatases variants [52]. A kinetic model was constructed in which only the lid-open form of the enzyme can be inactivated by peroxide while the lid-closed conformation is protected. By using the single-molecule rates for lid opening and closing and the peroxide inactivation rate of the lidless *M. tb* tyrosine phosphatases variant PtpA, this model is capable of quantitatively accounting for PtpB's rate of inactivation.

High-resolution single-molecule experiments on PtpB were able to provide the first experimental demonstration of large-amplitude conformational dynamics of the protein's novel two-helix lid. In addition, the measured rates help to understand the functional significance of the lid as a dynamic filter that protects the enzyme's active site from the host's chemical defenses. This structural adaptation illustrates the diverse strategies that enzymes have evolved to harness spontaneous conformational fluctuations to improve their fitness.

5.4 Summary

The study of proteins at the single-molecule level promises unique insights into the understanding of enzyme function at the molecular level. The ability to observe time-dependent conformational fluctuations of enzymes at biologically relevant timescales without relying on ensemble averaging complements available high-resolution structural techniques, such as X-ray crystallography and NMR, and helps to provide a complete view of these molecules in solution. The picture that emerges from this work is one in which enzyme structure is highly dynamic and has evolved to harness random thermal motions in ways that complement and enhance biological function. This level of microscopic molecular detail is required for a quantitative understanding of the forces that dictate enzyme function in solution and promises to push our basic understanding of the structure–function relationship in the future. Due to the technical nature of single-molecule fluorescence methods new advances in methodology, instrumentation, fluorescent probes, and data analysis will continue to push the limits of the technique in terms of both time and special resolution. Indeed, a widespread adaptation of single-molecule techniques to an ever-expanding set of biological questions can be expected in the years to come.

References

[1] Daniel, R. M.; Dunn, R. V.; Finney, J. L.; Smith, J. C. The role of dynamics in enzyme activity. *Annu. Rev. Biophys. Biomol. Struct.* 2003, 32, 69–92.

[2] Zaccai, G. Biochemistry – How soft is a protein? A protein dynamics force constant measured by neutron scattering. *Science* 2000, 288, 1604–1607.

[3] Bourgeois, D.; Schotte, F.; Brunori, M.; Vallone, B. Time–resolved methods in biophysics. 6. Time–resolved Laue crystallography as a tool to investigate photo-activated protein dynamics. *Photochem. Photobiol. Sci.* 2007, 6, 1047–1056.

[4] Callender, R.; Dyer, R. B. Advances in time-resolved approaches to characterize the dynamical nature of enzymatic catalysis. *Chem. Rev.* 2006, 106, 3031–3042.

[5] Carey, P. R. Raman crystallography and other biochemical applications of Raman microscopy. *Annu. Rev. Phys. Chem.* 2006, 57, 527–554.

[6] Cho, M. H. Coherent two-dimensional optical spectroscopy. *Chem. Rev.* 2008, 108, 1331–1418.

[7] Palmer, A. G.; Kroenke, C. D.; Loria, J. P. In *Nuclear Magnetic Resonance of Biological Macromolecules, Pt B*, Academic Press, USA: 2001; Vol. 339, pp. 204–238.

[8] Xie, X. S.; Trautman, J. K. Optical studies of single molecules at room temperature. *Annu. Rev. Phys. Chem.* 1998, 49, 441–480.

[9] Weiss, S. Fluorescence spectroscopy of single biomolecules. *Science* 1999, 283, 1676–1683.

[10] Barkai, E.; Jung, Y. J.; Silbey, R. Theory of single-molecule spectroscopy: Beyond the ensemble average. *Annu. Rev. Phys. Chem.* 2004, 55, 457–507.

[11] Lippitz, M.; Kulzer, F.; Orrit, M. Statistical Evaluation of Single Nano-Object Fluorescence. *ChemPhysChem* 2005, 6, 770–789.

[12] Barkai, E.; Brown, F. L. H.; Orrit, M.; Yang, H., *Theory and Evaluation of Single-Molecule Signals*. World Scientific: Singapore, 2008; p 1–35.

[13] Schuler, B.; Eaton, W. A. Protein folding studied by single-molecule FRET. *Curr. Opin. Struct. Biol.* 2008, 18, 16–26.

[14] Park, H.; Toprak, E.; Selvin, P. R. Single-molecule fluorescence to study molecular motors. *Q. Rev. Biophys.* 2007, 40, 87–111.

[15] Seidel, R.; Dekker, C. Single-molecule studies of nucleic acid motors. *Curr. Opin. Struct. Biol.* 2007, 17, 80–86.

[16] Kinosita, K.; Adachi, K.; Itoh, H. Rotation of F-1-ATPase: How an ATP-driven molecular machine may work. *Annu. Rev. Biophys. Biomol. Struct.* 2004, 33, 245–268.

[17] Zhuang, X. W. Single-molecule RNA science. *Annu. Rev. Biophys. Biomol. Struct.* 2005, 34, 399–414.

[18] Roy, R.; Hohng, S.; Ha, T. A practical guide to single-molecule FRET. *Nat. Methods* 2008, 5, 507–516.

[19] Walter, N. G.; Huang, C. Y.; Manzo, A. J.; Sobhy, M. A. Do-it-yourself guide: how to use the modern single-molecule toolkit. *Nat. Methods* 2008, 5, 475–489.

[20] Moerner, W. E.; Kador, L. Optical-Detection and Spectroscopy of Single Molecules in a Solid. *Phys. Rev. Lett.* 1989, 62, 2535–2538.

[21] Haran, G. Single-Molecule Raman Spectroscopy: A Probe of Surface Dynamics and Plasmonic Fields. *Acc. Chem. Res.* 43, 1135–1143.

[22] Lakowicz, J. R. *Principles of Fluorescence Spectroscopy*. 2nd ed.; Springer: New York, 1999; p 298–301.

[23] Michalet, X.; Weiss, S.; Jager, M. Single-molecule fluorescence studies of protein folding and conformational dynamics. *Chem. Rev.* 2006, 106, 1785–1813.

[24] Lu, H. P.; Xun, L. Y.; Xie, X. S. Single-molecule enzymatic dynamics. *Science* 1998, 282, 1877–1882.

[25] Yang, H.; Luo, G. B.; Karnchanaphanurach, P.; Louie, T. M.; Rech, I.; Cova, S.; Xun, L. Y.; Xie, X. S. Protein conformational dynamics probed by single-molecule electron transfer. *Science* 2003, 302, 262–266.

[26] Xie, X. S.; Dunn, R. C. Probing Single-Molecule Dynamics. *Science* 1994, 265, 361–364.

[27] Giepmans, B. N. G.; Adams, S. R.; Ellisman, M. H.; Tsien, R. Y. Review - The fluorescent toolbox for assessing protein location and function. *Science* 2006, 312, 217–224.

[28] Kapanidis, A. N.; Weiss, S. Fluorescent probes and bioconjugation chemistries for single-molecule fluorescence analysis of biomolecules. *J. Chem. Phys.* 2002, 117, 10953–10964.

[29] Waggoner, A. In *Biochemical Spectroscopy*, 1995; Vol. 246, pp. 362–373.

[30] Keppler, A.; Gendreizig, S.; Gronemeyer, T.; Pick, H.; Vogel, H.; Johnsson, K. A general method for the covalent labeling of fusion proteins with small molecules in vivo. *Nat. Biotechnol.* 2003, 21, 86–89.

Chapter 5

[31] Summerer, D.; Chen, S.; Wu, N.; Deiters, A.; Chin, J. W.; Schultz, P. G. A genetically encoded fluorescent amino acid. *Proc. Natl. Acad. Sci. USA* 2006, 103, 9785–9789.

[32] Yang, J. Y.; Yang, W. Y. Site-Specific Two-Color Protein Labeling for FRET Studies Using Split Inteins. *J. Am. Chem. Soc.* 2009, 131, 11644–11646.

[33] Griffin, B. A.; Adams, S. R.; Tsien, R. Y. Specific covalent labeling of recombinant protein molecules inside live cells. *Science* 1998, 281, 269–272.

[34] Rasnik, I.; McKinney, S. A.; Ha, T. Nonblinking and longlasting single-molecule fluorescence imaging. *Nat. Methods* 2006, 3, 891–893.

[35] Vogelsang, J.; Kasper, R.; Steinhauer, C.; Person, B.; Heilemann, M.; Sauer, M.; Tinnefeld, P. A reducing and oxidizing system minimizes photobleaching and blinking of fluorescent dyes. *Angewandte Chemie-International Edition* 2008, 47, 5465–5469.

[36] Stryer, L.; Haugland, R. P. Energy Transfer - a Spectroscopic Ruler. *Proc. Natl. Acad. Sci. USA* 1967, 58, 719–726.

[37] Yang, H. The Orientation Factor in Single-Molecule Forster-Type Resonance Energy Transfer, with Examples for Conformational Transitions in Proteins. *Isr. J. Chem.* 2009, 49, 313–321.

[38] Watkins, L. P.; Yang, H. Information bounds and optimal analysis of dynamic single molecule measurements. *Biophys. J.* 2004, 86, 4015–4029.

[39] Yang, H.; Xie, X. S. Probing single-molecule dynamics photon by photon. *J. Chem. Phys.* 2002, 117, 10965–10979.

[40] Watkins, L. P.; Chang, H.; Yang, H. Quantitative single-molecule conformational distributions: a case study with poly-(L-proline). *J. Phys. Chem. A* 2006, 110, 5191–5203.

[41] Axelrod, D. In *Biophysical Tools for Biologists, Vol 2: in Vivo Techniques*, 2008; Vol. 89, pp. 169–221.

[42] Pal, P.; Lesoine, J. F.; Lieb, M. A.; Novotny, L.; Knauf, P. A. A Novel Immobilization Method for Single Protein spFRET Studies. *Biophys. J.* 2005, 89, L11–L13.

[43] Groll, J.; Moeller, M. Star Polymer Surface Passivation for Single-Molecule Detection. *Methods in Enzymology, Vol 472: Single Molecule Tools, Pt A: Fluorescence Based Approaches* 2010, 472, 1–18.

[44] Talaga, D. S. Information theoretical approach to single-molecule experimental design and interpretation. *J. Phys. Chem. A* 2006, 110, 9743–9757.

[45] Cramér, H. *Mathematical Methods of Statistics*. Princeton University Press: Princeton, NJ, 1946; p 416–452.

[46] Chung, H. S.; Gopich, I. V.; McHale, K.; Cellmer, T.; Louis, J. M.; Eaton, W. A. Extracting rate coefficients from single-molecule photon trajectories and FRET efficiency histograms for a fast-folding protein. *The Journal of Physical Chemistry. A* 2011, 115, 3642–3656.

[47] Hanson, J. A.; Duclerstacit, K.; Watkins, L. P.; Bhattacharyya, S.; Brokaw, J.; Chu, J. W.; Yang, H. Illuminating the mechanistic roles of enzyme conformational dynamics. *Proc. Natl. Acad. Sci. USA* 2007, 104, 18055–18060.

[48] Kubo, R. Note on the Stochastic Theory of Resonance Absorption. *J. Phys. Soc. Jpn.* 1954, 9, 935–944.

[49] Anderson, P. W. A Mathematical Model for the Narrowing of Spectral Lines by Exchange or Motion. *J. Phys. Soc. Jpn.* 1954, 9, 316–339.

[50] Gopich, I. V.; Szabo, A. Single-macromolecule fluorescence resonance energy transfer and free-energy profiles. *J. Phys. Chem. B* 2003, 107, 5058–5063.

[51] Geva, E.; Skinner, J. L. Two-state dynamics of single biomolecules in solution. *Chem. Phys. Lett.* 1998, 288, 225–229.

[52] Flynn, E. M.; Hanson, J. A.; Alber, T.; Yang, H. Dynamic Active-Site Protection by the M. tuberculosis Protein Tyrosine Phosphatase PtpB Lid Domain. *J. Am. Chem. Soc.* 2010, 132, 4772–4780.

[53] Frauenfelder, H.; McMahon, B. H.; Austin, R. H.; Chu, K.; Groves, J. T. The role of structure, energy landscape, dynamics, and allostery in the enzymatic function of myoglobin. *Proc. Natl. Acad. Sci. USA* 2001, 98, 2370–2374.

[54] Smiley, R. D.; Hammes, G. G. Single molecule studies of enzyme mechanisms. *Chem. Rev.* 2006, 106, 3080–3094.

[55] Brender, J. R.; Dertouzos, J.; Ballou, D. P.; Massey, V.; Palfey, B. A.; Entsch, B.; Steel, D. G.; Gafni, A. Conformational dynamics of the isoalloxazine in substrate-free p-hydroxybenzoate hydroxylase: Single-molecule studies. *J. Am. Chem. Soc.* 2005, 127, 18171–18178.

[56] Kuznetsova, S.; Zauner, G.; Aartsma, T. J.; Engelkamp, H.; Hatzakis, N.; Rowan, A. E.; Nolte, R. J. M.; Christianen, P. C. M.; Canters, G. W. The enzyme mechanism of nitrite reductase studied at single-molecule level. *Proc. Natl. Acad. Sci. USA* 2008, 105, 3250–3255.

[57] Edman, L.; Foldes-Papp, Z.; Wennmalm, S.; Rigler, R. The fluctuating enzyme: a single molecule approach. *Chem. Phys.* 1999, 247, 11–22.

[58] Flomenbom, O.; Velonia, K.; Loos, D.; Masuo, S.; Cotlet, M.; Engelborghs, Y.; Hofkens, J.; Rowan, A. E.; Nolte, R. J. M.; Van der Auweraer, M.; de Schryver, F. C.; Klafter, J. Stretched exponential decay and correlations in the catalytic activity of fluctuating single lipase molecules. *Proc. Natl. Acad. Sci. USA* 2005, 102, 2368–2372.

[59] van Oijen, A. M.; Blainey, P. C.; Crampton, D. J.; Richardson, C. C.; Ellenberger, T.; Xie, X. S. Single-molecule kinetics of lambda exonuclease reveal base dependence and dynamic disorder. *Science* 2003, 301, 1235–1238.

[60] Henzler-Wildman, K. A.; Thai, V.; Lei, M.; Ott, M.; Wolf-Watz, M.; Fenn, T.; Pozharski, E.; Wilson, M. A.; Petsko, G. A.; Karplus, M.; Hubner, C. G.; Kern, D. Intrinsic motions along an enzymatic reaction trajectory. *Nature* 2007, 450, 838–844.

[61] Maragakis, P.; Karplus, M. Large amplitude conformational change in proteins explored with a plastic network model: Adenylate kinase. *J. Mol. Biol.* 2005, 352, 807–822.

[62] Miyashita, O.; Onuchic, J. N.; Wolynes, P. G. Nonlinear elasticity, proteinquakes, and the energy landscapes of functional transitions in proteins. *Proc. Natl. Acad. Sci. USA* 2003, 100, 12570–12575.

[63] Muller, C. W.; Schlauderer, G. J.; Reinstein, J.; Schulz, G. E. Adenylate kinase motions during catalysis: An energetic counterweight balancing substrate binding. *Structure* 1996, 4, 147–156.

[64] Yan, H. G.; Tsai, M. D. In *Advances in Enzymology*, 1999; Vol. 73, p 103.

[65] Jencks, W. P. Binding-Energy, Specificity, and Enzymic Catalysis - Circe Effect. *Adv. Enzymol. Relat. Areas Mol. Biol.* 1975, 43, 219–410.

[66] Kumar, S.; Ma, B. Y.; Tsai, C. J.; Wolfson, H.; Nussinov, R. Folding funnels and conformational transitions via hinge-bending motions. *Cell Biochem. Biophys.* 1999, 31, 141–164.

[67] Gunasekaran, K.; Ma, B. Y.; Nussinov, R. Is allostery an intrinsic property of all dynamic proteins? *Proteins: Struct. Funct., Bioinf.* 2004, 57, 433–443.

[68] Cole, S. T.; Brosch, R.; Parkhill, J.; Garnier, T.; Churcher, C.; Harris, D.; Gordon, S. V.; Eiglmeier, K.; Gas, S.; Barry, C. E.; Tekaia, F.; Badcock, K.; Basham, D.; Brown, D.; Chillingworth, T.; Connor, R.; Davies, R.; Devlin, K.; Feltwell, T.; Gentles, S.; Hamlin, N.; Holroyd, S.; Hornby, T.; Jagels, K.; Krogh, A.; McLean, J.; Moule, S.; Murphy, L.; Oliver, K.; Osborne, J.; Quail, M. A.; Rajandream, M. A.; Rogers, J.; Rutter, S.; Seeger, K.; Skelton, J.; Squares, R.; Squares, S.; Sulston, J. E.; Taylor, K.; Whitehead, S.; Barrell, B. G. Deciphering the biology of Mycobacterium tuberculosis from the complete genome sequence. *Nature* 1998, 393, 537–544.

[69] Beresford, N. J.; Mulhearn, D.; Szczepankiewicz, B.; Liu, G.; Johnson, M. E.; Fordham-Skelton, A.; Abad-Zapatero, C.; Cavet, J. S.; Tabernero, L. Inhibition of MptpB phosphatase from Mycobacterium tuberculosis impairs mycobacterial survival in macrophages. *J. Antimicrob. Chemother.* 2009, 63, 928–936.

[70] Singh, R.; Rao, V.; Shakila, H.; Gupta, R.; Khera, A.; Dhar, N.; Singh, A.; Koul, A.; Singh, Y.; Naseema, M.; Narayanan, P. R.; Paramasivan, C. N.; Ramanathan, V. D.; Tyagi, A. K. Disruption of mptpB impairs the ability of Mycobacterium tuberculosis to survive in guinea pigs. *Mol. Microbiol.* 2003, 50, 751–762.

[71] Grundner, C.; Ng, H.-L.; Alber, T. Mycobacterium tuberculosis protein tyrosine phosphatase PtpB structure reveals a diverged fold and a buried active site. *Structure (London, England: 1993)* 2005, 13, 1625–1634.

[72] Grundner, C.; Perrin, D.; Hooft van Huijsduijnen, R.; Swinnen, D.; Gonzalez, J. r.; Gee, C. L.; Wells, T. N.; Alber, T. Structural basis for selective inhibition of Mycobacterium tuberculosis protein tyrosine phosphatase PtpB. *Structure (London, England: 1993)* 2007, 15, 499–509.

[73] den Hertog, J.; Groen, A.; van der Wijk, T. Redox regulation of protein-tyro sine phosphatases. *Arch. Biochem. Biophy.* 2005, 434, 11–15.

[74] Rhee, S.; Bae, Y.; Lee, S.; Kwon, J. Hydrogen Peroxide: a key messenger that modulates protein phosphorylation through cysteine oxidation. *Sci. STKE* 2000, 2000, pe1.

Chapter 5

6. Small–Angle Scattering

Cy M. Jeffries
Jill Trewhella

Chapter 6

Quantitative Biology: From Molecular to Cellular Systems edited by Michael E. Wall © 2012 CRC Press / Taylor & Francis Group, LLC. ISBN: 978-1-4398-2722-2

6.1   The Power of Scattering

The small-angle scattering (SAS) of X-rays or neutrons is becoming an increasingly accessible and powerful complementary tool for structural analysis of biological macromolecules and the complexes or assemblies they form in solution. Unlike X-ray crystallography or nuclear magnetic resonance (NMR) spectroscopy, SAS does not answer questions with respect to *atomic resolution* detail. Rather, the power of SAS comes from its ability to provide precise and accurate *global* structural parameters related to the size and shape of biological macromolecules and, under the right circumstances, to do so relatively quickly. For students and early career researchers who are seeking to have impact in the field of structural biology, X-ray crystallography and NMR spectroscopy are the traditional techniques of choice, but if a target simply refuses to neatly package itself into diffraction quality crystals or is too big for NMR, SAS can provide meaningful structural results that can generate new insights into a system of interest.

Aside from revealing the overall dimensions and shapes of biological macromolecules in solution, SAS can be invaluable for investigating conformational changes [1,2], structural flexibility [3], unstructured systems or regions [4,5], the effects of changing pH or buffer conditions [6], and even folding pathways [7]. Of particular interest, if neutrons and contrast variation can be applied, the power of SAS can be extended from analyzing individual macromolecules to deducing the molecular organization of protein–protein [8,9], protein–RNA [10,11], and protein–DNA complexes [12–14] and investigating the structure of macromolecular assemblies as large as ribosomes and filamentous actin [15–17]. There are literally hundreds of examples where solution scattering has been applied to answer fundamental questions beyond the purview of crystallography and NMR (see, e.g., [18,19]). As long as caution is maintained in interpreting results and the correct checks and controls are employed [20], SAS can be a very powerful and productive technique for structural molecular biologists [21]. There are multiple approaches to SAS data analysis, but of particular interest to structural biologists are methods that provide 3-D shape reconstructions from 1-D scattering data (Figure 6.1; [22–24]).

By using straightforward computer programs [25–27] it is now relatively uncomplicated to obtain useful 3-D structural models based on SAS data that advance our understanding of biomacromolecular function. For example, the structures of the individual domains from the N-terminus of myosin binding protein C have been solved using either crystallography [28,29] or NMR [30,31]; however, the intact protein has been recalcitrant to crystallization and is too big for NMR. Small-angle X-ray scattering (SAXS) experiments quickly revealed that the domains of this protein are organized in an extended, tandem configuration similar to the giant muscle protein titin [32,33], and further small-angle neutron scattering (SANS) studies have suggested how this structural disposition may be important in modulating the calcium signals that control muscle contraction [17]. This chapter focuses on the use of SAXS and SANS for the extraction of this and other kinds of structural information from biological macromolecules in solution.

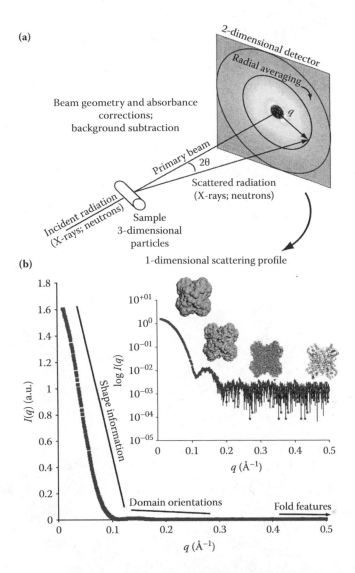

FIGURE 6.1 The basics of small-angle scattering. (a) A sample of monodisperse macromolecules is illuminated with incident radiation of known wavelength (X-rays or neutrons). Most of the radiation passes straight through (primary beam), while some is absorbed by the sample. Some of the radiation is scattered as a result of interacting with particles in solution. Measuring the intensities of the scattered radiation on a two-dimensional detector as a function of momentum transfer, q, (that relates to the scattering angle, 2θ, Equation (6.1)), and performing radial averaging of the intensity data produces a one-dimensional scattering profile ($I(q)$ vs. q). (b) The intensities of the scattered radiation at low angles are related to the shape of the macromolecules. (Inset) Scattering intensities are often placed on a log $I(q)$ scale to help observe weak scattering features in the mid and high-q region of a profile ($\sim$0.1–0.35 Å^{-1}).

6.2 What Is SAS?

The small-angle scattering of X-rays or neutrons is inherently linked to their properties as *plane waves*, like the waves generated on the surface of the ocean. Both X-rays and neutrons can be described in terms of wave functions that have a wavelength (λ), amplitude, and phase, even though neutrons are often thought of as particles (Figure 6.2).

Chapter 6

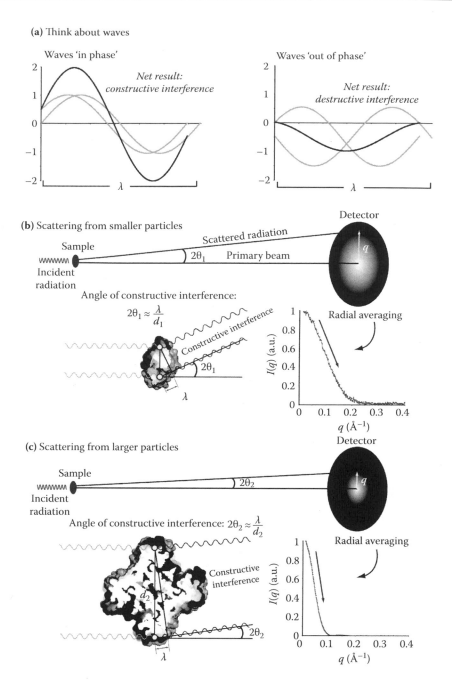

(a) Think about waves

(b) Scattering from smaller particles

(c) Scattering from larger particles

FIGURE 6.2 Think about waves. (a) Both X-rays and neutrons can be described as wave functions that have a wavelength, amplitude, and phase. The addition of two scattered waves (light gray) can produce constructive interference or destructive interference (black). In SAS, the measured intensities of constructively interfering scattered radiation through a scattering angle 2θ relate, in a reciprocal fashion, to the distribution of distances within a particle. (b) Larger angles are required to investigate smaller particles and (c) vice versa.

As an incident X-ray or neutron plane wave passes through a macromolecule, it can pass through unaffected, can be absorbed, or can interact with individual atoms to generate secondary scattered waves. X-rays are scattered by the orbiting electrons of an atom, while neutrons are scattered by the nucleus. As long as a macromolecule is supported in a medium that has a different scattering density compared with the macromolecule itself (e.g., a difference in electron density between a protein and its solvent) and as long as the scattering event is elastic (i.e., no change in energy and hence no change in the amplitude or wavelength), the net sum of the secondary scattered waves will produce a distribution of scattered intensity at small angles with respect to the direction of the incident radiation. This scattering intensity distribution will depend upon how many of the scattered waves constructively interfere with each other such that their amplitudes are additive. Additive constructive interference lies at the heart of why structural biologists can use small-angle X-ray or neutron scattering to probe macromolecular structures. The angles at which constructive interference occurs relate to the preserved averaged distance relationships between atoms in a target. Changing the shape of the target will alter the distribution of distances between atoms and thus the scattering intensity as a function of angle. More formally, if two atoms are separated by a distance d and they scatter a plane wave elastically, then the maximum additive constructive interference will occur when the coherently scattered waves travel at an angle that is $\sim\lambda/d$ from the incident wave. Thus, the intensity of the scattered radiation as a function of angle will be related in *reciprocal* scattering space to the atom-pair separation distances and their distribution within a macromolecule in *real* space [34,35]. This scattered intensity distribution, referred to as the *form factor*, is thus most directly related to the form or shape of the scattering particle expressed in terms of the distribution of distances between atoms within a particle's volume as well as the elemental composition internal to that volume. For X-rays, the scattering density is generally expressed in terms of the average number of electrons per unit volume (electron density); the more electrons orbiting an atom the more likely it will scatter an X-ray. For neutrons the scattering density can be described as a conceptual length that is a measure of the likelihood that a given nucleus will scatter (Section 6.9); the neutron scattering length of an atom must be determined experimentally as there is no simple dependence between it and atomic number and consequently there can be large isotope effects. It is from the scattering form factor that information about shape and volume of a macromolecule in solution is derived. With respect to investigating biological macromolecules using SAS, the shape of the target molecule is represented in a scattering profile at low angles, usually less than 3°—smaller angles are required to investigate larger systems and vice versa (Figures 6.2b and 6.2c).

6.3 Small-Angle Scattering Data

Small-angle scattering profiles are generally presented as $I(q)$ versus q, where I is the intensity of the scattered radiation, and q (Å^{-1}) is the amplitude of the momentum transfer (or scattering vector) of the scattered radiation. The momentum transfer is related to the scattering angle, 2θ, by

$$q = \frac{4\pi \sin\theta}{\lambda},$$

(6.1)

where λ is the wavelength of the incident radiation. A scattering profile has a number of q-regions, and the low to mid-q region (0–~0.3 Å^{-1}) corresponding to atom-pair distances from ~1500–6 Å is most relevant to defining the overall shape of the scattering particle. Higher-angle scattering ($q > $ ~0.3 Å^{-1}) arises from atom-pair distances within a macromolecule less than ~5 – 10 Å and, if measured accurately, may provide insights into internal density fluctuations within a molecular boundary. This region, however, is a challenge to measure as the scattering is generally very weak (Figure 6.1), with signal-to-noise ratios several orders of magnitude larger than that in the low-angle region. Nonetheless, a number of groups are looking to capitalize on the increased intensity of synchrotron X-ray sources to use this region to complement NMR data for improved solution structure refinement [36–38], while others are using wide-angle scattering (WAS, $q > 1.0$ Å^{-1}) to explore protein fold secondary structure changes [39,40].

Biological macromolecules investigated using SAS are usually studied in aqueous solution, and thus the measured scattering profile represents a time- and ensemble-average of the scattering from randomly oriented particles. So long as the individual particles are identical and remain *monosdisperse* (e.g., as defined monomers, dimers, trimers, etc.) and the conditions approximate a dilute solution such that there are no intermolecular distance correlations, the scattering profile will be proportional to the net scattering produced by all atom-pairs internal to a single particle averaged over all orientations; i.e., the form factor. In cases where Coulombic forces exist between particles that give rise to a mean distance of closest approach, interparticle distance correlations, or interparticle interference effects, can affect the scattering. This contribution to the scattering is termed the *structure factor* and, if present, will be convoluted with the form factor resulting in a suppression of the scattering at the smallest angles (Figure 6.3).

If not recognized and dealt with appropriately, interparticle interference will lead to apparent structural parameters that are systematically small. Interparticle interference is generally concentration dependent and can be alleviated by reducing macromolecular concentration or by changing solution conditions to modulate the net charge on particles (e.g., ionic strength or pH). As small interparticle interference effects scale linearly with concentration, it is also possible to use linear regression to extrapolate scattering data to infinite dilution to obtain a scattering profile free of these effects.

Nonspecific aggregation or sample heterogeneity causes significant and perhaps the most intractable of problems for small-angle scattering investigations. Due to the averaged nature of the information content of scattering data, samples that are aggregated, are not pure, or are a mixture in solution will produce scattering profiles that do not reflect the form-factor scattering derived from a single particle, but the proportionately weighted scattering profile produced by multiple species in solution (see also Supplementary Information). Thus, if the aim is to extract structural parameters and shape information about a macromolecule or complex, it is essential that samples are monodisperse, identical particles and preferable that the solution conditions approximate those of dilute solution (i.e., no interparticle interference). In summary, the measured scattering profile of $I(q)_{\text{total}}$ vs q can be described as

$$I(q)_{\text{total}} = I(q)S(C, q) \qquad (6.2)$$

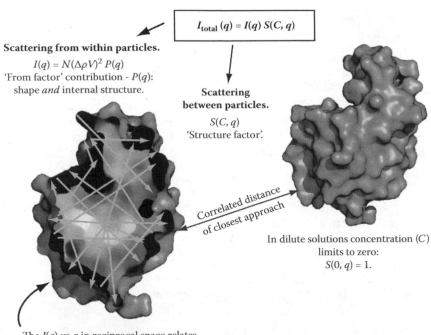

Scattering from within particles.

$$I(q) = N(\Delta\rho V)^2 P(q)$$
'From factor' contribution - $P(q)$:
shape *and* internal structure.

$$I_{total}(q) = I(q) S(C, q)$$

**Scattering
between particles.**

$S(C, q)$
'Structure factor'.

Correlated distance
of closest approach

In dilute solutions concentration (C)
limits to zero:
$S(0, q) = 1$.

The $I(q)$ vs q in reciprocal space relates
to the atom-pair distribution in real space, $P(r)$.

FIGURE 6.3 Form and structure factors. The total scattering intensity, $I(q)_{total}$, is derived from scattering from within particles ($I(q)$) and concentration dependent (C) scattering contributions made between particles (structure factors; $S(C,q)$). Under dilute monodisperse conditions, $S(C,q)$ limits to 1, and therefore $I(q)_{total}$ vs q will represent the form factor ($P(q)$) that reflects the distribution of atom-pair distances within a single particle (proportional to the number of noninteracting molecules (N) in solution, the contrast against the background solvent ($\Delta\rho$) squared and the volume of the particle (V) squared).

where $S(C,q)$ is the concentration (C) dependent structure factor contribution that when eliminated ($S(C,q) = 1$) will result in a measured $I(q)$ that is proportionate to the form factor $P(q)$:

$$I(q) \propto P(q) \tag{6.3}$$

6.4 Solvent Subtraction and Contrast

Practically speaking, SAS investigations are performed in solution, be it an aqueous buffer or sometimes other solvents. The supporting solvent is composed of atoms as is the cell containing the target of interest and as such will each contribute to the scattering measurement. Consequently, these contributions to the scattering profile must be subtracted to reveal the scattering derived from the target particle of interest, that is, SAS is a *subtractive* technique:

$$SAS_{(target)} = SAS_{(target+solvent+cell)} - SAS_{(solvent+cell)} \tag{6.4}$$

For optimal subtraction, one measures the (target + solvent + cell) and the (solvent + cell) in an identical cell in a fixed position, and the solvent must be precisely matched with the sample containing the target. Solvent matching becomes especially important within the very-low intensity higher-q region where the data will have proportionately large errors that can contribute to imprecise subtraction of the solvent contributions from the sample scattering. Obtaining matched solvent conditions (e.g., via sample dialysis) is a critical step in extracting global structural parameters and shape information from a target of interest (see also Supplementary Information).

Once the solvent scattering+cell contribution has been subtracted, it is relatively easy to intuit that if the mass of a particle of interest increases, then an increase in scattering intensity will result and that the number and distribution of distances between scattering centers within a target will relate to the rate at which the intensity drops off at ever-increasing angles. It also is easy to understand that increasing the number of particles in solution will increase the scattering intensity. What is less intuitive is that the scattering intensity in the small-angle region also depends, and indeed only arises, if there is a *difference* between the scattering density of the solvent and the particles of interest. This difference is known as contrast and is represented as $\Delta\rho = \rho - \rho_s$, where ρ and ρ_s are the mean scattering densities of the particle and the solvent, respectively. For example, in the case of X-ray scattering, upon increasing the number of electrons per unit volume in a solvent relative to a soluble macromolecule (e.g., by adding electron dense atoms like potassium, or sugars, glycerol, to a buffer), the difference in scattering density between the macromolecule and the solvent will decrease. If the scattering density of a soluble particle and the solvent become equal ($\rho = \rho_s$) (i.e., there is a homogeneous distribution of scattering density), then zero net scattering arises from constructive interferences between atom-pair scattering centers within the dissolved macromolecule (Figure 6.4).

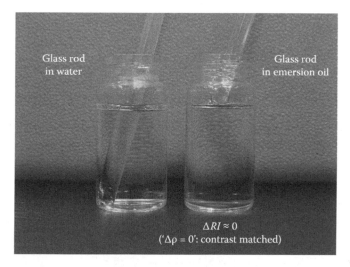

FIGURE 6.4   Contrast illustrated. When a glass rod is placed in emersion oil (left), it seems to disappear. This is because the refractive index (RI) of the glass and the oil are the same and the difference in refractive index (ΔRI) is zero—there is no contrast between the rod and the oil. In an analogous fashion, the scattering intensity of a macromolecule in solution depends on the difference between the scattering density of the macromolecule (ρ) and the background solvent (ρ_s). If $\Delta\rho$, or the mean contrast, is 0 then aside from potential minor contributions due to internal density fluctuations the scattering profile arising from the shape of the macromolecule will not be observed.

Under the conditions where $\Delta\rho = 0$ the scattering intensity arising from paired distance correlations within a target of interest will be essentially matched out from a scattering profile. In the case of neutron scattering, the differences in scattering power of atoms is influenced by their nuclear isotopic number (isotopes of the same element can have very different scattering properties.) As a result, the contrast between a macromolecule and its solvent can be impacted by isotopic substitutions that can be used very effectively to alter the contrast of a system to selectively solvent match scattering contributions made by components of macromolecular complexes. We will see later (Section 6.9) that it is the unique properties the isotopes of hydrogen that make neutron contrast variation with solvent matching relatively simple, thus providing a very powerful tool for studying biological macromolecular complexes. Contrast can be exploited in both SAXS and SANS experiments, but it is in SANS where this extra dimension can be exploited for structural studies (see Section 6.9 and also [8,16,17,22,41–46]).

6.4.1 Contrast Fundamentals

The fundamentals of scattering for a solution of identical, monodisperse particles as previously described are expressed mathematically as

$$I(q) = N \left\langle \left| \int_V (\rho(\vec{r}) - \bar{\rho}_s) e^{i\vec{q}\cdot\vec{r}} \, d\vec{r} \right|^2 \right\rangle_\Omega \tag{6.5}$$

where N is the number of dissolved scattering particles per unit volume, and $\langle \ \rangle$ indicates that the net scattered intensity emanating from all scattering centers within each particle is rotationally averaged over all orientations (Ω) due to the random tumbling of the particles in solution. Each atom-pair gives rise to a scattered circular wave whose form is expressed as $e^{i\vec{q}\cdot\vec{r}}$, where $\vec{r}$ is the vector between atom-pairs. The amplitude of the scattering from each atom-pair is proportional to the product of the contrast values at each atom center. The total scattering is the sum of all such contributions and here is expressed as an integral over the total particle volume (V). Measured scattering intensities are proportional to the square of the scattering amplitude and thus we see that $I(q) \propto |(\rho(\vec{r}) - \bar{\rho}_s)|^2$. For a uniform scattering density object, the term on the right-hand side of this equation is often designated simply as $\Delta\rho$ (the mean contrast) and consequently $I(q) \propto \Delta\rho^2$. Systems that have high contrast (large $\Delta\rho$) will produce more intense scattering signals and vice versa. The measured $I(q)$ can be simply expressed as

$$I(q) = N(\Delta\rho V)^2 P(q) S(C,q) \tag{6.6}$$

and we see, as long as there are no interparticle interference effects ($S(C,q) = 1$), that the scattering from a monodisperse system of particles in reciprocal space relates to the number of molecules as well as the square of its contrast with respect to the solvent and the molecular volume.

6.4.2   Playing with Contrast

When investigating macromolecular complexes composed of more than one macromo-lecular family member (e.g., a protein/RNA or protein/DNA complex), the contrast game becomes very interesting. Proteins and polynucleotides have different elemental ratios, and thus complexes comprising both have more than one region of average scattering density. Assuming that the internal density fluctuations within each component are small compared with the scattering density differences between each component and the solvent,* one can approximate the total scattering after solvent subtraction using

$$I(q_{total}) \propto \Delta\rho_1^2 I(q)_1 + \Delta\rho_1\Delta\rho_2 I(q)_{12} + \Delta\rho_2^2 I(q)_2 \tag{6.7}$$

where $\Delta\rho$ refers to the mean contrast of each component (designated by subscripts 1 or 2) and the solvent (sometimes termed as the background.) The $I(q)_{12}$ term is the cross-term that quantifies the proportion of the total scattering function arising from atom-pairs between the two regions of distinct scattering density, e.g., between the protein and RNA in the complex. The cross-term provides information on the separation of the centers of the two components within the complex and their orientations with respect to each other.

6.5   Before the Model Building Begins

A great strength of solution scattering is that you can always measure a scattering profile. A great weakness of solution scattering is that it can be hard to prove, unequivocally, that the measured profile is from what you think or hope it is. Critical to the success of a SAS experiment is sample preparation and assessing the quality of the data. Quality assess-ment can be time-consuming but is necessary to ensure that any derived structural parameters are accurate. It is critically important to determine that the macromolecules in the sample are monodisperse, free of nonspecific associations (aggregation) and with-out interparticle distance correlations that would give rise to interparticle interference. A careful analysis of the concentration dependence of the forward-scattering inten-sity at zero angle ($I(0)$) can be used to determine if there is interparticle interference or concentration dependent aggregation. $I(0)$ can also provide an estimate of the molecular weight of the scattering species—perhaps the most powerful indicator that the target is what you think it is. An $I(0)$ analysis should be done before more detailed structural modeling commences.

Extrapolating $I(q)$ to $q = 0$ yields $I(0)$, which is related to the volume of the scattering particle. For $q = 0$ Equation (6.6) becomes

$$I(0) = N(\Delta\rho V)^2 \tag{6.8}$$

which can be expressed as

$$I(0) = \frac{C\Delta\rho^2 v^2 M_r}{N_A}, \tag{6.9}$$

---

* The assumption of uniform scattering density within a macromolecule is only a first approximation. For most SAS applications based on averaged isotropic scattering, it is a reasonable assumption to make.

where N is the number of scattering particles per unit volume, V the volume of the scattering particle, C is mass per unit volume in g.cm^{-3}, v is the partial specific volume of the scattering particle in cm$^3 \cdot$g^{-1}, $\Delta\rho$, the contrast in cm^{-2} and N_A is Avagadro's number (47). By expressing $I(0)$ in these terms it is possible to estimate the molecular weight, M_r, for the scattering particle using readily measurable or known parameters (see Section 6.7.4–6). However, there is no way to directly measure $I(0)$ because the primary beam used to illuminate the sample is coincident with $q = 0$ Å^{-1}. So how do you perform an $I(0)$ analysis and what other structural parameters can be extracted from the data.

6.6 The Basic Structural Parameters from SAS

As the scattered intensity depends on the molecular weight and the concentration of a macromolecule, a right balance must be met between obtaining scattering data with good counting statistics versus collecting data that are affected by structure-factor contributions—interparticle interference—or by concentration dependent aggregation. Once the scattering from a matched solvent has been subtracted from a sample in solution, a careful analysis of the scattering intensity at zero angle can provide an estimate of the molecular weight of the scattering species.

6.6.1 Scaling Data for $I(0)$ Analysis and the Radius of Gyration

Different types of SAS instruments have different source characteristics and different beam geometries, so prior to analysis it is advisable to scale the data by either using (1) the scattering from water to place the data on an absolute scale [47] or (2) a secondary macromolecular standard [48,49] (see Section 6.7). Once the scattering data are scaled, structural parameters, including $I(0)$ and radius of gyration (R_g), can be interpreted.

The R_g value is a useful and simple indicator of the scattering density distribution in the particle volume; it equates to the contrast weighted root mean square (or quadratic mean) distance of all volume elements occupied by scattering centers with respect to the center of scattering density. When considering a SAXS experiment on a protein, the center of electron density will be approximately coincident with the protein's center of mass, and thus R_g will provide a measure of the mass distribution around this center. Larger proteins will have larger volumes and hence a larger R_g, but the shape of a protein will also affect things—proteins of the same volume that are increasingly asymmetric will have increased R_g values. Thus, both the size and shape (i.e., the distribution of atom-pairs within a protein) will affect the magnitude of R_g. In a system where two regions of scattering density are present (ρ_1, ρ_2), for example, a protein/RNA complex, the center of the scattering density across both components may not necessarily coincide with the centre of mass of the complex—the magnitude of R_g will be determined by the relative contribution made by each scattering density region (1 and 2):

$$R_g^2 = f_1' R_{g1}^2 + f_2' R_{g2}^2 + f_1' f_2' D^2 \tag{6.10}$$

Chapter 6

where

$$f_1' = \frac{\Delta\rho_1 V_1}{\Delta\rho_1 V_1 + \Delta\rho_2 V_2} \tag{6.11}$$

and $f_2' = 1 - f_1'$, D is the distance between the centers of mass of the components, and V is the volume of the individual components. These relations basically mean that the measured R_g of a two-component system is related to the contrast of each individual component with respect to the solvent and with respect to each other and is sensitive to the distribution of distances within the volumes occupied by the individual components and whole complex. These relations, in combination with that described in Equation (6.7), are the key to SANS contrast variation where, by selectively altering the contrast of the solvent or the components of a complex, the shapes of the individual components within a complex can be deduced.

$I(0)$ and R_g values can be determined using (1) the Guinier approximation or (2) indirect inverse Fourier transform methods. The latter yields $P(r)$, the contrast-weighted probable pair-wise distribution of scattering elements in real space. The determination of $I(0)$ and R_g and a careful evaluation of their concentration dependence allows an experimenter to assess the quality of the scattering data and the samples, including detecting levels of aggregation or interparticle interference.

6.6.2 The Guinier Approximation

Guinier [50] showed that for a system of monodisperse spheres the following approximation holds at very low scattering angles:

$$I(q) = I(0)e^{\frac{-q^2 R_g^2}{3}} \tag{6.12}$$

This relationship indicates that the scattering intensity drops off in the very low-q region of a scattering profile as a negative exponent of q^2 that is also dependent on the size of the particle as represented by the radius of gyration. A plot of the natural logarithm of the experimental $I(q)$ versus q^2 should produce a negative linear relationship that when extrapolated to the $\ln I(q)$ intercept at $q = 0$ (zero-angle) will yield the forward-scattering intensity ($I(0)$.) The slope of a Guinier plot is proportional to R_g^2. Most soluble, monodisperse proteins and protein complexes have a Guinier region in their scattering profiles that can be readily analyzed (defined by $qR_g < 1.3$ for globular particles) [51]. However, care must be taken when investigating particularly large proteins, where the number of data points on which to base the Guinier extrapolation decreases (that will introduce error) or where proteins significantly deviate from globularity. Highly extended or filamentous proteins have Guinier regions for values of $qR_g < 0.8$–1.0 depending upon the degree of asymmetry [52]. For highly asymmetric particles of uniform cross section, a modified Guinier analysis in the form of a plot of $\ln(I(q)q)$ versus q^2 will yield $I_c(0)$ which is related to the mass-per-unit length of the particle and its radius of gyration of cross section, R_c, which is the contrast-weighted root mean square of all area elements from the center of the scattering density with respect to the cross section [34,53].

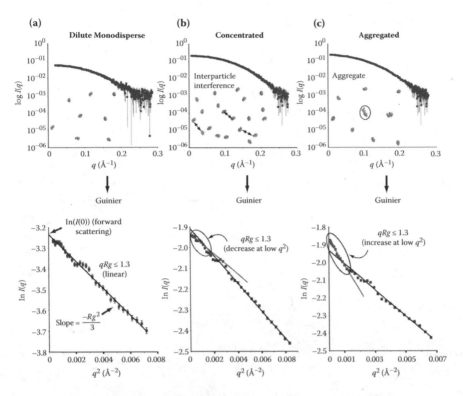

FIGURE 6.5 Guinier plots. To the untrained eye, there seems very little difference between the three scattering profiles collected from the same protein under different experimental conditions. However, plots of $\ln I(q)$ vs q^2 reveal that there are significant differences in the Guinier region of the scattering profiles. The dilute, monodisperse, protein sample (a) generates a well-defined negative linear correlation in the Guinier region with slope $-R_g^2/3$ and $\ln(I(q))$ intercept corresponding to the natural log of the forward-scattering intensity ($I(0)$). At higher concentrations (b) there is a depression in the intensities in the low-q region of the Guinier plot due to the presence of interparticle interference. The remaining sample (c) is affected by aggregation as noted by a significant upturn in the Guinier plot at very low-q.

Guinier analysis can also be used to qualitatively assess the presence of aggregation or interparticle interference (Figure 6.5). Observable upward deviations in $\ln I(q)$ away from linearity at low q is an indication that the sample has large molecular weight contaminates or contains non-specific aggregates. Interparticle interference, on the other hand, causes a decrease in intensity at low q values, resulting in a downward curvature at low q in Guinier plots. However, subtler aggregation or interparticle interference might not be readily detected using a Guinier plot, which puts greater importance in accurate assessments of parameters such as the molecular weight of the scattering particles from $I(0)$ (Section 6.7) determined either by Guinier or $P(r)$ analysis.

6.6.3 Real-Space Distance Distributions, $P(r)$

The $I(q)$ versus q dependence in reciprocal space is related to the probable distribution of contrast weighted atom-pair distances in real space, $P(r)$, by a Fourier transform

$$I(q) = 4\pi \int_0^{D_{max}} P(r) \frac{\sin qr}{qr} dr \qquad (6.13)$$

Chapter 6

where D_{max} is the maximum particle dimension or, conversely,

$$P(r) = \frac{1}{2\pi} \int_0^\infty I(q)q \cdot r \sin(qr)dq \qquad (6.14)$$

In practice, however, scattering data are measured at discreet intervals of q over a finite q-range bounded by a q_{min} and a q_{max}, so the integration in (6.14) cannot be done analytically. As a result, indirect transform methods must be used to calculate $P(r)$ from $I(q)$ [35,54]. The simplest indirect methods depend upon describing a scattering profile in terms of a sum of mathematical functions whose individual Fourier transforms are known. The fit to the scattering profile is optimized by the determination of coefficients that multiply each of these functions, which then can be applied to the corresponding Fourier transforms in real space to determine $P(r)$. The number of coefficients that can be accurately determined depends upon the q-range measured and the size of the particle (D_{max}). Criteria used to assess the quality of the fit include the smoothness of the transform, deviations from the data, and the stability of the solution [54]. The area under the modeled $P(r)$ profile gives $I(0)$:

$$I(0) = 4\pi \int_0^{D_{max}} P(r)dr \qquad (6.15)$$

while R_g corresponds to the second moment of $P(r)$:

$$R_g^2 = \frac{\int P(r)r^2\,dr}{2\int P(r)dr} \qquad (6.16)$$

Indirect methods are robust in extracting the atom-pair distance distribution from the scattering data. The process of choosing D_{max} and assessing the resulting $P(r)$ are made relatively straightforward using the regularization methods developed first by Glatter [35] and refined by Svergun [54]. These computational techniques provide a graphical output of $P(r)$ and determine R_g and $I(0)$ as well as assessing the agreement between the modeled $I(q)$ and the scattering data.

6.6.4 Fuzzy Boundaries, D_{max} and $P(r)$

Biological macromolecules are not solid objects; that is, they are not bounded by a hard or defined two-dimensional surface at the interface with the surrounding solvent. Their hydrated surfaces are neither two- nor three-dimensional but somewhere in between [55] and are affected by thermal fluctuations as well as by exchange with the bulk water or other solutes derived from the solvent. This hydration layer will have a different scattering density when compared with both the protein and the bulk solution [56]—there is a thin layer of contrast enveloping the already fuzzy surface. Consequently, there should be an observed gradual diminishment in $P(r)$ at long vector lengths as r limits to D_{max}. If the D_{max} chosen for the $P(r)$ solution is too short, $P(r)$ will descend quickly at a sharp

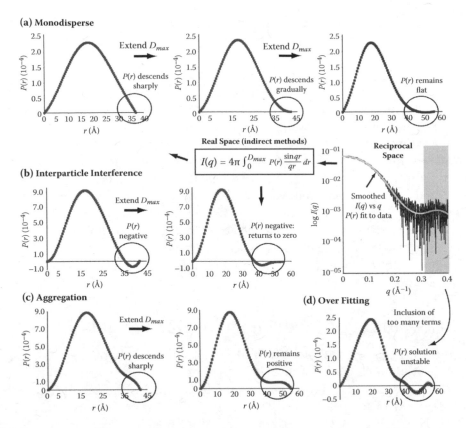

FIGURE 6.6 $P(r)$ profiles. The probable atom-pair distance distribution in real space, $P(r)$, is related to the scattering intensities in reciprocal space by a Fourier transformation. Using indirect methods it is possible to fit $P(r)$ models to the data. For monodisperse macromolecules in solution (a) $P(r)$ will gradually descend toward 0 as r approaches the particle's maximum dimension (D_{max}) and remain relatively flat at longer vector lengths. If the D_{max} chosen is too short, the profile will descend quickly and appear to be prematurely truncated. Interparticle interference (b) can produce negative values of $P(r)$ at long r, while aggregation (c) often produces persistent positive values of r irrespective of D_{max}. (d) $P(r)$ instability can occur when too many terms in the transform are used to fit noisy high-q data that can be affected by poor background subtraction.

angle, and $P(r)$ will appear to be prematurely truncated. Conversely, if the D_{max} chosen is too long, several outcomes can be noted in the shape of the profile (Figure 6.6).

If, after having descended smoothly to 0, $P(r)$ remains flat around $P(r) = 0$ as r increases, then the solution is stable and provides evidence that the sample is free of aggregation and interparticle interference and the solvent subtraction is accurate. The D_{max} is that r-value where the $P(r)$ first reaches zero at long vector lengths. If a $P(r)$ profile remains positive no matter what r-value is chosen for D_{max}, aggregation is likely present in the sample. Conversely, where a $P(r)$ goes negative at long vector lengths and then returns to steady at zero, interparticle interference and a structure factor term are likely affecting the scattering data. Finally, if the transform oscillates around 0 there may be overfitting of the data arising from having more terms in the transform that can be accurately determined. The number of terms used to solve the $P(r)$ is determined by the q-range and D_{max}. Increasing q_{max} increases not only the number of terms but also the fitting's sensitivity to how well the solvent scattering has been subtracted from the sample scattering at high-q. Due to the low scattering intensity and proportionately large errors in the high-q regions of the scattering profile, the inclusion of too much noisy high-q data that may be affected by imprecise solvent subtraction can lead to $P(r)$ instability.

To this point we have considered only $P(r)$ analyses of particles with only all positive or all negative contrast against a solvent. The existence of a mix of positive and negative contrast regions within a sample can generate real negative values in $P(r)$ and not necessarily at the longest vector lengths. In the case of X-rays a mix of positive and negative contrasts is unlikely to be observed—except perhaps for shell particles*—but during a neutron scattering contrast variation experiment the scattering density of the solvent is often altered and component contrasts can vary from positive to negative, in which case positive and negative $P(r)$ values can occur.

Once a stable $P(r)$ solution has been determined, the derived $I(0)$ and R_g values will be more accurate than those obtained by Guinier analysis. The shape of $P(r)$ will provide insights into the shape of the scattering particle and the distribution of contrast. Instabilities in $P(r)$ will indicate data quality issues.

6.7   How to Determine the Molecular Weight

Three methods can be used to determine the molecular weight of a scattering species in solution from SAXS data: (1) absolute scaling of $I(0)$; (2) comparison of $I(0)$ against a secondary standard; and (3) in the case of folded proteins the determination of the area under a plot of $I(q)q^2$ versus q that yields the invariant, Q, from which a molecular volume can be estimated. Ultimately all three derive the molecular weight (M_r) for the scattering particle, which can be subsequently compared with the expected molecular weight of a target calculated from its atomic composition (M_{rcalc}). Although exceptions may occur in cases for which some of the parameters used in the M_r calculation are not well known [57], if the experimentally determined molecular weight deviates by more than 10–15% of the expected value, the target under investigation—even if it has a linear Guinier region and transforms well to produce a reasonable $P(r)$ profile—one must consider the possibility there is aggregation, contamination, or inter-particle interference.

6.7.1   Watch Your Weight

It is important that the molecular weight determined from the scattering data is compared with an accurately calculated M_{rcalc}. For example, if a protein has an N-terminally linked 6-histidine tag that has not been removed (e.g., for affinity chromatography purposes) then the amino acid sequence comprising the tag must be included in the calculation of M_{rcalc} as do any bound metals, covalently linked cofactors, and glycosides, all of which contribute to the scattering. There are several excellent computationally based methods to determine the molecular weight of proteins and DNA based on their sequence, including PROTPARAM [58] and MULCh [41].

---

* A shell particle could be something like a lipoprotein complex where the polypeptide (contrast region 1) completely envelopes an internal hydrophobic core comprising of lipids (contrast region 2).

6.7.2   Concentrate on Concentration

The accurate determination of the concentration of a scattering species in a SAS experiment is important for reliable M_r estimates using $I(0)$. In practice it is more difficult to determine accurate biomolecular concentrations than is generally understood. For proteins, provided the samples are pure and their solutions devoid of ultraviolet absorbing agents that might change with time (e.g., dithiothreitol), the most convenient and precise method is based on the absorbance at 280 nm using extinction coefficients calculated from the primary amino acid sequence (e.g., using PROTPARAM [58]). Quantitative amino acid analysis can also be used; here the protein is digested into its component amino acids, and their concentrations are evaluated. The use of conjugating dyes, such as Bradford reagent, is more problematic, and unless a reliable standard curve for each specific protein has been standardized against a more accurate method a dye-based technique should be an option of last resort.

6.7.3   Concentration Series

Multiple SAXS experiments performed at various sample concentrations (C) can be used to identify whether a system is monodisperse or is influenced by aggregation or interparticle interference. If concentration is determined in units of in $mg \cdot mL^{-1}$ then

$$M_r \propto \frac{I(0)}{C} \tag{6.17}$$

Note that if the concentration were expressed in molar units, $I(0)/C$ would be proportional to the square of M_r due to the dependence of the scattering intensity on the square of the molecular volume (6.8). Keeping to units of concentration in terms of mass per unit volume ($mg \cdot mL^{-1}$), however, then $I(0)/C$ versus C for monodisperse particles should be a constant within experimental error (which will generally be dominated by errors in the concentration determination). If there is a positive correlation in $I(0)/C$ versus C, a concentration-dependent formation of higher molecular weight species is indicated. Conversely, if $I(0)/C$ decreases with increasing concentration, a structure-factor term is indicated due to interparticle interference (Figure 6.7).

6.7.4 M_r from $I(0)$ on an Absolute Scale

The M_r of a scattering particle can be determined from $I(0)$ providing the scattering data are on an absolute scale; that is, the data are in units of cm^{-1}, and the units for all of the parameters of the scattering particle that influence the scattered intensity ($I(0)$) are as described in Equation (6.9). The absolute scaling enables the calculation of the molecular weight of a scattering particle $g \cdot mol^{-1}$. To place the experimental data on an absolute scale, the angle-independent scattering from water is measured and corrected for empty sample cell scattering using the same experimental setup as the sample (normalizing to, e.g., exposure time, temperature). As the scattering from water has been standardized for a wide range of temperatures and pressure [47], it then becomes a case of deriving a multiplicative

Chapter 6

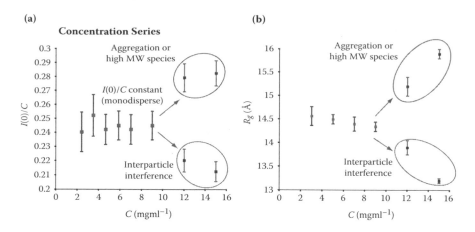

FIGURE 6.7 Forward-scattering intensity ($I(0)$) and concentration (C): small-angle X-ray scattering. (a) For pure macromolecular systems that are free of aggregation or interparticle interference, $I(0)/C$ vs C (in mg.mL⁻¹) should be relatively constant. If a persistent increase in $I(0)/C$ is observed as concentration increases then this is a sign of nonspecific aggregation or the formation of other high molecular weight species in solution. If a linear decrease in $I(0)/C$ is observed it is diagnostic of interparticle interference. (b) A similar pattern is observed in the experimentally determined R_g.

scale factor of the standardized water scattering to the experimentally determined water scattering,

$$sf = \frac{H_2O_{experiment}}{H_2O_{standard}} \tag{6.18}$$

which, when applied to the a sample scattering, will place the intensities on an absolute scale. Rearranging Equation (6.9), the molecular weight of the scattering particle can then be determined from $I(0)$ as long as the concentration has been estimated in $g \cdot cm^{-3}$ via

$$M_r = \frac{I(0)N_A}{C(\Delta\rho v)^2} \tag{6.19}$$

where N_A is Avogadro's number. Both $\Delta\rho$ and v can be calculated using MULCh [41], which accepts as inputs the molecular formulae of the solvent components and the amino acid or nucleotide sequences of the biological macromolecules under investigation.

6.7.5 M_r Determination Using a Secondary Standard

The M_r of a scattering species can be estimated by comparing $I(0)$ with a secondary standard of monodisperse particles of the same scattering density [48,49,59]. For example, dilute samples of lysozyme (up to 8 mg · ml⁻¹) are known to be monodisperse monomers in aqueous solutions containing 150 mM NaCl, 40 mM sodium acetate pH 3.8 [48] and can be used to standardize protein SAS data. If the particle concentration (C) is measured in mg · mL⁻¹

$$M_r \propto \frac{I(0)}{C} \tag{6.20}$$

and thus

$$K = \frac{I(0)}{C \cdot M_r} \tag{6.21}$$

where K is a constant for any monodisperse protein in solution. The M_r of the particle of interest is determined by

$$M_{r_{protein}} = \frac{I(0)_{protein} \cdot C_{lysozyme} \cdot M_{r_{lysozyme}}}{C_{protein} \cdot I(0)_{lysozyme}} \tag{6.22}$$

The advantage of this method is that there is no need to place the scattering data on an absolute scale to obtain $I(0)$, but a target is assumed to have a similar scattering length density as the secondary standard. For targets having a different scattering density or perhaps more than one region of scattering density, such as an RNA/protein complex, a correction factor due to the different contrasts must be determined.

6.7.6 M_r-Based on Volume

A method to accurately determine the M_r of a folded protein from SAXS profiles that is independent of knowing the protein concentration or scattering densities recently was published by Fischer and colleagues [60]. Assuming that a typical protein has an average mass density of 0.83 Da.$Å^{-3}$ [60], the M_r is simply

$$M_r = V \times 0.83 \text{ Daltons} \tag{6.23}$$

where V is the volume. In principle, the volume of a particle is related to the forward-scattering intensity $I(0)$ by

$$V = 2\pi^2 \frac{I(0)}{Q} \tag{6.24}$$

If the scattering profile is normalized so that $I(0) = 1$ then this further simplifies to

$$V = \frac{2\pi^2}{Q} \tag{6.25}$$

where Q is the absolute scattering invariant defined as

$$Q = \int_0^\infty I(q)q^2 \, dq \tag{6.26}$$

which equates to the area under a plot of $I(q)q^2$ versus q for q from 0 to ∞. This plot is known as a Kratky plot (Figure 6.8).

Chapter 6

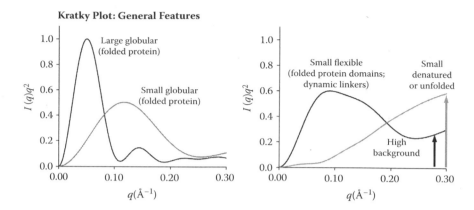

FIGURE 6.8 Kratky plots. A plot of $I(q)q^2$ versus q, or a Kratky plot, can be informative with respect to identifying whether scattering particles are folded and compact, conformationally flexible, or unfolded in solution. The area under the plot to infinite q is the invariant (Q) that relates to the volume of a particle in solution.

In practice, scattering data are collected over discrete, not continuous, intervals of q over a finite q-range, from q_{min} and q_{max} and as such it is possible to calculate only an apparent Q'

$$Q' = \sum_{q_{max}}^{q_{min}} I(q)q^2 \Delta q \tag{6.27}$$

and not the absolute invariant from SAS data.

Fischer et al. [60] employed an empirical approach (Figure 6.9) to solve the issue of estimating the absolute Q value from SAS data given the practical constraints of a scattering experiment. First, the smoothed $I(q)$ vs q profile, obtained from a $P(r)$ model fit to the data, is used for the integration between $0 < q < q_{max}$ where $I(0)$ is normalized to 1. Second, correction factors are employed that account for the effects truncating q_{max} by devising q_{max}-dependent volume correction coefficients, A and B, that relate the experimental volume V' to the absolute volume V via

$$V = A + V'B \tag{6.28}$$

where A and B were determined using simulated SAXS profiles of 1,148 unique high-resolution protein structures in the Protein Databank [60]. For folded proteins in solution, this approach works very well. However, for the analysis of proteins with regions of flexibility or that are unfolded or are atypical due to the presence of metals, cofactors, or binding partners with different mass densities (e.g., DNA), the method devised by Fischer et al. begins to incur errors in the M_r determination.

6.8 Molecular Modeling

Only after the Guinier, $P(r)$, $I(0)$, and M_r analyses indicate that the sample under investigation looks like a monodisperse solution of noninteracting particles (i.e., conditions adequately approximate those of a dilute solution) is it time to develop more sophisticated

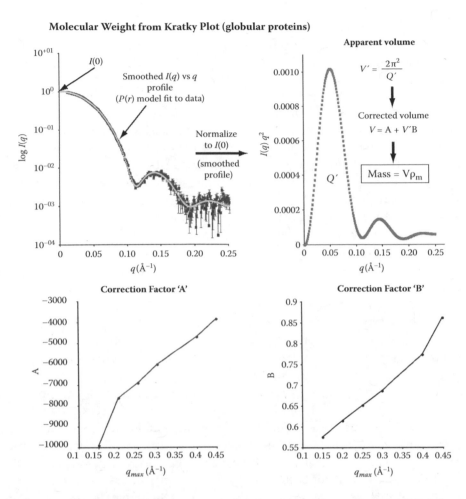

FIGURE 6.9 Molecular weight determination from Kratky plots. Fischer and colleagues [60] developed a concentration-independent method to determine the molecular weight of folded proteins from their scattering profiles using Kratky plots. A smoothed $P(r)$ model fit to the data is normalized to $I(0)$, then plotted on a Kratky plot. The apparent invariant Q' is calculated from the area under the plot up to a defined q_{max}. Q' relates to an apparent molecular volume V', which can be corrected at different values of q_{max} (using A and B) to obtain the actual volume V. The molecular weight of a protein can then be determined by multiplying V by the average mass density (ρ_m) of a protein.

structural models. What follows is a brief description of how to model scattering data using a few useful techniques and how far you might want push the (molecular) envelope.

6.8.1 *Ab Initio* Shape Restoration

Dmitri Svergun and his team produced a suite of free, widely accessible, and easy-to-use computer programs to model SAS data (offered in the ATSAS suite). One of the easiest to use is DAMMIN [27]—recently rereleased in a faster version as DAMMIF [61]—which generates possible 3-D dummy atom models (DAMs) of a scattering particle (assumed to be of a uniform scattering density) from a 1-D scattering profile. Briefly, DAMMIN takes the smoothed $I(q)$ versus q profile derived from the $P(r)$ model fit to the experimental data and optimizes the fit of a DAM to the smoothed profile using a simulated annealing procedure. The program starts by packing a sphere of maximum dimension, usually D_{max}, with smaller spheres (dummy atoms) that are assigned a phase such as solvent (0) or protein (1). The entire structure of the DAM is represented as a phase assignment vector, and the model $I(q)$ versus q scattering profile for the DAM can be calculated throughout the minimization. At the beginning of the procedure, the temperature of the system is held high so that for any one of the dummy atoms in the DAM the difference between its being assigned as either phase 0 or 1 is small; that is, there is a small energy difference between the two phases. As the system is cooled, the energy difference between the two phases at that position in the DAM increases depending on the overall minimization of the phase assignment vector and the DAM fit to the data. Eventually the phase at a position becomes fixed as either solvent or protein; the net result is a DAM that represents the shape of a macromolecule in solution and fits the data (those dummy atoms assigned as solvent are discarded in the final output file; Figure 6.10a).

Since more than one 3-D shape can theoretically fit a 1-D scattering profile, the procedure is usually run multiple (at least 10) times and the resulting solutions are aligned, averaged, and volume corrected to produce a model of the scattering particle in solution [62]. A measure of confidence in the final overall result is how similar each of the independent models used in the averaging are with respect to each other. The averaging procedure produces, among other parameters, a normalized spatial discrepancy value that is helpful in assessing how similar the individual DAMMIN or DAMMIF are and how unique the solution might be [62]. For systems composed of multiple components with different scattering densities or phases (e.g., a protein/RNA complex), the related program MONSA [16] can be used to model both phases and identify where these phases occur with respect to each other in a complex—that is, determine the orientations of subunits within the complex's molecular envelope. MONSA requires inputs for the volume fraction and contrasts of each phase and is run in parallel against multiple scattering data sets, such as X-ray scattering derived from a complex plus X-ray scattering derived from one or more of the individual components. MONSA is a particularly powerful tool for refining the molecular envelopes of biomolecular complexes from multiple SANS data sets collected in a contrast variation series (see Section 6.9.) Such SANS data sets also can be modeled in parallel with SAXS data sets [63] to further improve the determination of the shape of a macromolecular complex and the shapes and disposition of its components.

6.8.2   Atomic Models

One of the strengths of SAS is its ability to probe the biomolecular solution states of a range of targets. Where a high-resolution structure of a protein is available, such as a crystal, NMR, or homology model, a model SAXS profile can be calculated and evaluated against an experimental SAXS profile using CRYSOL (CRYSON for SANS) [64]. The goodness-of-fit of the atomic model against the experimental data is assessed using a χ^2 value

$$\chi^2 = (N-1)^{-1} \sum \left[\frac{I(q) - kI_m(q)}{\sigma(q)} \right]^2$$

(6.29)

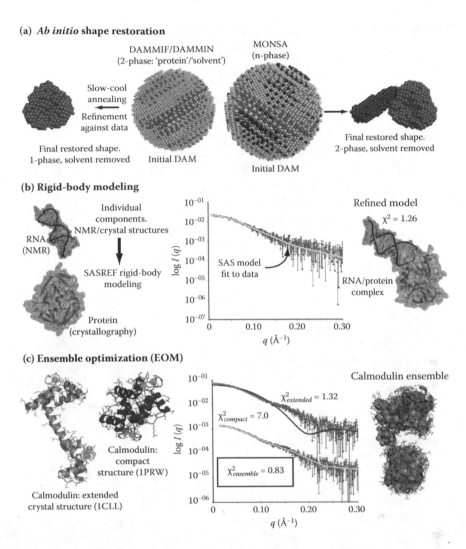

(a) *Ab initio* **shape restoration**

DAMMIF/DAMMIN
(2-phase: 'protein'/'solvent')

MONSA
(n-phase)

Slow-cool annealing

Refinement against data

Final restored shape.
1-phase, solvent removed   Initial DAM

Initial DAM

Final restored shape.
2-phase, solvent removed

(b) Rigid-body modeling

Individual components.
NMR/crystal structures

RNA (NMR)

SASREF rigid-body modeling

SAS model fit to data

Protein (crystallography)

Refined model

$\chi^2 = 1.26$

RNA/protein complex

(c) Ensemble optimization (EOM)

Calmodulin: compact structure (1PRW)

Calmodulin: extended crystal structure (1CLL)

Calmodulin ensemble

$\chi^2_{extended} = 1.32$

$\chi^2_{compact} = 7.0$

$\chi^2_{ensemble} = 0.83$

FIGURE 6.10 Shape restoration and molecular modeling. (a) The molecular shape and envelopes for a monodisperse system of particles in solution can be restored from SAS data using DAMMIN or MONSA. (b) Rigid-body modeling of individual subunits or domains of known structure can be performed using SASREF to determine the position of components within a complex, in this case a RNA/protein complex (10). (c) Ensemble optimization method (EOM) can be employed to obtain a series or family of structures that fit the SAS data. Calmodulin is a case where the two of the protein's primary crystal forms do not fit the solution scattering profile (indicated by elevated χ^2 values.) Upon applying EOM, there is a significant improvement in the fit to the data ($\chi^2 = 0.83$), suggesting that the state of calmodulin in solution is best represented as an ensemble of structures.

where N is the number of experimental points, $I(q)$ and $I_m(q)$ are the experimental and atomic model scattering intensities, respectively, k is a scaling factor, and $\sigma(q)$ the experimental error. For lab-based X-ray sources, if a correspondence exists between a high-resolution structure and a scattering profile and the statistical errors have been propagated correctly, χ^2 will approach 1.0 (a very good fit somewhere in the region of 0.8–1.15). It then becomes easy to identify whether a high-resolution structure does not fit the scattering data, and this is where the real power of SAS can come into play.

Chapter 6

In some cases, the crystal structure may not fully represent the solution state. A classic example is the calcium binding protein calmodulin where SAXS and NMR data revealed that it exists a conformationaly flexible ensemble of structures in solution that undergo conformational changes upon calcium binding—flexibility that turned out to be at the heart of how this protein functions as a regulator of calcium signaling [65–67]. In such instances where high-resolution structures fail to represent solution conditions, the programs SASREF and BUNCH [26] and EOM [68] can all be used to produce models that fit scattering data. SASREF can refine the positions of subunits within a complex or domains within a protein (Figure 6.10b); BUNCH can refine domain orientations as well as model regions of missing mass (using dummy atoms) for sequence segments of unknown structure. In the case of conformational dynamic proteins or proteins that have regions of significant flexibility (e.g., extended loops), EOM can generate ensembles of structures that fit the data (Figure 6.10c). A literal cache of structures in the Protein Databank can be exploited for model refinement against SAS data to provide insight into the overall solution states of biomacromolecules, their complexes, and how they might structurally reorganize in response to changes in environment, binding to small molecules, or other binding partners. However, aside from linking the $I(0)$ of a scattering experiment to the M_r of a species in solution, there is no way to validate SAS-refined models without data from other sources, unlike the more data-rich techniques of crystallography or NMR. Therefore, it is important to interpret the SAS-derived modeling solutions with care and incorporate other pieces of information to improve the confidence in the final analyses.

6.8.3 The Uniqueness of Structural Modeling

The information content of a scattering profile is essentially shape information, and commenting on atomic resolution details of a model that fits SAS data has to be approached cautiously. As with *ab initio* shape restoration, different atomic models may fit a SAS profile. Thus, multiple refinement calculations are often necessary to gauge whether a consensus of likely structures is available to aid in assessing the uniqueness of the predicted overall placement of domains or subunits within a biomacromolecule or complex. Importantly, the frequency with which a particular class of models is obtained is not a basis for discriminating among different classes of models. Even if dozens of refinements are performed and a minority class of differently organized models emerges that appear to fit the SAS data equally well as a majority consensus, these minority solutions maybe just as valid in terms of representing the shape and scattering density distribution within the experimental sample. In such circumstances a significance test on differences in χ^2 values may discriminate between alternate models or groups of models; however, more information is often needed in the modeling to reduce the number of possibilities.

As SAS data are really sensitive only to scattering density distribution, caution must be exercised when modeling proteins that consist of multiple modules of similar size and shape. For example, take a protein that consists of five domains of roughly equivalent mass whose crystal structures have been individually solved—A, B, C, D, E—that adopts a Y shape in solution based on *ab initio* modeling. Because of the shape and mass equivalency of the domains, any combination of domains A–D in the shape of a Y will produce equivalent fits to the data (Figure 6.11a).

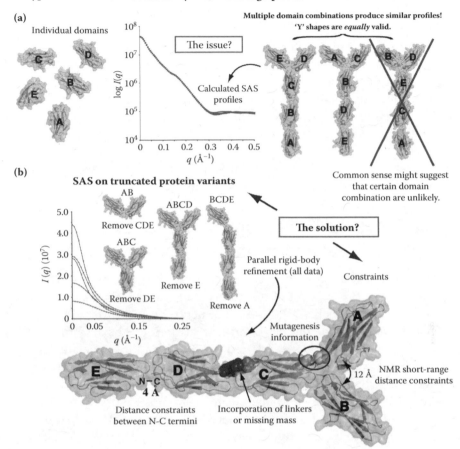

FIGURE 6.11 Reducing the number of model parameters. Rigid-body modeling of multiple high-resolution structures derived from NMR or crystallography against SAS data can pose a problem: there is often more than one shape that can fit a scattering profile. If a hypothetical protein Y contains five domains of approximately equal mass, almost any combination of domains within the Y-shaped envelope (a) will produce almost identical scattering profiles. Reducing the number of parameters in the modeling (b) by incorporating further SAS information (e.g., from truncated protein variants) or distance constraints from other methods (e.g., NMR or mutagenesis) improves the confidence in the final model.

Common sense might suggest that if the C- and N-termini of all the individual domains are separated by few residues in the primary amino acid sequence of the intact protein, it is unlikely that domain A would be positioned right next to domain D in the final Y-shaped structure or that domain E will end up neighboring domain B. These assumptions reduce the number of possible permutations in the modeling by discarding unlikely solutions. Codifying known constraints in the modeling reduces the number of degrees of freedom during refinement and can improve the confidence in the final consensus model.

The more constraints that can be introduced in to a refinement of an atomic model against SAS data, the better. Connectivity constraints that require the C- and N-termini of neighboring A, B, C, D, and E modules of the Y protein to be within a reasonable distance of each other is just one example. The incorporation of additional scattering data sets from truncated Y-protein variants for parallel SAS refinement, or short distance constraints derived from NMR, provide hard data to constrain the modeling (Figure 6.11b). NMR is very powerful for identifying which regions of a macromolecule interact with binding partners [10] and can be used very effectively in combination with SAS to constrain rigid-body refinement of macromolecular structures in solution (NMR can help SAS and vice versa [36–38,69]). Information from mutant screening experiments that alter binding interfaces also can yield contact information between components. Electron microscopy can provide structural information for very large macromolecular systems [70]. What is the take-home message? Bootstrapping information from other techniques is invaluable—as little as one distance constraint can transform an intractable modeling problem to a tractable one. One technique that is unique in its ability to increase the information content of a scattering experiment is small-angle neutron scattering with contrast variation.

6.9 The Powerful Advantage of SANS for Probing Macromolecular Complexes

SANS with contrast variation adds an extra dimension to scattering data that effectively increases the information content of the experiment particularly for multisubunit biomacromolecular complexes [24,71]. Recall that neutrons are scattered via interactions with the nuclei of atoms. The physics of nuclear scattering is beyond the scope of this chapter, and the interested reader is referred to [71,72] for a more detailed and formalized description of events. However, a few relevant items to structural analyses of biomacromolecules in solution can be explained simply. The key to understanding neutron scattering and how it can be used to investigate macromolecular systems is that, unlike small-angle X-ray scattering, the contrast of a SANS experiment can be easily and systematically manipulated by altering the neutron scattering density of a macromolecule or the solvent. This added power of contrast variation comes because the isotopes of hydrogen, ^{1}H and ^{2}H (deuterium), have very different neutron scattering properties.

6.9.1 Neutron Basics

Neutrons have zero charge, a spin of 1/2, a magnetic moment and are scattered primarily via interactions with the nuclei of atoms. While X-rays that can induce the free radicalization of electrons that might cause bond breakage and sample damage, neutrons are by-the-by less damaging to biomolecules, assuming that no elements in a system undergo neutron activation to release α, β, or γ radiation that can ionize the target (potassium, for example, is a γ emitter.) The scattering power of a nucleus depends upon the properties of the compound neutron–nucleus that forms in the scattering event. Simplistically, the ability of a nucleus to scatter a neutron is expressed in terms of a conceptual length, b, that relates to the radius of the area of a nuclear cross section; the greater the radius, the greater the area (cross section) and the more likely a nucleus will scatter.* There is no simple relationship between atomic number and neutron scattering

---

* X-ray scattering power, although often expressed in terms of electron density, also can be expressed as a length. For an atom with Z electrons the coherent scattering cross section is $4p(Zr_0)^2$, where $r_0 = e^2/m_ec^2$ ($= 0.28 \times 10^{-12}$ cm); e, electron charge; m_e, electron mass; and c, speed of light.

power, so the cross sections of elements and their isotopes have been determined empirically.

Neutron scattering theory parallels that for X-rays. If we think of a neutron in terms of its properties as a plane wave—as opposed to a particle—elastic, coherent neutron scattering from a macromolecule will produce a small-angle scattering profile just like X-rays, except the intensities will reflect atom-pair distance distributions of nuclei within a macromolecule, contrast weighted with respect to the neutron scattering density of a solvent. It happens that most elements commonly found in biomolecules (^{12}C, ^{16}O, ^{14}N, ^{31}P, and mostly ^{32}S) scatter neutrons coherently with approximately the same scattering power with one important exception: the naturally abundant isotope of hydrogen (^{1}H). The ^{1}H isotope is unusual for two reasons. First, the coherent scattering length for ^{1}H is about half the value and, more importantly, is negative with respect to other biological elements. This negative b-value basically means that the amplitude of the wave function of a neutron inverts as it scatters from a ^{1}H nucleus (the scattered wave will be exactly 180° out of phase with the incident wave.) Second, ^{1}H has a proportionately large incoherent scattering cross section compared with other elements. In the case of incoherent scattering, the phase relationships between the scattered neutron waves do not have a defined relationship between them. This incoherent scattering contributes to a significant background signal in SANS experiments in systems with high ^{1}H content; for example, buffers made up in regular light water [73]. The incoherent scattering background produced by ^{1}H$_2$O is so high that path length of the sample cell used for SANS in high ^{1}H$_2$O buffers is limited to 1 mm or less to avoid significant multiple scattering events. It is, however, the inversion of the phase of a coherently scattered neutron from a ^{1}H nucleus and the resulting negative scattering length wherein the real power of SANS and contrast variation comes into play.

The heavy isotope of hydrogen, deuterium (^{2}H or D), is like the other biological elements in that ^{2}H has an overall positive coherent scattering length, is of similar magnitude to that of ^{12}C, ^{16}O, ^{14}N, ^{31}P, and has a proportionately low incoherent scattering length—pretty much the opposite situation encountered with ^{1}H (^{2}H$_b^{coherent} = 0.667 \ 10^{-12}$ cm: ^{2}H$_b^{incoherent} = 0.404 \ 10^{-12}$ cm ^{1}H$_b^{coherent} = -0.374 \ 10^{-12}$ cm: ^{1}H$_b^{incoherent} = 2.527 \ 10^{-12}$ cm). Therefore, solutions composed of light water (H$_2$O or ^{1}H$_2$O) have increased incoherent scattering and have a very different (a negative) overall neutron scattering density when compared with heavy water (D$_2$O or ^{2}H$_2$O) ($\rho_{H_2O} = -0.6 \ 10^{10}$ cm^{-2}; $\rho_{D_2O} = 6.4 \ 10^{10}$ cm^{-2}). Since scattering density is simply the sum of the scattering powers of all the atoms in a volume divided by that volume ($\Sigma b_i / V$), changing the number of ^{1}H atoms in a volume will change the neutron scattering density. Hence, an experimenter can alter the contrast during a scattering experiment—that is, alter the difference in neutron scattering density between the solvent and a target biomacromolecule—by simply altering the H$_2$O to D$_2$O ratio in the solvent.

6.9.2   The Power of Neutron Scattering Contrast

In the case of a protein in solution, increasing the volume fraction of D$_2$O in the solvent from 0% v/v to ~43% v/v will result in an overall decrease in the positive difference in scattering density between the protein and solvent to a point where $\Delta\rho = 0$. While there may be small contributions to the scattering arising from internal fluctuations

Chapter 6

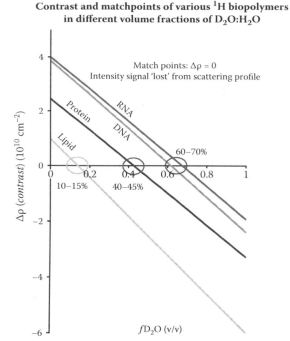

FIGURE 6.12 Contrast of different biomacromolecules in D_2O: small angle neutron scattering. Lipids, protein, and polynucleotides (RNA/DNA) all have different ratios of hydrogen (^1H) per unit volume and consequently have different natural contrasts and solvent match points in a small-angle neutron-scattering experiment.

in scattering density caused by regions of localized nonuniformity (e.g., from localized differences in hydrogen exchange with the solvent), the coherent scattering contribution from the protein will effectively disappear at the 43% v/v D_2O (i.e., its solvent match point). Upon further increasing the D_2O fraction from 43 to 100% v/v, the contrast between the protein and the solvent will change signs, resulting in the reappearance of a coherent small-angle scattering signal from the protein.

For a multisubunit complex in which the ratio of ^1H per unit volume is different between the individual subunits, each component should have a different D_2O solvent match point (Figure 6.12). For example, DNA and protein have different elemental and hydrogen compositions per unit volume, so within a protein/DNA complex two regions of naturally different neutron scattering length densities will be solvent matched at different D_2O concentrations: ~43% v/v for the protein; and ~63% v/v for the DNA (Figure 6.13a). After solvent subtraction, SANS data collected in 0% v/v D_2O will have scattering contributions derived from both components: at 43% v/v D_2O only the DNA will contribute; at ~63% v/v only the protein contributes; and at 100% D_2O, both components contribute again. Thus, a solvent-matching experiment can be designed that will provide shape information for the individual protein and DNA components as well as the whole complex in different D_2O concentrations.

A somewhat more sophisticated contrast variation experiment compared with simple solvent matching is one in which a full contrast series is measured, for example, measuring SANS data for a protein/DNA complex in 0, 10, 20, 30, 55, 80, 90, and

Protein/DNA complex - natural contrast in SANS

(a)

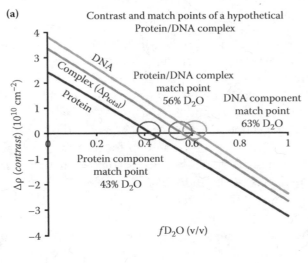

(b)

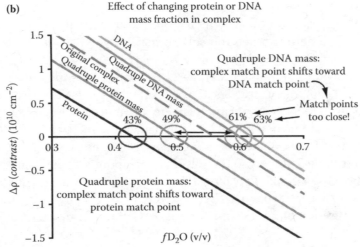

FIGURE 6.13   Contrast calculations from a protein/DNA complex. The scattering contributions made by individual components of a protein/DNA complex in a SANS investigation can be isolated depending on the solvent match points of the components. (a) The strength of the scattering signal will depend on the total contrast ($\Delta\rho_{total}$) of a protein/DNA complex at different D_2O concentrations. In 0% v/v D_2O the scattering profile will encode shape information from the whole complex. By placing the complex in ~43% v/v D_2O the contribution made to the scattering by the protein will be matched out and the scattering will represent that from the DNA ($\Delta\rho_{total}$ positive) In ~63% v/v D_2O the DNA contribution will be matched out and the scattering will encode shape information about the protein ($\Delta\rho_{total}$ negative) (b) The magnitude of the scattering signal for each of the individual components in a protein/DNA complex will depend on their mass ratios in the complex. Altering the ratio will shift the match points. If the individual component and whole complex solvent match points are too close, then it will be difficult to measure accurate SANS data due to a compromise in intensity as $\Delta\rho_{total}$ limits to zero.

100% v/v D_2O. In such a scenario, using the relationships described in Equations (6.7), (6.10), and (6.11) and generating a set of linear equations that must be simultaneously satisfied for each contrast point (from which the contrast coefficients can be calculated from chemical composition), it is possible to extract the scattering profiles of the entire complex ($I(q)_{total}$), each component ($I(q)_1$, $I(q)_2$), and the cross-term scattering arising from atom-pair distances between components ($I(q)_{12}$). Thus, a contrast series can reveal the shape of the overall complex as well as the individual components and their dispositions with respect to each other.

For our protein/DNA experiment, in between 43% v/v and 63% v/v D_2O there will be a point at which the contrast against the solvent for the entire complex is zero and there will be no measurable scattering. The overall complex match point depends on the scattering length densities of the individual components and their volume fractions. For example, a protein of 22 kDa in complex with DNA with an MW of 56 kDa has an overall solvent match point at 56% D_2O v/v. If the molecular weight of the protein is quadrupled, the match point for the whole complex shifts toward the match point of the protein (down to ~50% v/v D_2O.) Conversely, if the molecular weight of the DNA is quadrupled, the complex match point shifts in the other direction up to 61% v/v D_2O, perilously close to the 63% match point of the DNA by itself (Figure 6.13b). Consequently, in designing a neutron solvent-matching experiment it is important to ensure that the match point of the whole complex is well separated from the match points of the individual components; otherwise, it might be impossible to measure one of the individual component scattering profiles due to the lack of scattering intensity (made more acute if there is significant incoherent scattering from ^1H_2O in the solvent). The solution to the problem of poorly separated match points can be quite simple: artificially change the number of nonexchangeable hydrogens per unit volume on one of the components in the complex so the solvent match points of the components and the complex are better separated. Replacing ^1H with ^2H in one of the components not only aids the investigation of protein/DNA (or protein/RNA) complexes but also further extends the application of SANS with contrast variation to protein–protein complexes in solution.

6.9.3 Biodeuteration

Although SAXS can reveal the overall shape of a native protein–protein complex, it generally cannot discern the shapes and dispositions of the components because proteins have roughly equivalent electron densities. A similar situation holds for a SANS experiment. Aside from differences in amino acid composition that can alter the number of hydrogens in exchange with a solvent, which produces small fluctuations in neutron scattering density, the components of a native protein–protein complex will have roughly equivalent amounts of hydrogen per unit volume. Consequently, unlike a protein–DNA (or RNA) complex, in a protein–protein complex there isn't a major difference in natural contrast against a solvent. This situation, however, can be dramatically changed by selectively deuterating one of the protein components to decrease the number of nonexchangeable ^1H atoms per unit volume, which shifts the solvent match point of the deuterated component so that it is separated from that of the nondeuterated component as well as that of the whole complex. This kind of selective deuteration facilitates SANS experiments similar to those previously described for the protein/DNA complex, enabling structural characterization of the overall complex and the components within it. In this case, the important separation of the component and complex match points will again depend upon the level of deuteration in one of the proteins and its volume fraction within the complex.

Deciding on appropriate levels of deuteration required for a contrast variation experiment and estimates of the component and complex match points can be done prior to a SANS investigation using the *Contrast* module of the MULCh suite [41]. *Contrast* simply requires a user to input the molecular composition of the solvent (e.g., buffer components Tris, NaCl), the amino acid sequences of the proteins in each subunit, their stoichiometry (from which the volume fractions are derived), and the level of deuteration on one of the components. From this basic information, *Contrast* calculates the neutron scattering density and contrast $(\Delta\rho)$ of the individual components and the whole complex at various D_2O concentrations. As an example, a 1:1 protein complex forms between protein X and protein Y that each has a molecular weight of ~22 kDa (~27,000 Å^3). If neither protein is labeled with deuterium, the match point of the components and the complex will all be around 43% v/v D_2O, and it won't be possible to isolate the component scattering functions (Figure 6.14). However, if protein Y is labeled with deuterium so that on average 60% of all nonexchangeable [1]hydrogens have been replaced with ^2H, the solvent match point of the whole complex shifts to 67% v/v D_2O, the match point for Y increases to 93% v/v D_2O, and protein X (as it is unlabeled) stays the same at ~43% v/v D_2O (Figure 6.14b). With the solvent match points separated it is thus possible to collect SANS data from the whole complex (0% v/v D_2O), the deuterated protein Y component (at 43% v/v D_2O), and protein X (in high D_2O.) What if protein Y is deuterated even more, say on average to 75% ^2H? Will this improve the match point separation? *Contrast* calculates that the match point of the complex will increase (up to ~73% v/v D_2O) but will the match point of protein Y up to 105% v/v D_2O, making it impossible to solvent match. Deciding on the level of deuteration on one of the components of a multisubunit complex before performing a SANS with contrast variation can significantly boost an experiment's success. The *Contrast* module of MULCh enables an experimentalist to assess the effects of the volume fractions between components on the scattering signal and whether stand-alone experiments might be required to supplement a full contrast series, such as hyperdeuteration of a small component or a reversal of the component deuteration pattern (see also Supplementary Information).

6.9.4 Contrast, $I(0)$, and R_g Relations in a SANS Experiment

A full contrast series—0, 10, 20, 30, 55, 80, 90, and 100% v/v D_2O—will provide the most comprehensive structural information, including information on the relative dispositions of components within a complex [41,42,74]. As for any SAS experiment, the SANS scattering signal in a contrast variation experiment depends upon the contrast of the macromolecular system under investigation against the solvent. Thus, it is important to have an estimate of what the experimental contrasts are; if the contrasts are incorrect or a mistake is made in making up the contrast series (e.g., a volume error when making up %v/v D_2O buffers), then the component scattering functions extracted from the contrast variation data will not be accurate representations of the

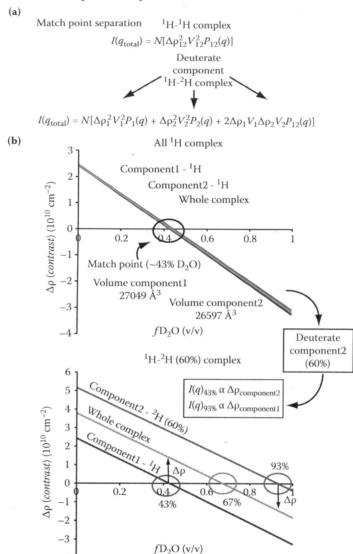

FIGURE 6.14 Protein–protein complexes and the affect of biodeuteration: SANS match point splitting. (a) Subunits of a 1:1 ^{1}H protein–^{1}H protein complex (designated as subscripts 1 and 2) will have approximately the same contrast against a background solvent, and the net average scattering intensity ($I(q_{total})$) will be proportionate to the typical parameters: the contrast ($\Delta\rho$) and volume (V) squared, the form factor ($P(q)$), and number of particles in solution (N). (b) However, if one of the proteins in the complex is deuterated, this causes the scattering function to split into contributions made by the individual components and between the components.

components (practical aspects of sample handling are discussed in the Supplementary Information). Aside from having quantified the average level of deuteration of any deuterated component used in a SANS investigation, it is also pertinent to measure the mass density of each buffer used in the contrast series (in $g \cdot cm^{-3}$), preferably using densitometry on the actual solvents to quantify the % v/v D_2O. By substituting the experimentally determined volume fractions of D_2O in the solvent into the calculated $\Delta\rho$ vs % v/v D_2O relationship from MULCh [41] it is possible to obtain reasonable estimates of the experimental contrasts. Assessment of the quality of the contrast variation data also requires careful R_g and $I(0)$ analyses (derived from Guinier or $P(r)$) that should adhere to specific patterns depending on the change contrast or D_2O concentration.

Ibel and Stuhrmann [75] showed that there is a quadratic dependency between the experimental radius of gyration squared and the inverse of the mean contrast across the whole complex against the solvent, $\Delta\bar{\rho}$ (cf. Equations (6.10) and (6.11), which express R_g^2 as functions of the mean contrasts of the individual components):

$$R_g^2 = R_m^2 + \frac{\alpha}{\Delta\bar{\rho}} - \frac{\beta}{\Delta\bar{\rho}^2}, \qquad (6.30)$$

where R_m is radius of gyration of the particle expressed in terms of a uniform scattering density at infinite contrast and α and β are coefficients that take into account scattering density fluctuations within the particle and provide information about their distribution. Plotting R_g^2 vs $1/\Delta\bar{\rho}$ should, within experimental error, adhere to a quadratic relation. Furthermore, the square root of the experimental forward-scattering intensities normalized to concentration and plotted against the fraction of D_2O in the solvent should produce a linear relationship leading up to and away from the match point of the whole complex, and the $I(0)$ should give a consistent experimentally determined M_r for all the contrasts measured (Figure 6.15; Equation (6.19)).

If the R_g and $I(0)$ analyses adhere to these relations, then it is possible to calculate the component scattering functions using the contrast variation data, from which the individual component shape parameters as well as the distribution of distances between the components of the complex can be determined. The MULCh suite of programs [41] can guide the experimenter through the basic neutron scattering and contrast variation analysis, including calculating the component and cross-term scattering functions, that is wise to complete as a guide and as an estimate of the quality of the data for 3-D modeling with MONSA and SASREF7 (Figure 16 and [16,26,63]).

The shape restoration program MONSA was designed for systems containing regions of multiple contrast [16] and will generate a low-resolution representation of such a complex that distinguishes the shapes of components having distinct scattering density. SASREF7 can be employed to rigid-body refine atomic models of components with respect to each other against a contrast variation series to build models that simultaneously fit all the data [26] and can be evaluated in terms of individual goodness-of-fit χ^2 values for each scattering contrast using CRYSON [56]. SANS with contrast variation opens up a whole field of new possibilities for investigating the structures of biomacromolecular complexes in solution [8,9,17,45,57,63,76–79].

Chapter 6

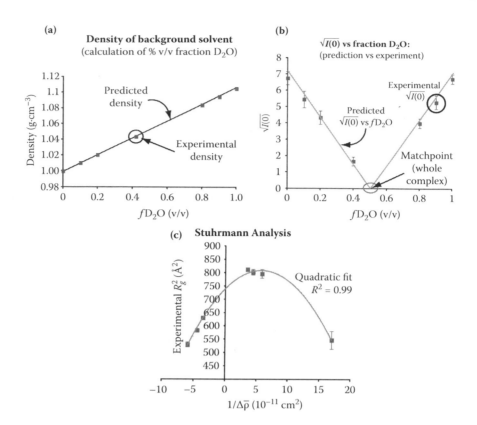

FIGURE 6.15 D_2O densities, $I(0)$ and Stuhrmann analyses for a SANS contrast variation experiment. (a) Experimental buffer (mass) densities containing different volume fractions of D_2O should be similar to those predicted from MULCh (41). (b) For a ^1H protein–^2H protein complex the $\sqrt{I(0)}$ normalized to protein concentration versus D_2O fraction should be linear leading up to and away from the complex match point. [In practice it is common to multiply the experimental values above the match point by –1. In doing so a linear relationship between "$\sqrt{I(0)}$" versus fD_2O should be obtained with a fD_2O-intercept corresponding to the match point of the complex.] (c) Ibel and Stuhrmann [75] showed that a quadratic relation exists between R_g and the inverse of the mean contrast across the whole complex ($\Delta\bar{\rho}$). It is expected that for a full contrast series from a macromolecular complex the experimentally determined R_g will adhere to closely to this relation.

6.10 The Future

Small-angle scattering has experienced a significant surge in interest from the structural biology community in the first decade of the 21st century. In this period there has been an approximate 300% increase in the number of publications that use SAXS or SANS as a complementary technique for characterizing the structures of macromolecules and their complexes in solution. The reason for this surge in interest is due to a confluence of a number of advances—most significantly in laboratory- and synchrotron-based X-ray instrumentation, computer software, and steady improvements in recombinant technologies—that have resulted in the expression and purification of increasingly diverse and interesting biomacromolecular systems. Complementing crystallography and NMR, SAS is revealing new and interesting aspects of structural biology (see Chapter 11). SAS is nonetheless very demanding experimentally, and it is important to understand what controls and checks are required to assess the quality of the data as well to appreciate the

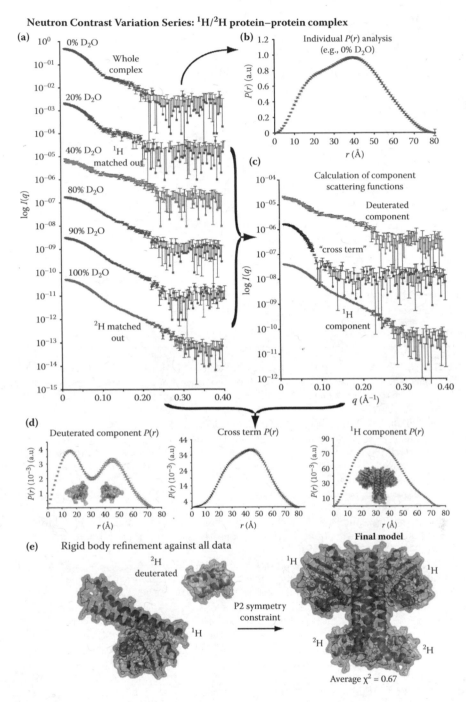

FIGURE 6.16 Neutron contrast variation data and modeling. SANS with contrast variation from a ^{1}H-^{2}H protein–protein complex in six D_2O concentrations is shown. From these data (a), the individual $P(r)$ profiles (b) can be calculated from every contrast point to extract $I(0)$ and R_g. The component scattering profiles (c) can be derived from the data from which shape information from the individual components can be extracted, and the cross-term can be calculated from which the distribution of distances between the components of the complex (d) can be determined. Parallel rigid-body refinement of the individual components in the complex produces a model that fits the data (e). The data presented here have been scaled for clarity.

limitations of the information present in the data when generating wonderful 3-D models. Hopefully this chapter has outlined some of the basics of scattering for the structural molecular biologist and highlighted certain aspects of what to look out for when performing a SAS experiment and how to get the most out of the data.

Acknowledgments

The authors wish to thank Dr. Robert Hynson of the School of Molecular Bioscience at the University of Sydney for help in preparing Figure 6.4 and John Chow for providing the calmodulin scattering data (Figure 6.10). The MULCh suite of modules written by Dr. Andrew Whitten (IMB, University of Queensland) and colleagues can be accessed online at (http://research.mmb.usyd.edu.au/NCVWeb/) and include a tutorial based on the data presented in Figure 6.16. The ATSAS package of programs (e.g., GNOM, CRYSOL, CRYSON, SASREF) developed by Dimitri Svergun and the Biological Small Angle Scattering Group that the EMBL (Hamburg) can be accessed online (http://www.embl-hamburg.de/biosaxs/software.html). Supplementary information for this Chapter is available at http://www.crcpress.com/product/isbn/9781439827222

References

1. Chan, J., A. E. Whitten, C. M. Jeffries, I. Bosanac, T. K. Mal, J. Ito, H. Porumb, T. Michikawa, K. Mikoshiba, J. Trewhella, and M. Ikura. 2007. Ligand-induced conformational changes via flexible linkers in the amino-terminal region of the inositol 1,4,5-trisphosphate receptor. *J Mol Biol* 373:1269–1280.
2. Wall, M. E., S. C. Gallagher, and J. Trewhella. 2000. Large-scale shape changes in proteins and macro-molecular complexes. *Annu Rev Phys Chem* 51:355–380.
3. Nemeth-Pongracz, V., O. Barabas, M. Fuxreiter, I. Simon, I. Pichova, M. Rumlova, H. Zabranska, D. Svergun, M. Petoukhov, V. Harmat, E. Klement, E. Hunyadi-Gulyas, K. F. Medzihradszky, E. Konya, and B. G. Vertessy. 2007. Flexible segments modulate co-folding of dUTPase and nucleocapsid proteins. *Nucleic Acids Res* 35:495–505.
4. VanOudenhove, J., E. Anderson, S. Krueger, and J. L. Cole. 2009. Analysis of PKR structure by small-angle scattering. *J Mol Biol* 387:910–920.
5. Wang, Y., J. Trewhella, and D. P. Goldenberg. 2008. Small-angle X-ray scattering of reduced ribonuclease A: effects of solution conditions and comparisons with a computational model of unfolded proteins. *J Mol Biol* 377:1576–1592.
6. Zhang, X., P. V. Konarev, M. V. Petoukhov, D. I. Svergun, L. Xing, R. H. Cheng, I. Haase, M. Fischer, A. Bacher, R. Ladenstein, and W. Meining. 2006. Multiple assembly states of lumazine synthase: a model relating catalytic function and molecular assembly. *J Mol Biol* 362:753–770.
7. Russell, R., X. Zhuang, H. P. Babcock, I. S. Millett, S. Doniach, S. Chu, and D. Herschlag. 2002. Exploring the folding landscape of a structured RNA. *Proc Natl Acad Sci U S A* 99:155–160.
8. Whitten, A. E., D. A. Jacques, B. Hammouda, T. Hanley, G. F. King, J. M. Guss, J. Trewhella, and D. B. Langley. 2007. The structure of the KinA-Sda complex suggests an allosteric mechanism of histidine kinase inhibition. *J Mol Biol* 368:407–420.
9. Jacques, D. A., D. B. Langley, C. M. Jeffries, K. A. Cunningham, W. F. Burkholder, J. M. Guss, and J. Trewhella. 2008. Histidine kinase regulation by a cyclophilin-like inhibitor. *J Mol Biol* 384:422–435.
10. Claridge, J. K., S. J. Headey, J. Y. Chow, M. Schwalbe, P. J. Edwards, C. M. Jeffries, H. Venugopal, J. Trewhella, and S. M. Pascal. 2009. A picornaviral loop-to-loop replication complex. *J Struct Biol* 166:251–262.
11. Bilgin, N., M. Ehrenberg, C. Ebel, G. Zaccai, Z. Sayers, M. H. Koch, D. I. Svergun, C. Barberato, V. Volkov, P. Nissen, and J. Nyborg. 1998. Solution structure of the ternary complex between aminoacyl-tRNA, elongation factor Tu, and guanosine triphosphate. *Biochemistry* 37:8163–8172.
12. Zhao, J., J. Wang, D. J. Chen, S. R. Peterson, and J. Trewhella. 1999. The solution structure of the DNA double-stranded break repair protein Ku and its complex with DNA: a neutron contrast variation study. *Biochemistry* 38:2152–2159.

13. Krueger, J. K., B. S. McCrary, A. H. Wang, J. W. Shriver, J. Trewhella, and S. P. Edmondson. 1999. The solution structure of the Sac7d/DNA complex: a small-angle X-ray scattering study. *Biochemistry* 38:10247–10255.

14. Rambo, R. P., and J. A. Tainer. 2010. Bridging the solution divide: comprehensive structural analyses of dynamic RNA, DNA, and protein assemblies by small-angle X-ray scattering. *Curr Opin Struct Biol* 20:128–137.

15. Funari, S. S., G. Rapp, M. Perbandt, K. Dierks, M. Vallazza, C. Betzel, V. A. Erdmann, and D. I. Svergun. 2000. Structure of free Thermus flavus 5 S rRNA at 1.3 nm resolution from synchrotron X-ray solution scattering. *J Biol Chem* 275:31283–31288.

16. Svergun, D. I., and K. H. Nierhaus. 2000. A map of protein–rRNA distribution in the 70 S *Escherichia coli* ribosome. *J Biol Chem* 275:14432–14439.

17. Whitten, A. E., C. M. Jeffries, S. P. Harris, and J. Trewhella. 2008. Cardiac myosin-binding protein C decorates F-actin: implications for cardiac function. *Proc Natl Acad Sci U S A* 105:18360–18365.

18. Neylon, C. 2008. Small angle neutron and X-ray scattering in structural biology: recent examples from the literature. *Eur Biophys J* 37:531–541.

19. Petoukhov, M. V., and D. I. Svergun. 2007. Analysis of X-ray and neutron scattering from biomacromolecular solutions. *Curr Opin Struct Biol* 17:562–571.

20. Jacques, D. A., and J. Trewhella. 2010. Small-angle scattering for structural biology-expanding the frontier while avoiding the pitfalls. *Protein Sci* 19:642–657.

21. Svergun, D. I. 2010. Small-angle X-ray and neutron scattering as a tool for structural systems biology. *Biol Chem* 391:737-743.

22. Svergun, D., and M. H. J. Koch. 2003. Small-angle scattering studies of biological macromolecules in solution. *Rep Progr Phys* 66:1735–1782.

23. Mertens, H. D., and D. I. Svergun. 2010. Structural characterization of proteins and complexes using small-angle X-ray solution scattering. *J Struct Biol*:doi:10.1016/j.jsb.2010.1006.1012

24. Lakey, J. H. 2009. Neutrons for biologists: a beginner's guide, or why you should consider using neutrons. *J R Soc Interface* 6 Suppl 5:S567–573.

25. Chacon, P., F. Moran, J. F. Diaz, E. Pantos, and J. M. Andreu. 1998. Low-resolution structures of proteins in solution retrieved from X-ray scattering with a genetic algorithm. *Biophys J* 74:2760–2775.

26. Petoukhov, M. V., and D. I. Svergun. 2005. Global rigid body modeling of macromolecular complexes against small-angle scattering data. *Biophys J* 89:1237–1250.

27. Svergun, D. I. 1999. Restoring low resolution structure of biological macromolecules from solution scattering using simulated annealing. *Biophys J* 76:2879–2886.

28. Fisher, S. J., J. R. Helliwell, S. Khurshid, L. Govada, C. Redwood, J. M. Squire, and N. E. Chayen. 2008. An investigation into the protonation states of the C1 domain of cardiac myosin-binding protein C. *Acta Crystallogr D Biol Crystallogr* 64:658–664.

29. Govada, L., L. Carpenter, P. C. da Fonseca, J. R. Helliwell, P. Rizkallah, E. Flashman, N. E. Chayen, C. Redwood, and J. M. Squire. 2008. Crystal structure of the C1 domain of cardiac myosin binding protein–C: implications for hypertrophic cardiomyopathy. *J Mol Biol* 378:387–397.

30. Ababou, A., M. Gautel, and M. Pfuhl. 2007. Dissecting the N-terminal myosin binding site of human cardiac myosin-binding protein C. Structure and myosin binding of domain C2. *J Biol Chem* 282:9204–9215.

31. Ababou, A., E. Rostkova, S. Mistry, C. Le Masurier, M. Gautel, and M. Pfuhl. 2008. Myosin binding protein C positioned to play a key role in regulation of muscle contraction: structure and interactions of domain C1. *J Mol Biol* 384:615–630.

32. Jeffries, C. M., A. E. Whitten, S. P. Harris, and J. Trewhella. 2008. Small-angle X-ray scattering reveals the N-terminal domain organization of cardiac myosin binding protein C. *J Mol Biol* 377:1186–1199.

33. von Castelmur, E., M. Marino, D. I. Svergun, L. Kreplak, Z. Ucurum-Fotiadis, P. V. Konarev, A. Urzhumtsev, D. Labeit, S. Labeit, and O. Mayans. 2008. A regular pattern of Ig super-motifs defines segmental flexibility as the elastic mechanism of the titin chain. *Proc Natl Acad Sci U S A* 105:1186–1191.

34. Glatter, O., and O. Kratky. 1982. *Small Angle X-ray Scattering*. Academic Press, New York.

35. Glatter, O. 1977. A new method for the evaluation of small-angle scattering data. *J Appl Cryst* 10:415–421.

36. Grishaev, A., J. Wu, J. Trewhella, and A. Bax. 2005. Refinement of multidomain protein structures by combination of solution small-angle X-ray scattering and NMR data. *J Am Chem Soc* 127:16621–16628.

Chapter 6

37. Grishaev, A., V. Tugarinov, L. E. Kay, J. Trewhella, and A. Bax. 2008. Refined solution structure of the 82-kDa enzyme malate synthase G from joint NMR and synchrotron SAXS restraints. *J Biomol NMR* 40:95–106.

38. Wang, J., X. Zuo, P. Yu, I. J. Byeon, J. Jung, X. Wang, M. Dyba, S. Seifert, C. D. Schwieters, J. Qin, A. M. Gronenborn, and Y. X. Wang. 2009. Determination of multicomponent protein structures in solution using global orientation and shape restraints. *J Am Chem Soc* 131:10507–10515.

39. Makowski, L. 2010. Characterization of proteins with wide-angle X-ray solution scattering (WAXS). *J Struct Funct Genomics* 11:9–19.

40. Cammarata, M., M. Levantino, F. Schotte, P. A. Anfinrud, F. Ewald, J. Choi, A. Cupane, M. Wulff, and H. Ihee. 2008. Tracking the structural dynamics of proteins in solution using time-resolved wide-angle X-ray scattering. *Nat Methods* 5:881–886.

41. Whitten, A. E., S. Z. Cai, and T. J. 2008. MULCh: modules for the analysis of small-angle neutron contrast variation data from biomolecular assemblies. *J Appl Cryst* 41:222–226.

42. Goodisman, J., and H. Brumberger. 1971. Scattering from a multiphase system. *J Appl Cryst* 4:347–351.

43. Heller, W. T., E. Abusamhadneh, N. Finley, P. R. Rosevear, and J. Trewhella. 2002. The solution structure of a cardiac troponin C-troponin I-troponin T complex shows a somewhat compact troponin C interacting with an extended troponin I-troponin T component. *Biochemistry* 41:15654–15663.

44. Kuzmanovic, D. A., I. Elashvili, C. Wick, C. O'Connell, and S. Krueger. 2006. The MS2 coat protein shell is likely assembled under tension: a novel role for the MS2 bacteriophage A protein as revealed by small-angle neutron scattering. *J Mol Biol* 355:1095–1111.

45. Olah, G. A., S. E. Rokop, C. L. Wang, S. L. Blechner, and J. Trewhella. 1994. Troponin I encompasses an extended troponin C in the Ca(2+)-bound complex: a small-angle X-ray and neutron scattering study. *Biochemistry* 33:8233–8239.

46. Zhao, J., E. Hoye, S. Boylan, D. A. Walsh, and J. Trewhella. 1998. Quaternary structures of a catalytic subunit-regulatory subunit dimeric complex and the holoenzyme of the cAMP-dependent protein kinase by neutron contrast variation. *J Biol Chem* 273:30448–30459.

47. Orthaber, D., A. Bergmann, and O. Glatter. 2000. SAXS experiments on absolute scale with Kratky systems using water as a secondary standard. *J Appl Cryst* 33:218–225.

48. Krigbaum, W. R., and F. R. Kugler. 1970. Molecular conformation of egg-white lysozyme and bovine alpha-lactalbumin in solution. *Biochemistry* 9:1216–1223.

49. Kozak, M. 2005. Glucose isomerase from *Streptomyces rubiginosus*—potential molecular weight standard for small-angle X-ray scattering. *J Appl Cryst* 38:555–558.

50. Guinier, A. 1938. The diffusion of X-rays under the extremely weak angles applied to the study of fine particles and colloidal suspension. *Comptes Rendus Hebdomadaires Des Seances De L Acad Des Sci* 206:1374–1376.

51. Konarev, P. V., V. V. Volkov, A. V. Sokolova, M. H. J. Koch, and D. I. Svergun. 2003. PRIMUS: a Windows PC-based system for small-angle scattering data analysis. *J Appl Cryst* 36:1277–1282.

52. Hjelm, R. P. 1985. The small-angle approximation of X-ray and neutron scatter from rigid rods of non-uniform cross section and finite length. *J Appl Cryst* 18:452–460.

53. Guinier, A., and G. Fournet. 1955. *Small-angle scattering of X-rays (structure of matter series)*. Wiley, New York.

54. Svergun, D. I. 1992. Determination of the regularization parameter in indirect-transform methods using perceptual criteria. *J Appl Cryst* 25:495–503.

55. Timchenko, A. A., O. V. Galzitskaya, and I. N. Serdyuk. 1997. Roughness of the globular protein surface: analysis of high resolution X-ray data. *Proteins* 28:194–201.

56. Svergun, D. I., S. Richard, M. H. Koch, Z. Sayers, S. Kuprin, and G. Zaccai. 1998. Protein hydration in solution: experimental observation by X-ray and neutron scattering. *Proc Natl Acad Sci U S A* 95:2267–2272.

57. Jacques, D. A., M. Streamer, S. L. Rowland, G. F. King, J. M. Guss, J. Trewhella, and D. B. Langley. 2009. Structure of the sporulation histidine kinase inhibitor Sda from *Bacillus subtilis* and insights into its solution state. *Acta Crystallogr D Biol Crystallogr* 65:574–581.

58. Gasteiger, E., C. Hoogland, A. Gattiker, S. Duvaud, M. R. Wilkins, R. D. Appel, and A. Bairoch. 2005. *Protein identification and analysis tools on the ExPASy server*. Humana Press, Totowa.

59. Mylonas, E., and D. Svergun. 2007. Accuracy of molecular mass determination of proteins in solution by small-angle X-ray scattering. *J Appl Cryst* 40:s245–s249.

60. Fischer, H., M. O. Neto, H. B. Napolitano, I. Polikarpov, and A. F. Craievich. 2010. The molecular weight of proteins in solution can be determined from a single SAXS measurement on a relative scale. *J Appl Cryst* 43:101–109.

61. Franke, D., and D. Svergun. 2009. DAMMIF, a program for rapid ab-initio shape determination in small-angle scattering. *J Appl Cryst* 42:342–346.
62. Volkov, V., and D. Svergun. 2003. Uniqueness of *ab initio* shape determination in small-angle scattering. *J Appl Cryst* 36:860–864.
63. Petoukhov, M. V., and D. I. Svergun. 2006. Joint use of small-angle X-ray and neutron scattering to study biological macromolecules in solution. *Eur Biophys J* 35:567–576.
64. Svergun, D., C. Barberato, and M. H. Koch. 1995. CRYSOL—a program to evaluate X-ray solution scattering of biological macromolecules from atomic coordinates. *J Appl Cryst* 28:768–773.
65. Trewhella, J., and J. K. Krueger. 2002. Small-angle solution scattering reveals information on conformational dynamics in calcium-binding proteins and in their interactions with regulatory targets. *Methods Mol Biol* 173:137–159.
66. Trewhella, J., D. K. Blumenthal, S. E. Rokop, and P. A. Seeger. 1990. Small-angle scattering studies show distinct conformations of calmodulin in its complexes with two peptides based on the regulatory domain of the catalytic subunit of phosphorylase kinase. *Biochemistry* 29:9316–9324.
67. Trewhella, J. 1992. The solution structures of calmodulin and its complexes with synthetic peptides based on target enzyme binding domains. *Cell Calcium* 13:377–390.
68. Bernado, P., E. Mylonas, M. V. Petoukhov, M. Blackledge, and D. I. Svergun. 2007. Structural characterization of flexible proteins using small-angle X-ray scattering. *J Am Chem Soc* 129:5656–5664.
69. Gabel, F., B. Simon, M. Nilges, M. Petoukhov, D. Svergun, and M. Sattler. 2008. A structure refinement protocol combining NMR residual dipolar couplings and small angle scattering restraints. *J Biomol NMR* 41:199–208.
70. Comoletti, D., M. T. Miller, C. M. Jeffries, J. Wilson, B. Demeler, P. Taylor, J. Trewhella, and T. Nakagawa. 2010. The macromolecular architecture of extracellular domain of alphaNRXN1: domain organization, flexibility, and insights into trans-synaptic disposition. *Structure* 18:1044–1053.
71. Krueger, J. K., and G. D. Wignall. 2006. *Small-angle neutron scattering from biological molecules.* Springer, New York.
72. Squires, G. L. 1978. *Introduction to the theory of thermal neutron scattering.* Dover, New York.
73. Rubinson, K. A., C. Stanley, and S. Krueger. 2008. Small-angle neutron scattering and the errors in protein structures that arise from uncorrected background and intermolecular interactions. *J Appl Cryst* 41:456–465.
74. Svergun, D. I., and M. H. Koch. 2002. Advances in structure analysis using small-angle scattering in solution. *Curr Opin Struct Biol* 12:654–660.
75. Ibel, K., and H. B. Stuhrmann. 1975. Comparison of neutron and X-ray scattering of dilute myoglobin solutions. *J Mol Biol* 93:255–265.
76. Jacques, D. A., D. B. Langley, R. M. Hynson, A. E. Whitten, A. Kwan, J. M. Guss, and J. Trewhella. 2011. A novel structure of an antikinase and its inhibitor. *J Mol Biol* 405:214–226.
77. Comoletti, D., A. Grishaev, A. E. Whitten, I. Tsigelny, P. Taylor, and J. Trewhella. 2007. Synaptic arrangement of the neuroligin/beta-neurexin complex revealed by X-ray and neutron scattering. *Structure* 15:693–705.
78. Howarth, J. W., J. Meller, R. J. Solaro, J. Trewhella, and P. R. Rosevear. 2007. Phosphorylation-dependent conformational transition of the cardiac specific N-extension of troponin I in cardiac troponin. *J Mol Biol* 373:706–722.
79. Heller, W. T., N. L. Finley, W. J. Dong, P. Timmins, H. C. Cheung, P. R. Rosevear, and J. Trewhella. 2003. Small-angle neutron scattering with contrast variation reveals spatial relationships between the three subunits in the ternary cardiac troponin complex and the effects of troponin I phosphorylation. *Biochemistry* 42:7790–7800.

Chapter 6

7. Subcellular Signaling Dynamics

Jeffrey J. Saucerman
Jin Zhang

7.1   Introduction

Cellular signaling networks exhibit enormous complexity, not only in the diversity of molecules involved but also in the way that molecular signals vary spatially and temporally. This chapter describes some of the experimental and theoretical approaches used

Quantitative Biology: From Molecular to Cellular Systems edited by Michael E. Wall © 2012 CRC Press / Taylor & Francis Group, LLC. ISBN: 978-1-4398-2722-2

Chapter 7

to dissect the subcellular signaling processes that underlie many cellular decisions. A general theme is that an iterative cycle between experiments and modeling will ultimately be necessary to develop a quantitative understanding of cellular information processing.

7.2 Biochemical Messages Varying in Space and Time

7.2.1 Information Processing by Signaling Networks

Cells are faced with a wide range of tasks, and they must be constantly making decisions about whether and how to respond to changes in the external environment. Indeed, responsiveness to external stimuli is a fundamental aspect of the definition of life. A cell's environment can change in many ways, such as exposure to lipids, hormones, nutrients, and metabolites. In response to these environmental signals, cell decisions must be made that will determine the appropriate phenotype of the cell. Even the most basic cells constantly make a wide range of choices due to the changing environment: should it pick survival or death? Proliferation or quiescence? Migration? Differentiation? Cells must also detect and respond to changes in their internal states, such as the DNA damage response and cell cycle checkpoints. Malfunctioning cellular decisions can lead to disease. For example, perturbation of many of the aforementioned basic cell fate processes can lead to cancer.

More specialized cells must make cell decisions appropriate for their constituent tissue or organ. Cardiac myocytes make decisions about whether to increase or decrease their contractility, growth, and death or survival. Indeed, all of these specific cell decisions can be regulated by the same β-adrenergic signaling pathway. This allows for cellular contingency plans; the heart enhances contractile function in response to transient stress, but if stress continues alternate decisions may be made such as growth or death. For this reason, while β-adrenergic signaling is initially beneficial, sustained signaling will ultimately lead to heart failure [1]. Smooth muscle is surprisingly plastic, allowing substantial phenotypic switching between the more typical contractile phenotype to a proliferative/migratory phenotype. This phenotypic switch is a central element in the development of atherosclerosis, where smooth muscle cells migrate to and proliferate at the site of injury [2]. Phenotypic switching is also performed by pathogens, often switching from a dormant phenotype to virulent depending on environmental conditions [3]. Thus, it is not sufficient for biology to just catalog the list of functions carried out by a given cell; it is critical to understand precisely how a cell selects a particular behavior for a given situation.

Although it was once argued that the genome was the central processing unit (CPU) of the cell, it is increasingly clear that cell decisions are in fact made by distributed signaling and regulatory networks throughout and between cells [4]. Receptors, ion channels, and adhesions act as sensors for these networks. After making a *cell decision*, the signaling network needs to communicate with the protein machinery of the cell to execute these decisions, often by posttranslational modification such as phosphorylation. However, we are just beginning to understand the design principles of how the actual cellular decisions are made. One cellular strategy is to integrate multiple pathways through cross-talk so that the cell can integrate multiple sources of information.

But even a single pathway can make surprisingly complex decisions. Calcium is a widely used second messenger, controlling phenotypes such as fertilization, learning, and contraction while also being a common trigger of necrosis and apoptosis [5]. Cyclic AMP (cAMP) has a similar wide range of signaling roles, acting through protein kinase A, Epac and cyclic nucleotide-gated channels to regulate gene transcription, contraction, and metabolism [6]. NFκB is another central signaling hub, with hundreds of known activators and the ability to either enhance or suppress tumor development depending on the context [7].

7.2.2   Signaling Compartmentation

One of the main ways a major signaling hub may perform complex cell decisions is by signaling compartmentation. By processing the spatiotemporal variation of a biochemical signal, the cell may be able to establish a more complete view of the cell's environment, not unlike a radio antenna. Brunton et al. [8] initially invoked the signaling compartmentation hypothesis to explain why certain G-protein coupled receptors, all signaling through cAMP, resulted in different functional consequences. But direct visualization of these cAMP gradients in intact cells took other 20 years [9].

In Section 7.3, live-cell imaging techniques for direct measurement of signaling compartmentation will be described. These new experimental strategies provide the opportunity to directly measure both normal signaling compartmentation and how spatial signaling is affected by pharmacologic or genetic perturbations, revealing the mechanistic basis for compartmentation. In Section 7.4, examples of integrated experimental and mathematical modeling approaches are provided. Finally, Section 7.5 provides a detailed case study of imaging and modeling of signaling compartmentation to further highlight the trade-offs in various experimental and modeling approaches in the context of addressing a particular biological question.

7.3   Live-Cell Imaging of Subcellular Signaling Dynamics

7.3.1   Fluorescence Imaging Techniques

Fluorescence microscopy is a fundamental technique of cell biology, particularly valuable for the study of cellular signaling dynamics or compartmentation. By labeling small molecules or proteins of interest, the dynamic, rich behaviors exhibited by cells can be directly observed. This section provides a brief overview of the different fluorophores available for labeling and also some of the main fluorescence microscopy methods available for using these fluorophores to image subcellular signaling dynamics. For further reading, there are several excellent practical guides for fluorescence microscopy and live-cell imaging [10–12].

7.3.1.1   Fluorophores

The most established class of fluorophores is organic fluorescent dyes. Fluorescein and rhodamine have long been used as complementary green- and red-emitting dyes, respectively, which are quite valuable when conjugated to specific antibodies for immunofluorescent labeling of endogenous proteins. Reasonably good spatial information is available in such images, although this is generally limited to fixed cells. A wide range of variants have been produced; for example, the Alexa range of dyes includes 17 distinct colors (Invitrogen). Through careful organic chemistry, a number of developed derivatives exhibit either excitation or emission spectra that are sensitive to biological signals such as calcium, sodium, pH, nitric oxide, and membrane potential (Invitrogen). In addition, a series of methodologies has been developed to label cellular recombinant proteins with fluorescent probes using genetically encodable tags, such as the tetracysteine/ biarsenical system, fluorogen activating proteins, and SNAP- and HALO tags [13,14]. These systems have the potential for targeting a diverse set of fluorescent indicators to specific subcellular locations. In a recent example, the tetracysteine/biarsenical system was successfully employed to target a Ca^{2+}–sensitive dye for monitoring rapid changes in [Ca^{2+}] dynamics generated by discrete microdomains located at the cell surface [15].

Another widely used class of fluorophores is fluorescent proteins. Since the original cloning of green fluorescent protein (GFP) from jellyfish, GFP and fluorescent proteins subsequently isolated from other organisms have been extensively reengineered to obtain a wide range of fluorescent colors and improve other fluorophore characteristics. Common color variants of GFP include cyan, yellow, and blue. Fluorescent proteins from corals have enabled a large number of additional colors particularly at longer wavelengths, providing orange, red, and more recently infrared-emitting fluorescent proteins [16]. This range of colors facilitates multiplexing of fluorophores and opens up new possibilities such as noninvasive in vivo imaging [17]. In terms of photophysical properties, these proteins have been engineered to enhance brightness while minimizing sensitivity to photobleaching, environmental variables (e.g., pH) and dimerization [16]. Some fluorescent proteins have also been engineered to be photosensitive—either increasing brightness or changing color when stimulated, allowing new protein tracking applications.

7.3.1.2   Live-Cell Imaging Approaches

The most widely used application of fluorescent proteins has been labeling a protein of interest for imaging of either static localization or to track protein translocation between cellular compartments. For example, a yeast library has been developed where in each strain a different gene product is tagged with GFP [18], allowing comprehensive analysis of subcellular protein localization [19]. Protein translocation can often be a readout of protein activity, such as the Akt PH domain-GFP fusion, which translocates to the plasma membrane with a rise in PIP_3 [20]. If multiple proteins are labeled with complementary fluorescent proteins, one can perform colocalization analysis, although this is generally limited to the resolution of the light microscope. However, recently several approaches have been developed to achieve super-resolution fluorescence microscopy at the nanometer scale [21]. There is always some concern that labeling a protein may alter its function or localization or the fact that an exogenous protein may not act the same as the endogenous protein. Therefore, it is important to perform control experiments to verify the correct localization and targeting of these fusion proteins.

If the timescales or spatial scales of interest are smaller, other approaches may be needed. Fluorescence resonance energy transfer (FRET) involves using complementary pairs of fluorophores and allows detection of protein interactions at < 10 nm resolution and will be discussed further in Section 7.3.2 (see also Chapter 5). If diffusion coefficients or rate constants are of interest, fluorescence recovery after photobleaching (FRAP) may be useful. In a FRAP experiment, the protein of interest is labeled with a fluorophore, and then a laser is used to rapidly photobleach the labeled proteins within a small region of the cell. A sequence of images is then obtained to measure the timecourse of recovery in fluorescence as labeled proteins from other regions of the cell diffuse into the bleached region. When fit to mathematical models, FRAP experiments can be used to estimate diffusion coefficients, rate constants, and the fraction of mobile molecules [12] as discussed further in Chapter 8. Recently developed cross-correlation approaches complement translocation, FRET, and FRAP approaches by providing intermediate resolution and a variety of quantitative readouts. Cross-correlation analysis has been used in analysis of cell adhesions to detect the dynamics and localization of protein complexes, rate constants, and stoichiometry within a protein complex [22,23]. Further, these cross-correlation approaches are accessible using many commercially available confocal microscopes.

7.3.2   FRET Reporter Development

As described already, FRET is sensitive to the distance and relative orientation of complementary fluorophores at the ~10 nm scale. This makes FRET well suited for the measurement of protein–protein interactions and conformational changes within a protein. The phenomenon of FRET is the nonradiative energy transfer from a donor fluorophore to a spectrally compatible acceptor fluorophore. The amount of energy transfer (FRET) is dependent on the distance between fluorophores (proportional to 1/radius^6), and their relative orientation and spectral overlap [24]. Maximal spectral overlap occurs when the emission spectrum of the donor molecule matches the excitation spectrum of the acceptor molecule.

7.3.2.1   Intermolecular FRET

The most common FRET application is intermolecular FRET, where two proteins are labeled with complementary fluorophores to assess protein–protein interaction. The most convenient readout of FRET is to excite the donor fluorophore and record the fluorescence emission from both donor and acceptor. The emission of the acceptor when exciting the donor is a measure of FRET termed sensitized emission. Therefore, while exciting the donor, an increase in FRET should cause an increase in acceptor fluorescence and a decrease in donor fluorescence.

However, the sensitized emission approach is sensitive to a variety of artifacts, particularly for intermolecular FRET. For example, photobleaching, environmental sensitivity, and the stoichiometry of the two fluorophores can all cause changes in apparent FRET. Several correction strategies have been proposed to deal with these to some degree. Gordon et al. [25] developed corrections for (1) spectral bleedthrough from donor fluorescence to acceptor emission channels; (2) direct excitation of the acceptor at the donor excitation wavelengths; and (3) dependence of apparent FRET on the concentrations of the donor and acceptor [25]. This approach was subsequently extended to include intensity range-based correction factors, which avoid over- and underestimation of spectral bleedthrough at various fluorescence intensities that are particularly important in confocal and multiphoton FRET [26].

However, intermolecular FRET measured by sensitized emission may still need to be confirmed by alternative approaches. The simplest alternative is acceptor photobleaching, which causes an increase in donor fluorescence if FRET was occurring [27]. A disadvantage of acceptor photobleaching is the single time-point nature, preventing imaging of protein interaction dynamics. Another technique, fluorescence lifetime imaging (FLIM), can provide a very reliable measure of FRET but requires specialized equipment [24]. FLIM involves either nanosecond pulses of laser excitation where the decay rate of fluorescence is measured (time-domain FLIM) or an oscillating laser excitation where the phase delay between excitation source and fluorescence emission (frequency-domain FLIM) is measured.

7.3.2.2 Intramolecular FRET

Although intermolecular FRET allows measurement of protein–protein interactions, intramolecular FRET allows measurement of conformational change in a given protein (see also Chapter 5). One approach is to take an endogenous protein or protein domain and label distinct regions with FRET pairs (often cyan and yellow fluorescent protein). For example, this approach has been used to measure cAMP by exploiting cAMP-dependent conformational changes in the cAMP-binding domain of Epac [28]; IP$_3$ using IP$_3$-dependent conformational changes in the IP$_3$-binding domain of the IP$_3$ receptor [29]; and β$_1$-adrenergic receptor conformations in response to ligand binding [30]. While intramolecular FRET is not subject to the aforementioned stoichiometry artifacts for intermolecular FRET, in some cases conformational changes in endogenous proteins may not be sufficient to obtain a detectable change in FRET. Thus, it may be necessary to optimize the location and identity of the fluorescent proteins involved.

Another approach to intramolecular FRET is to engineer a *molecular switch*, which undergoes a conformational change in response to a specific cellular analyte. One of the first examples of this approach was the development of cameleons, genetically encoded Ca sensors incorporating calmodulin and the M13 peptide that binds Ca/CaM complex [31]. Subsequently, this approach was extended to serine/threonine [32] and tyrosine [33] kinase/phosphatase activities, where sensors are based on a substrate peptide for the kinase/phosphatase and a phosphoamino binding domain. When the substrate is phosphorylated, it is sequestered by the phosphoamino binding domain causing a significant conformational change and thus a change in FRET signal. These approaches can take advantage of the wide range of available peptide substrates and optimization of both phosphoamino binding domains and linker regions to maximize the dynamic range of the FRET reporter. This general strategy is also applicable to monitoring dynamics of other posttranslational modifications [34] as well as activating signaling proteins such as small G proteins.

7.3.3 Subcellular Targeting of FRET–Based Biosensors

Unless the recombinant protein biosensor already incorporates a targeting motif, one would expect that the biosensor would be fairly uniformly expressed. Therefore, the signaling readouts, while providing excellent temporal resolution in live cells, may reflect whole-cell signaling activity rather than particular local signals. To measure specific subcellular signaling dynamics, targeting motifs can be added to send the reporter to a particular compartment. Targeting motifs are generally derived from those motifs found in endogenous proteins and have been used to send biosensors to a wide range of cellular compartments including the nucleus, plasma membrane, lipid rafts, mitochondria, endoplasmic reticulum, and even particular protein complexes [35]. For example, simultaneous imaging plasma membrane-targeted cAMP sensor with a nuclear PKA sensor has been used to reveal the propagation of signal through different compartments of this pathway [28]. FRET biosensors of protein kinase D activity have been targeted to NHERF protein scaffolds using PDZ domains, allowing measurement of a unique PKD activity signature within this protein complex [36].

7.4 Mathematical Modeling of Subcellular Signaling Dynamics

In engineering and the physical sciences, mathematical modeling has always been an integral part of the discovery and design process. Mathematical models help in evaluating whether a certain set of assumptions (e.g., molecular mechanisms) are quantitatively sufficient to explain the behavior of a complex system. In this sense, the fact that models are abstract and simplified is completely necessary—otherwise, new conceptual understanding could not be obtained. Specific approaches for mathematical modeling of signaling networks have been reviewed elsewhere [37,38], and there are also now good introductory textbooks on the topic [39,40]. This section reviews some of the key design principles of cell signaling that have been elucidated by combining experimental approaches with mathematical models.

Modeling of signaling compartmentation goes back almost 60 years. The mathematical models of Hodgkin and Huxley [41] predicted not only the action potential at a single location but also the self-propagating electrical wave along a squid axon. At about the same time, Alan Turing [42] published a landmark theoretical study titled "Chemical Basis of Morphogenesis." In this paper, he demonstrated how a simple reaction–diffusion model could generate a wide variety of stable spatial patterns such as spots and stripes. The underlying dynamical basis for these patterns was a short-range positive feedback morphogen interlinked with a longer-range negative feedback morphogen. Note that the Hodgkin–Huxley and Turing models both used interlinked positive and negative feedbacks, illustrating the versatility of this network motif.

Despite these early successes, a lack of appropriate experimental tools made it difficult to obtain a quantitative understanding of signaling compartmentation. With the advent of fluorescent dyes, fluorescent proteins, and advances in light microscopy as described in Section 7.3, a new generation of data-driven mathematical models became possible. Mathematical models are now being combined with a variety of live-cell imaging approaches to elucidate numerous aspects of signaling compartmentation. In particular, integrated imaging and modeling studies have investigated mechanisms underlying biochemical gradients, roles of cell shape, and molecular-scale compartmentation on protein scaffolds (Table 7.1).

7.4.1 Modeling Subcellular Signaling Gradients

By combining live-cell imaging and spatially explicit modeling, Fink et al. [43] examined the spatiotemporal interplay between Ca^{2+} waves and IP_3 in the neuron. They began by imaging Ca^{2+} (using the fluorescent dye fura-2) in response to flash photolysis of caged IP_3, finding that IP_3 induces stronger Ca signals in the neurite than the

Table 7.1 Examples of Mathematical Models Used to Study Subcellular Signaling Gradients

Category	Topic	Equations Type	Reference
Signaling gradients, idealized geometry	Action potential propagation	1-D PDE	[41]
	Morphogen patterns	2-D PDE	[42]
	Membrane-cytosol gradients	1-D PDE	[53]
	cAMP gradients	Compartmental ODE or 1-D PDE	[48]
	Nuclear Ran transport	Compartmental ODE	[45]
	PDGF gradient sensing	1-D PDE	[47]
	Cell shape and phosphorylation gradients	3-D PDE	[54]
	cAMP/PKA gradients	Compartmental ODE	[69]
	cAMP gradients	Compartmental ODE	[50]
Signaling gradients, image-based geometry	IP3/Ca signaling	2-D PDE	[43]
	Nuclear Ran transport	3-D PDE	[45]
	PDGF gradient sensing	2-D PDE	[47]
	cAMP/PKA gradients	2-D PDE	[69]
	cAMP/MAPK gradients	2-D PDE	[56]
Protein complexes	MAPK scaffolds	ODE	[61]
	AKAPs	ODE	[62], [63]
	FCεRI signaling	ODE	[68]
	EGFR	ODE	[67]

Notes: ODE, ordinary differential equation. PDE, partial differential equation.

soma, while bradykinin affects Ca^{2+} globally. To better understand the mechanisms of nonuniform IP_3 and Ca signaling in their experiments, they developed a 2-D partial-differential equation (PDE) mathematical model incorporating molecular mechanisms of Ca^{2+} uptake and release along with, importantly, the geometric distribution of receptors and cell geometry obtained in their experiments. These spatial details were critical for their study of how local geometry influences subcellular signaling. Their model was then validated against their measurements of IP_3-induced Ca^{2+} waves in individual cells, and then the model was used to make the new prediction that local bradykinin would stimulate Ca^{2+} more strongly at the neurite than the soma or growth cone of the neuron. They subsequently validated this prediction as well using local bradykinin applied with a micropipette. This study elegantly illustrates how models can be used to integrate a variety of experimental data (Ca^{2+} receptors/fluxes, cell geometry) to provide quantitative understanding of how a cell signaling system works. Further, this group's Virtual Cell software provided an analysis platform that enabled a number of subsequent spatial analyses of cell signaling [44].

Smith et al. [45] extended this integrated modeling and live-cell imaging approach to examine the signaling network regulating nucleocytoplasmic Ran transport. Fluorescently labeled Ran-GTP was microinjected into the cell, and the rate of Ran transport into the nucleus was imaged by confocal microscopy. Smith et al. also used Ran mutants to assess the contribution of passive versus active Ran transport. To assess transport from the nucleus to the cytosol, the cytosolic fluorescent Ran was repeatedly bleached, causing a decrease in nuclear Ran indicating the extent and rate of nucleus-to-cytosol transport. This approach is called fluorescence loss in photobleaching and is related to the FRAP approach described in Section 7.2.1. Together with available biochemical rate constants from the literature, this imaging data was used to develop a 3-D PDE mathematical model of Ran transport, which could reproduce the kinetics and localization of Ran transport seen experimentally. Due to the large computational requirements of this 3-D model, Smith et al. developed a simplified compartmental ordinary-differential equation (ODE) model, which was also able to reproduce the overall experimental findings. This indicated that nuclear transport rather than cytosolic diffusion was rate-limiting.

Smith et al. [45] then used their compartmental ODE model to perform new *in silico* experiments to identify key regulatory mechanisms. The model predicted that while the system was quite robust to perturbation of many expected key regulators, it was surprisingly sensitive to perturbation of the Ran exchange factor RCC1. They subsequently validated this prediction by repeating the RanGTP microinjection experiments using cells that express a temperature-sensitive RCC1. The model was also used to provide quantitative estimates of bidirectional flux rates through the nuclear pore. These studies were later extended with a more comprehensive mathematical model and additional mechanistic experiments, which found that the nuclear import machinery provides increased dynamic range of nuclear import rates at the expense of overall transport efficiency [46].

7.4.1.1 Cell Migration

Integrated modeling and live-cell imaging studies have also yielded important advances in our understanding of biochemical gradients involved in cell migration. Schneider and Haugh [47] combined live-cell measurements of PI 3-kinase signaling with mathematical models to understand how fibroblasts detect gradients of PDGF. They began with a simplified 1-D steady state PDE model to illustrate the hypothesis that, in these cells, PI 3-kinase recruitment is simply proportional to the local density of activated receptors, an absolute gradient-sensing mechanism. To test this hypothesis, they imaged a CFP-tagged pleckstrin homology domain of Akt (CFP-AktPH) as a biosensor of 3' PI by TIRF microscopy. By applying controlled spatial gradients of PDGF, they used this biosensor of PI 3-kinase signaling to provide dynamic and quantitative measurements of gradient sensing. These experiments validated the model predictions that strong gradient sensing requires a strong gradient with midpoint PDGF concentrations in a certain range.

To more directly assess the mechanisms of gradient sensing in the TIRF experiments, Schneider and Haugh [47] extended their mathematical models to incorporate image-based cell geometry, the actual PDGF concentration field, and recruitment of the CFP-AktPH biosensor. Model parameters were determined from their previous experimental and modeling studies of fibroblast responses to uniform PDGF. These 2-D-PDE equations were solved using the finite-element approach in FEMLAB (Comsol, Burlington, MA). Using this more detailed model, they were able to predict the gradients in PI 3-kinase signaling observed in individual fibroblasts and also predict the dynamic response to pulses in PDGF, which was then validated experimentally.

Overall, the model predictions indicated that the degree of PDGF-gradient sensing found in fibroblasts did not require some of the feedback loops found in other chemotactic cells, because PI 3-kinase signaling was not amplified relative to the external PDGF gradient. This helped explain the slower and less sensitive chemotaxis seen in fibroblasts. Further, hot spots of PI 3-kinase signaling at the leading edge of migrating cells were predicted to provide a spatial bias that may enable persistence of migration toward PDGF gradients [47].

7.4.1.2   cAMP Compartmentation

Several combined experimental and modeling studies have now examined the original cAMP compartmentation hypothesis already described: that functional selectivity of cAMP signaling is explained by compartmentation. Rich et al. [48] engineered cAMP-gated ion channels to measure rapid cAMP kinetics near the plasma membrane using patch clamping. They found that forskolin, a direct activator of adenylyl cyclase (cAMP source), induced large submembrane [cAMP] despite > 10-fold lower global [cAMP]. Indeed, forskolin stimulated large submembrane [cAMP] signals even during rapid dialysis of the bulk cytosol, suggesting that cAMP may be generated in diffusionally limited microdomains. To understand the mechanisms underlying the measured cAMP gradients, Rich et al. developed a simple steady state 1-D PDE model of cAMP diffusion. This model predicted that based on the cell size and previously measured rates of cAMP synthesis and the hypotheses of free cAMP diffusion, cAMP gradients should be neglible.

Rich et al. [48] then used their experimental data to develop a more molecularly detailed compartmental ODE model including submembrane and cytosolic compartments and a third compartment representing the patch pipette. Experiments using changes in ionic concentrations and controlled [cAMP] were used to characterize the diffusional exchange rates between the three compartments. Using their model, the authors found that they could quantitatively reproduce the experimentally observed cAMP gradients only when diffusional barriers were present between the submembrane and cytosolic compartments. While the precise nature of this submembrane barrier is not clear, it could potentially be due to the extensive endoplasmic reticulum in these cells. Subsequent studies used this approach to examine the response to prostaglandin-E1, finding that whereas submembrane cAMP rose transiently, global cAMP rose more slowly to a sustained level [49]. Experimental perturbations and modeling indicated that these dynamics were regulated by phosphodiesterases (cAMP sink) and passive diffusional barriers.

Iancu et al. [50] used a conceptually similar approach to examine receptor-dependent cAMP compartmentation in adult ventricular myocytes. They developed a compartmental ODE model including cytosolic, caveolar, and extra-caveolar membrane spaces and regulation of cAMP signaling by both β_1-adrenergic receptors and M_2 muscarinic receptors. Model parameters were obtained either directly from the biochemical literature or by adapting previously developed models. Iancu et al. validated their model predictions against data from cAMP FRET sensors expressed in adult ventricular myocytes [51], including the dose-response and kinetics of cAMP biosensors in response to combinations of β-adrenergic agonist isoproterenol and muscarinic agonist acetylcholine. These model validations provided a measure of confidence that the model contained many of the key features of this signaling system. Iancu et al. then showed how the model could predict the intriguing experimentally observed *rebound effect*, where following transient M_2-receptor-dependent inhibition of weak β_1-adrenergic receptor activation, cAMP rises to a higher level than before M_2 stimulation. Using their model, the authors were able to investigate potential molecular mechanisms that would be difficult to evaluate directly with experiments. They found that caveolar compartmentation was a critical determinant of the rebound effect, which relied on the differential targeting of G proteins between membrane compartments and subcaveolar diffusional barriers. Their subsequent study combined additional FRET sensors of cytosolic cAMP with mathematical models to predict that caveolae maintain much lower resting cAMP levels than the rest of the cell, allowing a large dynamic range in [cAMP] with receptor stimulation [52].

7.4.2   Cell Shape as a Regulator of Subcellular Signaling

A variety of theoretical studies have specifically examined the role of cell shape in the production of subcellular signaling gradients. Brown and Kholodenko developed a simple yet elegant 1-D PDE model in which a protein kinase is located at the plasma membrane and a protein phosphatase is either distributed homogeneously or on a second parallel membrane [53]. In both cases, they found that large gradients of substrate phosphorylation were predicted due to spatial segregation of kinase and phosphatase. While they were not modeling a particular kinase/phosphatase system, they used rate constants and concentrations that are typical of experimentally measured values. These types of models were later extended by Meyers et al. [54] to examine how changes in cell size or shape would affect the predicted phosphorylation gradients. The authors developed 3-D PDE models using idealized cell geometries (e.g., spheres, cuboids) and representative rate constants that were not intended to represent a particular system. Their models illustrated how in cells with membrane-bound kinase and cytosolic phosphatase, cell protrusions such as neurites would exhibit high phosphorylation whereas the cell body would be less phosphorylated. Analogously, flattening an initially spherical cell while maintaining constant volume was predicted to increase substrate phosphorylation, again due to increased surface area-to-volume ratio. While these models were intentionally abstract for generality, the predictions were used to further explain published experimental data showing highest Cdc42 GTPase activity in fibroblasts near the leading edge, where it is thinnest [55].

Neves et al. [56] took this idea a step further, integrating mathematical modeling with FRET imaging and immunofluorescence to examine compartmentation of cAMP/MAPK signaling in hippocampal neurons. The authors began by integrating biochemical rate constants, expression levels, and cell geometry into a 2-D PDE model of β-adrenergic signaling and downstream cAMP/PKA activation. Parameters were obtained from the biochemical literature where available. Analogous to the results of Meyers et al. [54], their model predicted that when combining a plasma membrane-bound cAMP source (adenylyl cyclase) with a cytosolic cAMP sink (phosphodiesterase), high cAMP levels were localized to the dendrites where there is a high local surface area/volume ratio. Next, they used FRET sensors of cAMP to validate these model predictions, confirming that a β-adrenergic receptor agonist increased cAMP in elongated dendrites but not in the cell body. These cAMP gradients in signaling were predicted to be particularly sensitive to the diameter of the dendrite. They then asked whether these patterns of localized signaling would extend downstream to the Raf/MAPK pathway. The 2-D mathematical model was extended to incorporate PKA-mediated phosphorylation of b-Raf and downstream MAPK activation, containing a negative feedback loop and a positive feed-forward motif. While gradients in PKA and MAPK activity were predicted to arise as well, they were less sensitive to changes in dendrite diameter, suggesting that additional mechanisms may be contributing to these gradients. Model analysis suggested that a negative feedback involving phosphodiesterase activation may help sustain downstream MAPK gradients, and this prediction was validated experimentally by immunofluorescence of neurons in a cultured tissue slice, treated with β-adrenergic agonist with and without phosphodiesterase inhibition. Thus, they found that while cell geometry was a key regulator of cAMP gradients, localized Raf/MAPK activity also required additional negative feedback and feed-forward loops.

7.4.3   Signaling on Protein Scaffolds

There is increasing appreciation that many signaling reactions take place at an even smaller level of spatial compartmentation, within protein scaffolds rather than by free diffusion. For example, > 50 varieties of A-kinase anchoring proteins (AKAPs) have been identified from yeast to man, although the functional role for many of these is not fully clear [57]. While quantitative experiments of signaling within these scaffolds has been limited to select examples [32,58], it has been hypothesized that anchoring proteins help enhance the rate, efficiency, and selectivity of biochemical information processing [59].

In an early example of mathematical modeling of signaling complexes, Levchenko et al. [60] adapted previous ODE models of MAPK signaling to incorporate scaffolding proteins [61]. In their two-member scaffold model, they modeled the enhanced interaction of MEK and MAPK when both were bound to the protein scaffold. Model analysis indicated that in addition to expected results such as increased speed and efficacy of signaling, there was a surprising biphasic dependence of MAPK activity on scaffold concentration. At low scaffold levels, there was insufficient scaffold to enhance signaling. At high scaffold levels, individual MEK and MAPK proteins were often isolated on separate scaffold molecules, ultimately inhibiting their interaction. Thus, this study illustrated how the same protein scaffold can either enhance or diminish signaling depending on expression state of the cell.

Saucerman et al. [62] developed an ODE model of β-adrenergic signaling in cardiac myocytes, including downstream PKA-mediated phosphorylation and consequences to changes in cellular Ca^{2+} dynamics. Previous FRET experiments had shown that tethering PKA to its substrate can increase phosphorylation rates by about 10-fold [32]. Therefore, Saucerman et al. modeled PKA tethering to certain substrates in known AKAPs using a phenomenological model where the effective concentration of the substrate was increased 10-fold. This modification was essential for predicting quick and efficient phosphorylation of low-abundance PKA substrates such as the L-type Ca channel, as seen experimentally. In a subsequent study, this approach was extended to simulate the consequences of a clinically observed mutation in KCNQ1, the pore-forming subunit of the I_{Ks} channel [63]. This mutation disrupts the interactions of KCNQ with its AKAP (yotiao), PKA, and PP1 [64]. Their model predicted how this gene mutation would prevent KCNQ1 phosphorylation during β-adrenergic signaling, leading to stress-induced lengthening of the action potential, cellular arrhythmia, and ultimately abnormal electrocardiograms in a 3-D PDE ventricular wedge model.

While the signaling complexes modeled by Levchenko et al. [60] and Saucerman et al. [62] contained rather few components, others have identified a challenge to modeling protein scaffolds of increasing complexity [65,66]. As the number of proteins in a complex increases, the number of signaling states available to the complex increases exponentially: a simple scaffold with two docking sites was shown to have 18 microstates and 33 state transitions [67]. A model of FcεRI receptor signaling, consisting of four main components, was shown to require 354 chemical species and 3680 reactions [68]. One approach to deal with combinatorial complexity is to autogenerate this system of equations using rule-based algorithms [68]. Another approach is to assume independence of protein docking to the complex, allowing one to drastically simplify prototypic models of signaling complexes [67]. While there have been only been a limited number of mathematical studies of signaling within protein complexes to date, improved experimental techniques will make this form of nanocompartmentation increasingly accessible and important to understand quantitatively.

7.5 Case Study: Integrated Imaging and Modeling of PKA Activity Gradients

In this section, the authors provide a detailed case study where live-cell imaging and computational modeling were used together to quantitatively characterize the dynamics and compartmentation of PKA activity in cardiac myocytes [69]. This study was composed of three parts, each addressing a particular question:

Section 7.5.1: What are the rate-limiting steps in the kinetics of β-adrenergic/PKA signaling?
Section 7.5.2: Can we explain differences in PKA signaling at the plasma membrane and the cytosol?
Section 7.5.3: Do localized gradients in cAMP generate downstream gradients in PKA-mediated phosphorylation?

While all of these questions are focused on PKA dynamics and compartmentation in cardiac myocytes, the subtle differences lead to quite different experimental and computational approaches needed to address a particular question. In Section 7.5.4, these individual projects are compared to draw new insights into the mechanisms of cell signaling and how one may choose appropriate experimental and theoretical methods for future projects.

7.5.1 Rate-Limiting Steps in β-adrenergic/PKA Signaling

The kinetics of β-adrenergic signaling are an important determinant of how quickly the heart can respond to stress. So how are the kinetics of this pathway determined? In many complex systems, individual components have a wide range of time constants,

Chapter 7

with the result that only a subset of the components is actually rate-limiting. So we asked, "What are the rate-limiting steps in the kinetics of β-adrenergic signaling to PKA activity in cardiac myocytes?" [69]

We first performed FRET experiments to measure the rate of increased PKA activity with the β-adrenergic receptor agonist isoproterenol (ISO). We found that PKA activity increased with a $t_{1/2}$ of 33 sec, similar to the rate of change in Ca^{2+} transients and myocyte shortening downstream of PKA. Observing similar kinetics in a single compartment ODE model of this pathway (adapted for neonatal myocytes from [62]) indicated that the overall behavior was consistent with experimental measurements. By looking at more detailed simulated timecourses of all species in the model, we found that while β-adrenergic receptor and Gsα activity increased rapidly with ISO, cAMP accumulated much more slowly and was in phase with the downstream PKA activity. This led to the hypothesis that cAMP accumulation was rate limiting in the β-adrenergic receptor/PKA pathway. To experimentally test this hypothesis, we performed flash photolysis of chemically caged cAMP (DMNB-cAMP), which allowed a light-activated rapid release of cAMP in the cell. We found very rapid PKA activity in this experiment, confirming the model predictions [69].

We next examined the overall off rate of this pathway when the β-adrenergic receptor antagonist propranolol was applied. In both model and simulation PKA activity decreased but more slowly than the on rate of the pathway, indicating that additional mechanisms may be involved. Indeed, the experimental response to propranolol was even more delayed than predicted in the simulations. With further model simulations, we found a potential explanation for the delayed response to propranolol. In conditions of high ISO, the model predicted that cAMP levels would be much higher than needed to saturate PKA. Therefore, when the β-adrenergic receptor is inhibited, it takes some time for cAMP to drop below saturating levels. Model analysis suggested a direct experimental approach for testing this hypothesis- repeating the DMNB-cAMP uncaging experiments at higher, PKA-saturating conditions. Indeed, these model predictions were validated experimentally, indicating that cAMP-mediated saturation of PKA creates a novel mechanism for sustaining PKA activity for some period when the upstream receptor is inhibited.

7.5.2   PKA Activity Gradients between Plasma Membrane and Cytosol

Previous studies have indicated a potential for cAMP gradients between the plasma membrane and the cytosol. Indeed, cAMP gradients have been suggested to be responsible for differential functional consequences of β-adrenergic and prostaglandin signaling on contractility and ion currents despite both generating cAMP signals. cAMP has been found to be particularly elevated in particulate fractions in β-adrenergic receptor stimulated myocytes, whereas myocytes treated with prostaglandin have elevated cAMP only in the soluble fraction. More recent studies with real-time cAMP indicators using either electrophysiologic or fluorescent proteins have directly measured distinct cAMP signals in the membrane that are dependent upon the stimulus. Here, we asked, "Are there gradients of PKA activity between plasma membrane and the cytosol? If so, how are they formed?" [69]

To address these questions, we expressed either the untargeted PKA activity reporter (AKAR) or plasma membrane-targeted reporter in cardiac myocytes. The plasma membrane targeted AKAR (pmAKAR) used a lysine sequence and a lipid modification domain from the small G protein Rho (KKKKKSGCLVL) at the C-terminal end of the reporter. In a manner quite consistent with previous biochemical assays [8], prostaglandin E_1 (PGE1) stimulated significantly higher PKA activity in the cytosol than in the plasma membrane. In contrast, the rate of increase in PKA activity was faster at the plasma membrane than the cytosol. On the other hand, we found that the β-adrenergic receptor agonist ISO stimulated similar magnitudes of PKA activity in cytosol and at plasma membrane, although plasma membrane PKA activity responded significantly faster.

To understand the mechanistic basis for the differences in PKA signaling between the plasma membrane and cytosol, we adapted our previous single-compartment computational model (described in Section 7.5.1) to include both plasma membrane and cytosolic compartments and to allow cAMP diffusion (a compartmental ODE model). To our surprise, initial versions of the model were not able to predict the experimentally measured delay in cytosolic PKA activity. Only when the apparent diffusion coefficient of cAMP was reduced by 100X (from 200 to 2 μm^2/sec) were appropriate cytosolic PKA delays predicted. These simulations predicted that there may be significant submembrane diffusional barriers to cAMP. Indeed, electron microscopy has identified a meshwork of actin filaments just below the plasma membrane. However, it is not clear whether such a cytoskeletal network would be quantitatively sufficient to reduce diffusion by such an extent. Future experiments will be required to resolve this issue. However, it should be noted that the initial discrepancy between experimental data and model predictions allowed us to focus on the potential importance of submembrane diffusional barriers.

7.5.3   PKA Activity Gradients within Cytosol

One limitation of the plasma membrane/cytosol PKA activity gradient measurements described in Section 7.5.2 was that an individual cell was expressing only the plasma membrane or the cytosolic PKA reporter. Due to biological variability, it was somewhat difficult to identify cytosolic PKA activity delays relative to the plasma membrane. Therefore, we aimed to more directly image PKA activity gradients within a single cell. We initially took our original model of β-adrenergic/PKA signaling in the neonatal cardiac myocyte (described in Section 7.4.1) and adapted it to the 2-D geometry of an individual cardiac myocyte. Using the Virtual Cell software [44], we tested several potential experimental designs for inducing cytosolic PKA activity gradients in a single cell. One rather robust protocol was a rapid uncaging of DMNB-cAMP on just one end of the cell (localized ultraviolet flash photolysis). We tested this protocol experimentally, finding that we could indeed measure global responses to global stimuli and a localized, transient PKA phosphorylation gradient in response to local DMNB-cAMP uncaging (Figure 7.1a). Two important internal controls were added as well. Preceding the local cAMP uncaging, we performed global UV uncaging to demonstrate that functional responses could be measured from the entire cell. Following the local DMNB-cAMP uncaging, Iso was added as an independent test that we could still measure functional responses from the entire cell.

With direct experimental evidence of subcellular PKA phosphorylation gradients, we asked whether quantitatively similar PKA phosphorylation gradients would be

Chapter 7

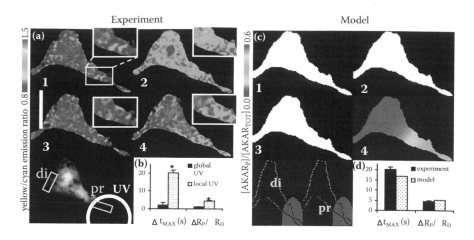

FIGURE 7.1 (See color insert.) FRET imaging and computational modeling of subcellular PKA activity gradients in cardiac myocytes. At time (2), DMNB-cAMP is uncaged globally, inducing a global increase in PKA-mediated phosphorylation that gradually subsides (time 3). With subsequent local DMNB-cAMP uncaging in the region denoted by a circle, both experiments (a) and model (c) showed subcellular PKA activity gradients that were limited primarily to the region stimulated (time 4). (b) Quantification of phosphorylation propagation time (Δt_{MAX}) and phosphorylation gradient magnitude ($\Delta R_P / \Delta R_D$). (d) Quantitative comparison of propagation time and gradient magnitude between experiment and model. (From Saucerman et al. [69]. With permission.)

predicted by the model when using the same cell geometries used in the imaging experiments. Indeed, using an apparent diffusion coefficient of $D_{cAMP} = 200$ μm^2/sec, similar to that seen in simple cells, we could predict quantitatively similar propagation times of the PKA-mediated *phosphorylation wave* and magnitude of the phosphorylation gradient (Figure 7.1b). The model was then perturbed in various ways to identify the molecular mechanisms that may contribute to these PKA phosphorylation gradients. We found that an apparent diffusion coefficient of $D_{cAMP} = 200$ μm^2/sec was critical for predicting the experimentally measured phosphorylation gradient magnitudes. Lower D_{cAMP} produced phosphorylation gradients that were too large, whereas larger D_{cAMP} produced phosphorylation gradients that were too small. Model predictions also highlighted the key role of phosphodiesterases in enhancing signaling gradients by degrading cAMP before it could diffuse across the cell. Interestingly, model predictions also suggested that cAMP buffering, or PKA-binding to cAMP that protects it from PDE-mediated degradation, also contributed significantly to phosphorylation gradients. Thus model analysis identified three main mechanisms for PKA-mediated phosphorylation gradients: restricted diffusion of cAMP, degradation of cAMP, and buffering of cAMP. Future experimental studies will be required to validate these model predictions.

7.5.4 Insights from Comparison of Models and Experiments at Each Scale

Sections 7.5.1–7.5.3 provided a particular case study of how live-cell FRET imaging and mechanistic computational modeling have been combined to quantitatively characterize the dynamics and compartmentation of PKA signaling in the cardiac myocyte. By comparing each of these sections, additional insights may provide practical guidance for future quantitative analyses of subcellular signaling dynamics. The general theme is that appropriate experimental and theoretical methods must be chosen based on the particular biological question to be asked and that repeated iteration between experiment and theory is needed to quantitatively understand signaling networks.

In each study of overall signaling dynamics, plasma membrane versus cytosolic PKA activity gradients, or intracytosolic PKA activity gradients, experimental and theoretical methods were carefully chosen to match the biological question of interest. In Section 7.5.1, the study was focused on mechanisms for rate-limiting PKA signaling. Therefore, full spatial simulations would have been overkill and may have made it more difficult to extract insights from the computational model. Similarly, cytosolic measurements of PKA activity were sufficient to get a basic idea of the dynamic pattern of PKA activation and deactivation with β-adrenergic receptor agonists or antagonists.

On the other hand, cAMP compartmentation could be a critical determinant of PKA phosphorylation kinetics. Indeed, comparison of the conclusions from Sections 7.5.1 (cAMP accumulation is rate-limiting) and 7.5.2 (submembrane barriers to cAMP diffusion explain delays in cytosolic PKA activity) reveal an apparent contradiction. How can both mechanisms be so important? One potential explanation is that the rate-limiting step in PKA signaling is not cAMP synthesis alone but the combination of cAMP synthesis and its diffusion to cytosolic PKA. Therefore, the cytosolic cAMP accumulation is rate-limiting to β-adrenergic stimulation of cytosolic PKA activity. Thus, the rate-limiting step for PKA activity in other cellular compartments may be somewhat different, and further study is required.

In Section 7.5.2, the focus was on measuring and modeling differences in PKA activity at the plasma membrane and the cytosol. Given the limited optical resolution of the epi-fluorescence microscope, we cannot reliably discriminate between similarly labeled reporters at the plasma membrane and the cytosol when expressed in the same cell. Therefore, in these studies we expressed either the plasma membrane reporter or the cytosolic reporter and compared the PKA activity kinetics from different cells. While this was sufficient for the study described, future studies may be able to take advantage of innovative new approaches for multiplexing FRET sensors [70]. Based on the limited ability to resolve the plasma membrane versus cytosol compartments optically, the experiments were designed to use sensors genetically targeted to the plasma membrane. Targeting via genetic localization sequences can be achieved at the molecular level, which was a strong advantage in these studies.

For the corresponding model described in Section 7.5.2, a two-compartment model was most appropriate because the experimental data were coming from two distinct physical locations with cAMP diffusion between these locations. While we could have developed a more complex model that explicitly showed the spatial distribution of cAMP as a function of distance from the plasma membrane (like in Section 7.5.3), the data were not available to support a model at that level. Indeed, while the model indicated the potential presence of subplasma membrane diffusional barriers, there is relatively little direct experimental data of such a physical structure to support this argument. Thus, the choice to use a two-compartment model here was based on the available experimental data and the particular biological question, which was focused on understanding overall differences between plasma membrane and cytosol PKA activity kinetics.

In the study on cytosolic PKA activity gradients (Section 7.5.3), higher spatial resolution was clearly required in both the imaging experiments and the computational modeling. In the experimental studies, spatial resolution was required to characterize the speed of the phosphorylation wave and the spatial extent of the gradients. Sufficient temporal resolution was also required to capture the transient spike in PKA phosphorylation (see Figure 7.1). Generally, imaging was performed at 0.2 Hz to minimize photobleaching over the entire 20 minute experiment (including the two internal controls). However, in a subset of experiments with only local uncaging, imaging was performed at 1 Hz. This ensured that the images recorded at 0.2 Hz had sufficiently captured the rapid PKA activity kinetics following DMNB-cAMP uncaging.

The biological questions and available experimental data also helped shape the appropriate level of modeling for cytosolic PKA activity gradients. Clearly, a spatially explicit model was required for comparison with the spatially detailed experimental data. However, there was still an important decision about whether to model PKA activity gradients in one, two, or three dimensions. While a 1-D model would have been simpler, there were concerns that this would not be directly comparable to the experimental data due to necessary assumptions about the boundary conditions. In particular, the 2-D model was able to capture the same 2-D geometry seen in the images, which would automatically account for any contribution of the specific cell geometry in these experiments. On the other hand, it could be argued that the cell is actually 3-D and therefore a 3-D model is necessary. A 3-D model was not developed at this time for several reasons. First and most importantly, the specific biological question and the corresponding experimental data were not truly 3-D. Second, these neonatal myocytes are fairly thin, and the uncaging of DMNB-cAMP was done uniformly in the z-direction. Finally, from practical considerations a 2-D model was much easier to build than 3-D, and we had not collected full 3-D geometry for each cell measured in the experiments.

Another interesting area for comparison of the models and experiments in Sections 7.5.1–7.5.3 is the different biological mechanisms that were important and each length scale. In the nonspatial study of β-adrenergic/PKA signaling dynamics (Section 7.5.1), cAMP synthesis and degradation rates were key drivers of the overall pathway kinetics. cAMP buffering also played an important role in stabilizing [cAMP] near a point of high PKA sensitivity, helping PKA to respond more quickly to a β-adrenergic stimulus. But as described already, when considering plasma membrane phosphorylation delays (Section 7.5.2), restricted cAMP diffusion is an important aspect as well. Interestingly, cAMP buffering did not seem to contribute much to plasma membrane/cytosol phosphorylation delays, and PDE-mediated cAMP degradation was important but only when also combined with restricted diffusion. In contrast, cytosolic PKA phosphoryaltion gradients (Section 7.5.3) did not require diffusion to be more restricted than expected in the cytosol, but both cAMP buffering and PDE-mediated cAMP degradation were key determinants of PKA activity gradients. This comparison shows that there is a wide range of mechanisms contributing to shape signaling dynamics and compartmentation, and their quantitative contribution will depend highly on the particular experimental conditions examined.

One common element to the studies in Sections 7.5.1–7.5.3 is that all of these relied on an iterative cycle between models and experiments. In the study of PKA dynamics (Section 7.5.1), initial experiments were followed by a model that was tuned slightly to exhibit similar activation/deactivation kinetics with isoproterenol and propranolol. Next, the model made new predictions about the rate-limiting steps and helped us develop experimental designs to test whether cAMP accumulation was rate-limiting (using DMNB-cAMP uncaging) and whether cAMP saturation of PKA was responsible for delayed PKA deactivation (by varying the level of DMNB-cAMP uncaging). Both of these model-guided designs were subsequently validated experimentally.

Study of plasma membrane/cytosol PKA activity gradients (Section 7.5.2) again started with an experimental observation of different kinetics and magnitudes of signaling in response to PGE1 or Iso. This was followed by a model where D_{cAMP} was tuned to match the observed phosphorylation delay. From there, additional model simulations were performed to examine how diffusional barriers, PDE-mediated degradation, cAMP buffering and other mechanisms contribute to plasma membrane/cytosolic PKA gradients.

Finally, the study on cytosolic PKA gradients (Section 7.5.3) was initially motivated by limitations of the experiments in the previous studies and began with a simple model used to design a robust experiment where cytosolic PKA gradients could be reliably measured. Once these measurements were obtained, the more detailed computational model was developed and validated against the experimental measurements. After successful validation, the model was used to make new predictions about the molecular mechanisms underlying cytosolic PKA activity gradients. These predictions provide an outline for future experimental validations of molecular mechanisms shaping cytosolic PKA activity gradients.

In conclusion, the case study in Section 7.5 illustrated three different approaches for combining computational modeling and FRET imaging to examine subcellular signaling dynamics of the β-adrenergic/PKA pathway. Importantly, the selection of experimental and theoretical methods for each of these sections was driven by the particular biological question to be asked. While a variety of molecular mechanisms contributed to subcellular PKA dynamics and compartmentation, the quantitative contribution of these mechanisms varied greatly depending on the particular experimental context. Finally, the quantitative analysis of this signaling system relied greatly on an interactive hypothesis-testing cycle of modeling and experiments.

References

1. Saucerman, J. J., and A. D. McCulloch. 2006. Cardiac beta-adrenergic signaling: from subcellular microdomains to heart failure. *Ann N Y Acad Sci* 1080:348–361.
2. Owens, G. K., M. S. Kumar, and B. R. Wamhoff. 2004. Molecular regulation of vascular smooth muscle cell differentiation in development and disease. *Physiol Rev* 84:767–801.
3. Fries, B. C., C. P. Taborda, E. Serfass, and A. Casadevall. 2001. Phenotypic switching of Cryptococcus neoformans occurs in vivo and influences the outcome of infection. *J Clin Invest* 108:1639–1648.
4. Noble, D. 2008. Genes and causation. *Philos Transact A Math Phys Eng Sci* 366:3001–3015.
5. Berridge, M. J., P. Lipp, and M. D. Bootman. 2000. The versatility and universality of calcium signalling. *Nat Rev Mol Cell Biol* 1:11–21.
6. Fimia, G. M., and P. Sassone-Corsi. 2001. Cyclic AMP signalling. *J Cell Sci* 114:1971–1972.
7. Perkins, N. D., and T. D. Gilmore. 2006. Good cop, bad cop: the different faces of NF-kappaB. *Cell Death Differ* 13:759–772.
8. Brunton, L. L., J. S. Hayes, and S. E. Mayer. 1981. Functional compartmentation of cyclic AMP and protein kinase in heart. *Adv Cyclic Nucleotide Res* 14:391–397.
9. Zaccolo, M., and T. Pozzan. 2002. Discrete microdomains with high concentration of cAMP in stimulated rat neonatal cardiac myocytes. *Science* 295:1711–1715.
10. Lichtman, J. W., and J. A. Conchello. 2005. Fluorescence microscopy. *Nat Methods* 2:910–919.
11. Brown, C. M. 2007. Fluorescence microscopy—avoiding the pitfalls. *J Cell Sci* 120:1703–1705.

Chapter 7

12. Frigault, M. M., J. Lacoste, J. L. Swift, and C. M. Brown. 2009. Live-cell microscopy—tips and tools. *J Cell Sci* 122:753–767.

13. Chen, I., and A. Y. Ting. 2005. Site-specific labeling of proteins with small molecules in live cells. *Curr Opin Biotechnol* 16:35–40.

14. O'Hare, H. M., K. Johnsson, and A. Gautier. 2007. Chemical probes shed light on protein function. *Curr Opin Struct Biol* 17:488–494.

15. Tour, O., S. R. Adams, R. A. Kerr, R. M. Meijer, T. J. Sejnowski, R. W. Tsien, and R. Y. Tsien. 2007. Calcium Green FlAsH as a genetically targeted small-molecule calcium indicator. *Nat Chem Biol* 3:423–431.

16. Shaner, N. C., P. A. Steinbach, and R. Y. Tsien. 2005. A guide to choosing fluorescent proteins. *Nat Methods* 2:905–909.

17. Shu, X., A. Royant, M. Z. Lin, T. A. Aguilera, V. Lev-Ram, P. A. Steinbach, and R. Y. Tsien. 2009. Mammalian expression of infrared fluorescent proteins engineered from a bacterial phytochrome. *Science* 324:804–807.

18. Huh, W. K., J. V. Falvo, L. C. Gerke, A. S. Carroll, R. W. Howson, J. S. Weissman, and E. K. O'Shea. 2003. Global analysis of protein localization in budding yeast. *Nature* 425:686–691.

19. Chen, S. C., T. Zhao, G. J. Gordon, and R. F. Murphy. 2007. Automated image analysis of protein localization in budding yeast. *Bioinformatics* 23:i66–71.

20. Andjelkovic, M., D. R. Alessi, R. Meier, A. Fernandez, N. J. Lamb, M. Frech, P. Cron, P. Cohen, J. M. Lucocq, and B. A. Hemmings. 1997. Role of translocation in the activation and function of protein kinase B. *J Biol Chem* 272:31515–31524.

21. Huang, B., M. Bates, and X. Zhuang. 2009. Super-resolution fluorescence microscopy. *Annu Rev Biochem* 78:993–1016.

22. Digman, M. A., P. W. Wiseman, C. Choi, A. R. Horwitz, and E. Gratton. 2009. Stoichiometry of molecular complexes at adhesions in living cells. *Proc Natl Acad Sci U S A* 106:2170–2175.

23. Digman, M. A., P. W. Wiseman, A. R. Horwitz, and E. Gratton. 2009. Detecting protein complexes in living cells from laser scanning confocal image sequences by the cross correlation raster image spectroscopy method. *Biophys J* 96:707–716.

24. Chen, Y., J. D. Mills, and A. Periasamy. 2003. Protein localization in living cells and tissues using FRET and FLIM. *Differentiation* 71:528-541.

25. Gordon, G. W., G. Berry, X. H. Liang, B. Levine, and B. Herman. 1998. Quantitative fluorescence resonance energy transfer measurements using fluorescence microscopy. *Biophys J* 74:2702–2713.

26. Chen, Y., and A. Periasamy. 2006. Intensity range based quantitative FRET data analysis to localize protein molecules in live cell nuclei. *J Fluoresc* 16:95–104.

27. Miyawaki, A., and R. Y. Tsien. 2000. Monitoring protein conformations and interactions by fluorescence resonance energy transfer between mutants of green fluorescent protein. *Methods Enzymol* 327:472–500.

28. DiPilato, L. M., X. Cheng, and J. Zhang. 2004. Fluorescent indicators of cAMP and Epac activation reveal differential dynamics of cAMP signaling within discrete subcellular compartments. *Proc Natl Acad Sci U S A* 101:16513–16518.

29. Remus, T. P., A. V. Zima, J. Bossuyt, D. J. Bare, J. L. Martin, L. A. Blatter, D. M. Bers, and G. A. Mignery. 2006. Biosensors to measure inositol 1,4,5-trisphosphate concentration in living cells with spatiotemporal resolution. *J Biol Chem* 281:608–616.

30. Rochais, F., J. P. Vilardaga, V. O. Nikolaev, M. Bunemann, M. J. Lohse, and S. Engelhardt. 2007. Real-time optical recording of beta1-adrenergic receptor activation reveals supersensitivity of the Arg389 variant to carvedilol. *J Clin Invest* 117:229–235.

31. Miyawaki, A., J. Llopis, R. Heim, J. M. McCaffery, J. A. Adams, M. Ikura, and R. Y. Tsien. 1997. Fluorescent indicators for Ca^{2+} based on green fluorescent proteins and calmodulin. *Nature* 388:882–887.

32. Zhang, J., Y. Ma, S. S. Taylor, and R. Y. Tsien. 2001. Genetically encoded reporters of protein kinase A activity reveal impact of substrate tethering. *Proc Natl Acad Sci U S A* 98:14997–15002.

33. Ting, A. Y., K. H. Kain, R. L. Klemke, and R. Y. Tsien. 2001. Genetically encoded fluorescent reporters of protein tyrosine kinase activities in living cells. *Proc Natl Acad Sci U S A* 98:15003–15008.

34. Aye-Han, N. N., Q. Ni, and J. Zhang. 2009. Fluorescent biosensors for real-time tracking of post-translational modification dynamics. *Curr Opin Chem Biol* 13:392–397.

35. Gao, X., and J. Zhang. 2010. FRET-based activity biosensors to probe compartmentalized signaling. *Chembiochem* 11:147–151.

36. Kunkel, M. T., E. L. Garcia, T. Kajimoto, R. A. Hall, and A. C. Newton. 2009. The protein scaffold NHERF-1 controls the amplitude and duration of localized protein kinase D activity. *J Biol Chem* 284:24653–24661.

37. Ma'ayan, A., R. D. Blitzer, and R. Iyengar. 2005. Toward predictive models of mammalian cells. *Annu Rev Biophys Biomol Struct* 34:319–349.

38. Janes, K. A., and D. A. Lauffenburger. 2006. A biological approach to computational models of proteomic networks. *Curr Opin Chem Biol* 10:73–80.

39. Fall, C. P. 2002. *Computational Cell Biology*. Springer, New York.

40. Alon, U. 2007. *An Introduction to Systems Biology: Design Principles of Biological Circuits*. Chapman & Hall/CRC, Boca Raton, FL.

41. Hodgkin, A. L., and A. F. Huxley. 1952. A quantitative description of membrane current and its application to conduction and excitation in nerve. *J Physiol* 117:500–544.

42. Turing, A. M. 1952. The Chemical Basis of Morphogenesis. Philosophical Transactions of the Royal Society of London. *Series B, Biological Sciences* 237:37–72.

43. Fink, C. C., B. Slepchenko, Moraru, II, J. Schaff, J. Watras, and L. M. Loew. 1999. Morphological control of inositol-1,4,5-trisphosphate-dependent signals. *J Cell Biol* 147:929–936.

44. Moraru, II, J. C. Schaff, B. M. Slepchenko, M. L. Blinov, F. Morgan, A. Lakshminarayana, F. Gao, Y. Li, and L. M. Loew. 2008. Virtual Cell modelling and simulation software environment. *IET Syst Biol* 2:352–362.

45. Smith, A. E., B. M. Slepchenko, J. C. Schaff, L. M. Loew, and I. G. Macara. 2002. Systems analysis of Ran transport. *Science* 295:488–491.

46. Riddick, G., and I. G. Macara. 2007. The adapter importin-alpha provides flexible control of nuclear import at the expense of efficiency. *Mol Syst Biol* 3:118.

47. Schneider, I. C., and J. M. Haugh. 2005. Quantitative elucidation of a distinct spatial gradient-sensing mechanism in fibroblasts. *J Cell Biol* 171:883–892.

48. Rich, T. C., K. A. Fagan, H. Nakata, J. Schaack, D. M. Cooper, and J. W. Karpen. 2000. Cyclic nucleotide-gated channels colocalize with adenylyl cyclase in regions of restricted cAMP diffusion. *J Gen Physiol* 116:147–161.

49. Rich, T. C., K. A. Fagan, T. E. Tse, J. Schaack, D. M. Cooper, and J. W. Karpen. 2001. A uniform extracellular stimulus triggers distinct cAMP signals in different compartments of a simple cell. *Proc Natl Acad Sci U S A* 98:13049–13054.

50. Iancu, R. V., S. W. Jones, and R. D. Harvey. 2007. Compartmentation of cAMP signaling in cardiac myocytes: a computational study. *Biophys J* 92:3317–3331.

51. Warrier, S., A. E. Belevych, M. Ruse, R. L. Eckert, M. Zaccolo, T. Pozzan, and R. D. Harvey. 2005. Beta-adrenergic- and muscarinic receptor-induced changes in cAMP activity in adult cardiac myocytes detected with FRET-based biosensor. *Am J Physiol Cell Physiol* 289:C455–461.

52. Iancu, R. V., G. Ramamurthy, S. Warrier, V. O. Nikolaev, M. J. Lohse, S. W. Jones, and R. D. Harvey. 2008. Cytoplasmic cAMP concentrations in intact cardiac myocytes. *Am J Physiol Cell Physiol* 295:C414–422.

53. Brown, G. C., and B. N. Kholodenko. 1999. Spatial gradients of cellular phospho-proteins. *FEBS Lett* 457:452–454.

54. Meyers, J., J. Craig, and D. J. Odde. 2006. Potential for control of signaling pathways via cell size and shape. *Curr Biol* 16:1685–1693.

55. Nalbant, P., L. Hodgson, V. Kraynov, A. Toutchkine, and K. M. Hahn. 2004. Activation of endogenous Cdc42 visualized in living cells. *Science* 305:1615–1619.

56. Neves, S. R., P. Tsokas, A. Sarkar, E. A. Grace, P. Rangamani, S. M. Taubenfeld, C. M. Alberini, J. C. Schaff, R. D. Blitzer, Moraru, II, and R. Iyengar. 2008. Cell shape and negative links in regulatory motifs together control spatial information flow in signaling networks. *Cell* 133:666–680.

57. Smith, F. D., L. K. Langeberg, and J. D. Scott. 2006. The where's and when's of kinase anchoring. *Trends in Biochemical Sciences* 31:316–323.

58. Dodge-Kafka, K. L., J. Soughayer, G. C. Pare, J. J. Carlisle Michel, L. K. Langeberg, M. S. Kapiloff, and J. D. Scott. 2005. The protein kinase A anchoring protein mAKAP coordinates two integrated cAMP effector pathways. *Nature* 437:574–578.

59. Pawson, T., and J. D. Scott. 1997. Signaling Through Scaffold, Anchoring, and Adaptor Proteins. *Science* 278:2075–2080.

60. Huang, C. Y., and J. E. Ferrell, Jr. 1996. Ultrasensitivity in the mitogen-activated protein kinase cascade. *Proc Natl Acad Sci U S A* 93:10078–10083.

61. Levchenko, A., J. Bruck, and P. W. Sternberg. 2000. Scaffold proteins may biphasically affect the levels of mitogen-activated protein kinase signaling and reduce its threshold properties. *Proc Natl Acad Sci U S A* 97:5818–5823.

62. Saucerman, J. J., L. L. Brunton, A. P. Michailova, and A. D. McCulloch. 2003. Modeling beta-adrenergic control of cardiac myocyte contractility *in silico*. *J Biol Chem* 278:47997–48003.

Chapter 7

63. Saucerman, J. J., S. N. Healy, M. E. Belik, J. L. Puglisi, and A. D. McCulloch. 2004. Proarrhythmic consequences of a KCNQ1 AKAP-binding domain mutation: computational models of whole cells and heterogeneous tissue. *Circ Res* 95:1216–1224.
64. Marx, S. O., J. Kurokawa, S. Reiken, H. Motoike, J. D'Armiento, A. R. Marks, and R. S. Kass. 2002. Requirement of a macromolecular signaling complex for beta adrenergic receptor modulation of the KCNQ1-KCNE1 potassium channel. *Science* 295:496–499.
65. Hlavacek, W. S., J. R. Faeder, M. L. Blinov, A. S. Perelson, and B. Goldstein. 2003. The complexity of complexes in signal transduction. *Biotechnol Bioeng* 84:783–794.
66. Bray, D. 2003. Genomics. Molecular prodigality. *Science* 299:1189–1190.
67. Borisov, N. M., N. I. Markevich, J. B. Hoek, and B. N. Kholodenko. 2005. Signaling through receptors and scaffolds: independent interactions reduce combinatorial complexity. *Biophys J* 89:951–966.
68. Faeder, J. R., W. S. Hlavacek, I. Reischl, M. L. Blinov, H. Metzger, A. Redondo, C. Wofsy, and B. Goldstein. 2003. Investigation of early events in Fc epsilon RI-mediated signaling using a detailed mathematical model. *J Immunol* 170:3769–3781.
69. Saucerman, J. J., J. Zhang, J. C. Martin, L. X. Peng, A. E. Stenbit, R. Y. Tsien, and A. D. McCulloch. 2006. Systems analysis of PKA-mediated phosphorylation gradients in live cardiac myocytes. *Proc Natl Acad Sci U S A* 103:12923–12928.
70. Clapp, A. R., I. L. Medintz, H. T. Uyeda, B. R. Fisher, E. R. Goldman, M. G. Bawendi, and H. Mattoussi. 2005. Quantum dot-based multiplexed fluorescence resonance energy transfer. *J Am Chem Soc* 127:18212–18221.

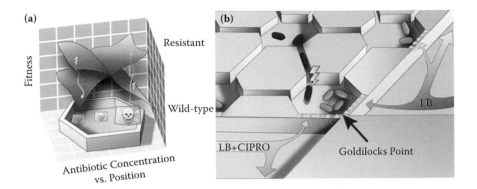

FIGURE 1.6 (a) The blue surface and the red surface are the fitness of the wild-type and resistant bacteria, respectively, versus space in a combination of antibiotic gradient and population gradients. Mutation from the wild-type to the resistant genome is represented by vertical transitions between the two fitness surfaces. (b) The basic design of our microecology creates high-population gradients using constriction of nutrient flow via nanoslits in the presence of an antibiotic gradient. Here the apex of the device is shown, where the gradients are highest. Channels allow movement of motile mutant bacteria along the population gradient. The nutrients and nutrient + Cipro streams are circulated by syringe pumps as shown.

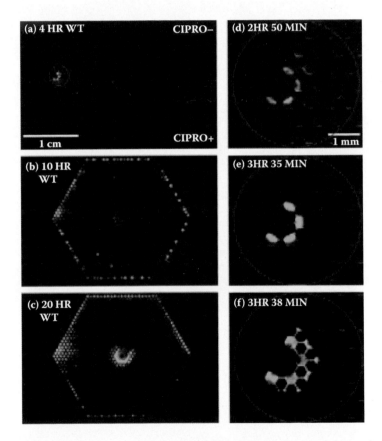

FIGURE 1.7 (a) Ignition of resistance to Cirpro at the Goldilocks point 4 hours after inoculation. (b) Spread of resistant bacteria around the periphery of the microecology at 10 hours after inoculation. (c) Return of Cipro-resistant bacteria to the center after 20 hours. The dynamics of the resistant bacteria at the Goldilocks point is shown with higher time resolution in (d)–(f) during the ignition of resistance over a 1 hour period.

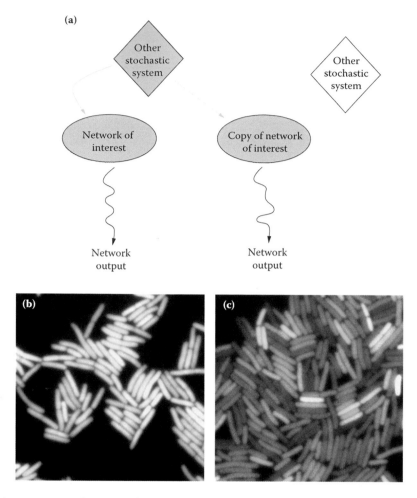

FIGURE 3.1 Intrinsic and extrinsic fluctuations. (a) Extrinsic fluctuations are generated by interactions of the system of interest with other stochastic systems. They affect the system and a copy of the system identically. Intrinsic fluctuations are inherent to each copy of the system and cause the behavior of each copy to be different. (b) Strains of *Escherichia coli* expressing two distinguishable alleles of Green Fluorescent Protein controlled by identical promoters. Intrinsic noise is given by the variation in color (differential expression of each allele) over the population and extrinsic noise by the correlation between colors. (c) The same strains under conditions of low expression. Intrinsic noise increases. (From Elowitz et al. [12]. With permission.)

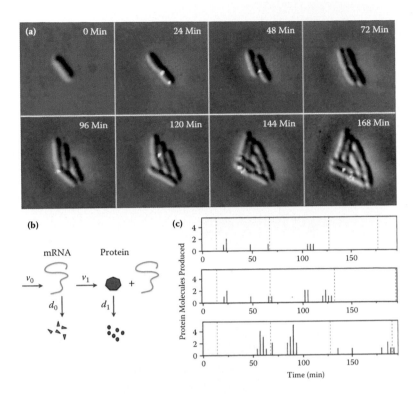

FIGURE 3.2 Stochasticity is generated by bursts of translation. (a) Time series showing occasional expression from a repressed gene in *E. coli*, which results in the synthesis of a single mRNA but several proteins. Proteins are membrane proteins tagged with Yellow Fluorescent Protein (yellow dots). (b) The two-stage model of constitutive gene expression. Transcription and translation are included. (c) Quantization of the data from (a) showing exponentially distributed bursts of translated protein. (From Yu et al. [19]. With permission.)

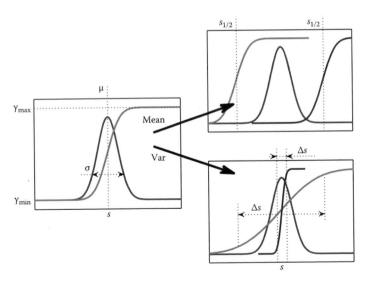

FIGURE 4.2 Parameters characterizing response to a signal. Left panel: the probability distribution of the signal, $P(s)$ (blue), and the best-matched steady state dose-response curve r_{ss} (green). Top right: mismatched response midpoint. Bottom right: mismatched response gain. (From [75]. With permission.)

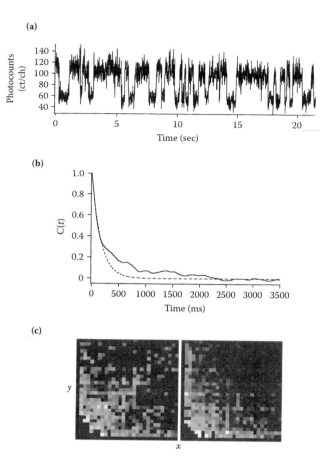

FIGURE 5.2 Single-molecule study of FAD fluorescence in cholesterol oxidase (COx). (a) Sample intensity versus time trajectory showing characteristic on and off blinking behavior as the FAD cofactor is oxidized and reduced during its catalytic cycle. (b) Intensity autocorrelation function for trajectory in (a) (black line). The data have been fit to a single exponential model (dashed line). (c) 2-D conditional probability distribution for the length of subsequent on times showing a diagonal feature (left) and with a lag of 10 on times showing no correlation in lengths (right). (From Lu et al. [24]. With permission.)

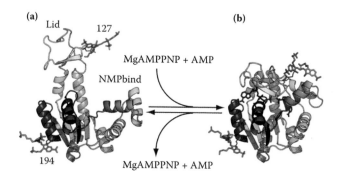

FIGURE 5.3 Adenylate kinase (AK) crystal structures. (a) Substrate-free structure with lids in an open state. Dyes have been modeled in at the labeling positions used in single-molecule experiments (A127C and A194C). (b) AMP-PNP and AMP bound AK crystal structure with both lids in the closed state. (From Hanson et al. [47]. With permission.)

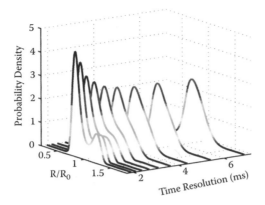

FIGURE 5.5 Time-dependent deconvoluted probability density functions for AK with AMPNP and AMP showing motional narrowing. (From Hanson et al. [47]. With permission.)

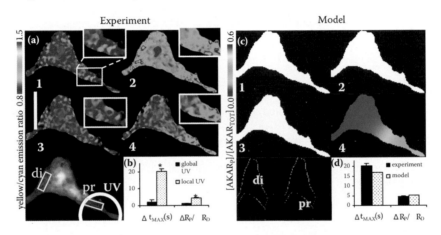

FIGURE 7.1 FRET imaging and computational modeling of subcellular PKA activity gradients in cardiac myocytes. At time (2), DMNB-cAMP is uncaged globally, inducing a global increase in PKA-mediated phosphorylation that gradually subsides (time 3). With subsequent local DMNB-cAMP uncaging in the region denoted by a circle, both experiments (a) and model (c) showed subcellular PKA activity gradients that were limited primarily to the region stimulated (time 4). (b) Quantification of phosphorylation propagation time (Δt_{MAX}) and phosphorylation gradient magnitude ($\Delta R_P / \Delta R_D$). (d) Quantitative comparison of propagation time and gradient magnitude between experiment and model. (From Saucerman et al. [69]. With permission.)

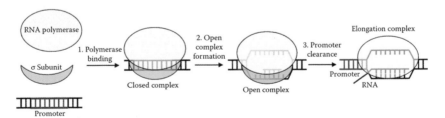

FIGURE 9.2 Transcription initiation consists of three major steps. (1) In polymerase binding, RNA polymerase and the σ subunit assemble onto the promoter. (2) In open complex formation, the promoter DNA is melted by RNA polymerase in preparation for transcribing single-stranded mRNA. (3) In promoter clearance, RNA polymerase slides downstream from the promoter, releases σ, and continues to transcribe as an elongation complex.

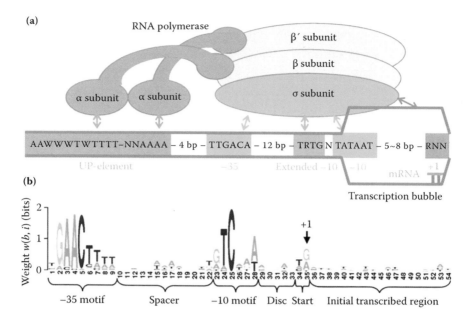

FIGURE 9.3 The consensus sequence of a promoter can be represented by a position weight matrix (PWM). (a) The interaction between RNA polymerase and promoter is illustrated. The consensus sequence of the σ^{70} promoter includes a –35, a –10, and sometimes an extended –10 region that interacts with the σ subunit of RNA polymerase. Another AT-rich region called the UP-element interacts with the α subunit. (b) An example sequence for the σ^E promoter is represented by a position weight matrix in sequence logo form. Bases that occur more frequently in promoters are represented by larger sequence logos. The unit, bits, is adapted from information theory as a measure of information content. (From Rhodius and Mutalik [44]. With permission.)

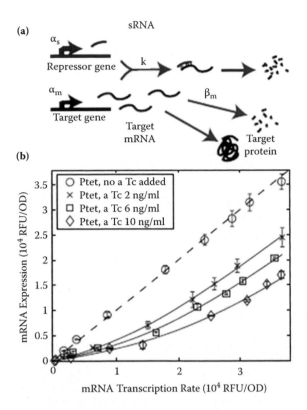

FIGURE 9.12 The quantitative behavior of sRNAs can be modeled using a kinetic model. (a) The sRNA and the target mRNA genes are shown in red and black, respectively. Only the target mRNAs that are not bound by sRNAs can translate into the target protein. (b) The expressions of several sRNA–mRNA pairs show the predicted behavior. The expression of the target gene, *rhyB*, increases linearly with its transcription rate after a threshold. This threshold is determined by the transcription rate of the sRNA, which is regulated by the small molecule aTc for the P_{tet} promoter that drives the sRNA expression. (From Levine et al. [78]. With permission.)

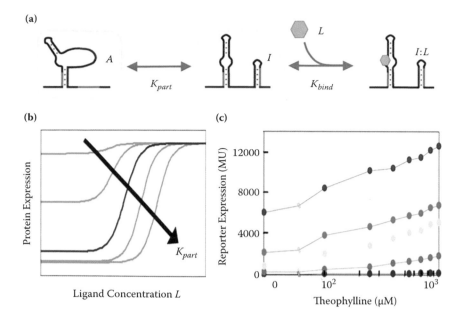

FIGURE 9.13 A mathematical model guides the design of riboswitches. (a) A riboswitch can fold into two secondary structures that are either active (A) or inactive (I) for the downstream regulation. A ligand (L) binds to the inactive form and stabilizes the riboswitch in the inactive form. (b) The transfer function of the riboswitch regulation changes with the equilibrium constant K_{part} between the active and inactive forms as determined by their secondary structure stability. (c) The protein expression of the regulated gene with respect to ligand (theophylline) responds differently according to the different secondary structure stabilities as predicted by the model. The colors red, orange, yellow, green, and blue denote riboswitch designs that each stabilizes the riboswitch in the inactive form more than the previous riboswitch, leading to elevated basal expression as well as dynamic range in the target protein expression as predicted by the model. Reporter expression is measured by a β-galactosidase assay in Miller units. (From Beisel and Smolke [80]. With permission.)

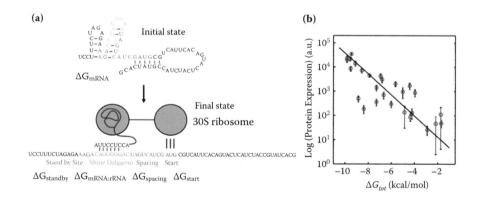

FIGURE 9.14 The strength of a ribosome binding site can be predicted by a thermodynamic model. (a) The ribosome binding site interacts with the 30S ribosomal subunit to initiate translation. To interact with the 16S rRNA in the 30S ribosomal subunit, several regions in the RBS must be unfolded, including the standby site, the Shine-Dalgarno sequence, the start codon, with a defined spacing between the Shine-Dalgarno sequence and the start codon. (b) Using a thermodynamic model, the protein expression level of a given RBS sequence can be predicted within two folds from the total free energy accounting for each of the sequence contributions. The expression level of reporter fluorescence proteins is measured by flow cytometry. (From Salis et al. [101]. With permission.)

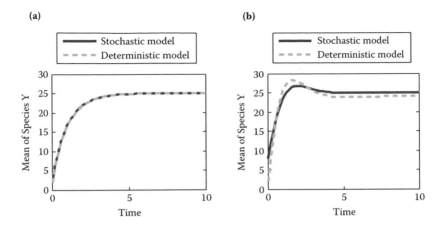

FIGURE 10.6 The effect of feedback in gene regulation. (a) Trajectory for the mean level of species Y in the absence of feedback. (b) Trajectory for the mean level of species Y with autoregulatory feedback. The dashed lines correspond to the solution to the deterministic ODE model, and the solid lines correspond to the mean of the stochastic model.

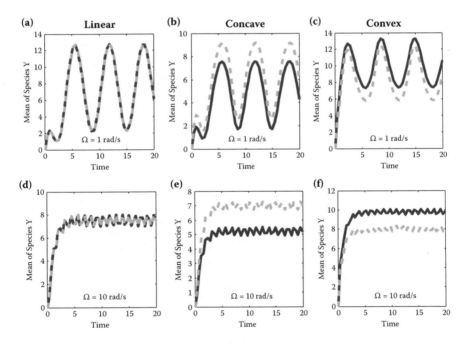

FIGURE 10.7 The effects of nonlinearities and stochasticity on signal transduction. (a) Trajectory for the mean level of species Y with when species X activates Y through linear regulation. (b) Trajectory for the mean level of species Y when species X activates Y with a concave function. (c) Trajectory for the mean level of species Y when species X represses Y with a convex function. (a)–(c) correspond to the same systems as (a)–(c), but where the external signal varies with a frequency of 1 rad/s. (d)–(f) correspond to a system where the external signal varies with a frequency of 10 rad/s. In all plots, the dashed lines correspond to the solution to the deterministic ODE model, and the solid lines correspond to the mean of the stochastic model.

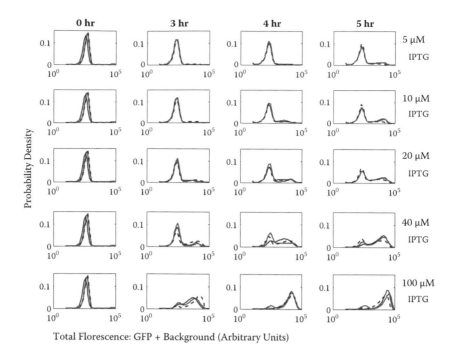

FIGURE 10.13 Measured (solid lines) and computed (dashed lines) histograms of GFP under the control of the *lac* operon and induced with IPTG. The columns correspond to different measurement times {0,3,4,5} hr after induction. The rows correspond to different levels of extracellular IPTG induction {5,10,20,40,100} μM. Experimental data are reproduced from Munsky et al. [52] (with permission), but a different model is used to fit this data as described in the text.

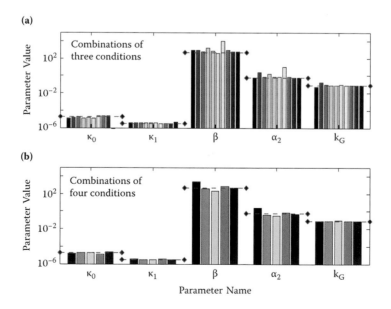

FIGURE 10.14 Identified parameters for the induction of *lac* with IPTG. (a) Parameters identified with every possible combination of three different IPTG concentration from {5,10,20,40,100} μM. (b) Parameters identified with every possible combination of four different IPTG concentrations. In each set of bars, the diamonds and the horizontal dashed lines correspond to the parameter set identified from all five IPTG levels.

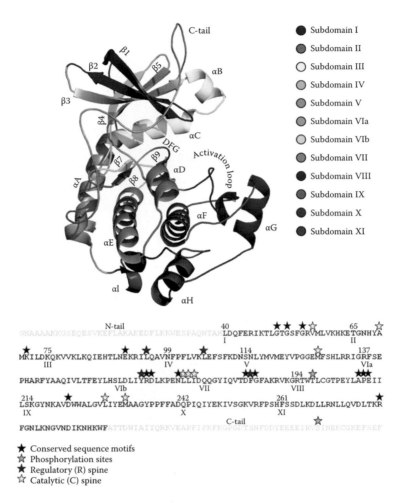

FIGURE 11.2 Subdomain assignments based on sequence conservation according to Hanks and Hunter (1995) shown on the PKA structure. Positions of the major conserved residues are shown as stars.

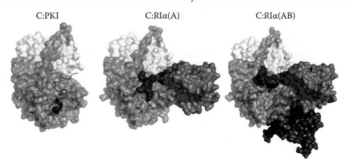

FIGURE 11.10 The catalytic subunit serves as a scaffold that interacts with multiple regulatory proteins. On the left is a representation of the catalytic subunit bound to a peptide from the heat-stable protein kinase inhibitor (PKI). The middle panel shows a catalytic subunit bound to a deletion mutant of RIα that contains a single nucleotide binding domain (CNB-A). On the right is the catalytic subunit bound to a deletion mutant of RIα that contains both single nucleotide binding domains (CNB-A and CNB-B). The catalytic subunit is shown as a space filling model with the N-lobe in white and the C-lobe in tan. PKI 1 to 24 is shown as a red ribbon. The R-subunits are also shown as space-filling models. The Inhibitor site that docks to the catalytic site is shown in red. The CNB-A domain in dark teal, CNB-B domain is in turquoise.

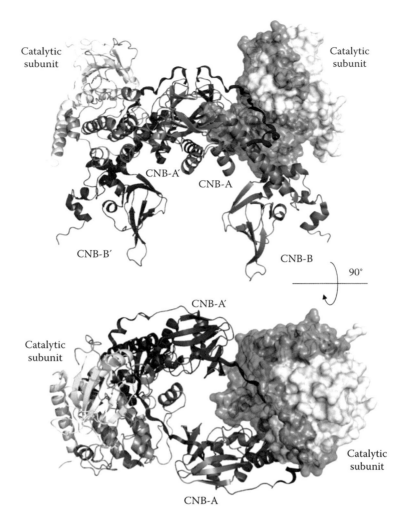

FIGURE 11.11 Model of the tetrameric RIα holoenzyme. Crystallization of a deletion mutant of the RIα subunit that contains an extended linker segment bound to a catalytic subunit revealed a tetrameric configuration of the two RC dimers. The model of the full length protein based on this tetrameric configuration is consistent with small-angle X-ray scattering and small-angle neutron scattering data. It is also consistent with single particle image reconstruction models (unpublished results). The upper panel shows the relative position of the dimerization docking domain and the well-separated locations of the catalytic subunits. Rotation of this model by 90 degrees reveals the incredible symmetry of the two active sites where the extended linker of one R-subunit is docked onto the CNB-A domain of the symmetry related dimer. Coloring of the catalytic subunit is as described in Figure 11.10, the R-subunits are shown as ribbons. In red is the linker region that becomes ordered only in the complex. The regulatory subunit of one domain is shown in turquoise, while the regulatory subunit from the symmetry related dimer is shown in dark teal.

8. Single-Cell Behavior

Philippe Cluzel

8.1 Introduction

Studying random fluctuations to characterize the properties of dynamical systems has been a classic approach of condensed matter physics and has more recently been extended to economics and biology. Historically, this field of research was prompted by the observation of what is known today as Brownian motion. In 1827, botanist Robert Brown used an optical microscope to observe that, when pollen from *Clarckia pulchella* was suspended in water, individual grains within the pollen displayed jittery movements [1]. He further demonstrated that the observed random motion was not due to the presence of "living" animalcules within the pollen, because the motion was also observed in microscopic

Quantitative Biology: From Molecular to Cellular Systems edited by Michael E. Wall © 2012 CRC Press / Taylor & Francis Group, LLC. ISBN: 978-1-4398-2722-2

Chapter 8

examinations of fossilized wood and dust [1]. Since this early observation of Brownian motion, there has been a steady stream of studies on this subject, but the first key conceptual advancement came from Albert Einstein in a 1905 paper [2] predicting that the mean square displacement of a Brownian particle was proportional to the time of observation. Einstein assumed that the motion of a Brownian particle is governed by the temperature of the fluid, which produces a random force and the friction of the fluid on the particle. These assumptions yield the classic formula of diffusion that gives the mean square distance covered by a Brownian particle during a given time interval as $\langle \delta x^2 \rangle_N \propto D\tau$, where D is the diffusion constant that depends on the temperature and the friction of the fluid. This formula emphasizes that some relevant information about the stochastic variable $\delta x(t)$ defined with $x(t) = \langle x \rangle_{Time} + \delta x(t)$ is obtained by characterizing its second moment (i.e., variance).

Here we bring Brownian motion back to biology. Most of the themes discussed in this chapter illustrate how the concept of the Brownian particle and the measure of its associated mean square displacement can be used directly to extract intracellular biochemical parameters from individual living cells. The literature on this subject is vast, and only a few examples are selected intentionally for the sake of simplicity and cohesiveness, focusing exclusively on bacteria. The intent of this chapter is not to present an exhaustive review of the subject but rather to discuss several methods that may be useful for characterizing the behavior of individual cells.

First, the principles and latest developments of fluorescence correlation spectroscopy (FCS) are described. FCS is based on monitoring the fluctuations of the fluorescence signal associated with dyes that diffuse in and out a small volume of detection [3–5]. The specific application developed to measure coding and noncoding RNAs in individual living bacteria [6,7] will be discussed.

Second, the well-known stochastic Langevin equation will be introduced; it describes the motion of a Brownian particle in the limit of low Reynolds number. A discussion will outline how the same phenomenological stochastic equation has been successfully used to model and characterize cell-to-cell variability in the expression of a single gene in bacteria [8]. Thus far, most of the studies have focused on the static distribution of the cell-to-cell variability within an isogenic population [8,9]. However, the discussion will be extended to the dynamic aspect of variability, and a description will be provided of how the analysis of temporal fluctuations of cellular behavior can be used to determine intracellular biochemical parameters [10]. Finally, the Langevin equation makes predictions about the linear response of dynamical systems to a small external perturbation. While this phenomenological equation has been used to describe physical systems at thermodynamic equilibrium, this framework may also be extended to living cells. This extension is possible when the system exhibits a well-defined steady state and has Markovian dynamics [11]. Under Markovian dynamics, the system has a short memory and can relax to a well-defined steady state. However, this framework is particularly relevant to study energy consuming mechanisms that are present in the living cells [12]. This chapter discusses the usefulness of this approach to characterize the existence of a fundamental interdependence between spontaneous fluctuations in living cells and their response to a small external stimulus. For nonequilibrium systems, such as living cells, this relationship between fluctuation and response has been recently formalized in a fluctuation-response theorem by Prost et al. [11].

While this theorem predicts the existence of a coupling between fluctuation and response, it does not reveal how this coupling varies as a function of the system parameters, such as gene expression or reaction rates. Therefore, it would be interesting to discuss under what circumstances biological systems would exhibit similar coupling. This approach predicts constraints between variability and response to environmental changes in specific classes of biological systems. The usefulness of this theorem will be illustrated using chemotaxis in *E. coli* as experimental system (see Chapter 13 for a review of chemotaxis in *E. coli*). The fluctuation-response theorem and the Langevin equation will be reinterpreted to highlight the existence of this fundamental relationship between behavioral variability and the response to a small chemotactic stimulus in single bacteria. Subsequently, the design principle in chemotaxis shared by other biological systems will be reported, which determines the coupling between response and fluctuations.

8.2 Fluorescence Correlation Spectroscopy

In characterizing the diffusion of fluorescent molecules in solution, Magde, Elson, and Webb [3–5] pioneered and developed a new quantitative technique for biology called FCS. The principle of this technique consists of measuring the fluorescence signal emitted from molecules diffusing freely in and out of a confocal volume of detection (see Chapters 5 and 7 for discussions of fluorescent probes). The fluorescence intensity fluctuates because individual molecules are entering and leaving the volume of detection defined by a diffraction–limited laser spot. A normalized autocorrelation function is used to analyze the fluctuations δn of emitted photons (i.e., fluorescence intensity)

$$G(t) = \frac{\langle \delta n(t') \delta n(t'+t) \rangle}{\langle n \rangle^2} \tag{8.1}$$

The function $G(t)$ depends on the shape of the illumination profile that is generally Gaussian. This section focuses only on 2-D diffusion. (For a detailed review of the mathematical derivations of the function $G(t)$ see [13]). While 2-D diffusion involves fewer parameters and can be a good approximation for 3-D diffusion, it is also well-suited to examine the diffusion of fluorescent markers in rod-like bacterial cells, such as *E. coli*. If we neglect the diffusion of molecules along the z-axis, $G(t)$ can be simplified to

$$G(t) = \frac{1}{N} \left(\frac{1}{\left(1 + t/\xi_{free}\right)} \right) \tag{8.2}$$

where N is the number of fluorescent particles per detection volume, and ξ_{free} is the effective diffusion time of these particles through the detection volume. Fitting the experimental autocorrelation function with $G(t)$ yields the number of molecules present on average in the volume of detection and the diffusion coefficient of the molecules, which depends on their size and the viscosity of their environment.

The power of this approach lies in the fact that the measurements are self-calibrated because they characterize the fluctuations of fluorescence relative to the mean fluorescence signal. Thus, the result does not depend on the intensity of the

Chapter 8

excitation light. Of course, there are several important limitations to this approach. First, fluctuations in the emitted fluorescence signal must be large enough relative to the mean. This condition restricts the technique to a low concentration of fluorescent markers. For example, the definition of the autocorrelation function becomes poor when there are more than 1,000 molecules per effective volume of detection of ~0.1 fl. Second, the technique is sensitive to bleaching, which introduces correlations in the fluctuations of intensity and introduces artifacts in the measurements. Third, if molecules are localized and cannot diffuse freely, then FCS does not work. Despite these limitations, FCS has proven to be a powerful tool in single-cell measurements [14–16].

8.2.1   Measuring Intracellular Protein Diffusion in Single Bacteria

While FCS was initially developed to measure small traces of dyes in solution [4], it has more recently been applied to living single cells [17]. This technique became less invasive with the use of genetically encoded fluorescent proteins [6,7]. Using FCS, one of the first examples of the direct quantification of protein concentration within single living cells revealed that the bacterial motor in *E. coli* exhibits an extremely steep input–output sigmoid relationship with a Hill coefficient of about 10 [16]. This result is specific to the single-cell approach because when the same bacterial motor was characterized at the population level, it had an input–output relationship that was much smoother with a Hill coefficient of ~3 [18–20]. To understand this discrepancy, it is important to note that that there is a strict relationship between the motor behavior and the concentration of a specific signaling protein that controls the motor behavior. When standard ensemble techniques are used to evaluate protein concentration, such as immunoblotting, they ignore the inherent cell-to-cell variability. As a result, ensemble average effectively smoothed out the typical motor characteristics, leading to lower values of the Hill coefficient.

8.2.2   Measuring Intracellular Diffusion of RNA in Single Bacteria

In addition to measuring protein concentration, FCS has been extended to quantify coding and noncoding RNA in single cells [6,7]. To label RNA, it is common to use the MS2 labeling system originally developed for yeast by Bertrand and colleagues [21] and subsequently adapted to *E. coli* [22]. This system employs two gene constructs encoded on plasmids: (1) a fusion of the RNA-binding MS2 coat protein and GFP (MS2-GFP); and (2) a 23-nucleotide *ms2* RNA binding site (*ms2*-binding site) located downstream of a gene on an RNA transcript. MS2-GFP can either diffuse freely through the cell or bind to the *ms2*-binding sites on the transcript. There are two *ms2*-binding sites, and each site binds an MS2-GFP homodimer. To monitor temporal variations in RNA concentration in real time, it is essential to account for the slow maturation time of GFP and thus to preexpress MS2-GFP. Free MS2-GFP fusion proteins diffuse through the detection volume (diffraction limited laser spot) with a typical time of ~1 ms. The sensitivity of detection of noncoding RNA is increased by fusing a ribosomal binding site to the tandem of *ms2*-binding sites, which produces an RNA/MS2-GFP/ribosome complex that diffuses with a typical diffusion time of ~30 ms, 30-fold slower than free MS2-GFP in the cytoplasm. These measurements have high temporal resolution, requiring only a 2 second acquisition time to obtain a reliable autocorrelation function.

To determine the concentration of bound GFP molecules and, therefore, the concentration of the RNA transcripts labeled with MS2 coat proteins, the original autocorrelation function (Equation (8.2)) is extended to Equation (8.3a) [13]. This new formula takes into account the fact that there is a mixture of bound and free molecules in the detection volume and that their diffusion time is much faster than the dynamics of binding and unbinding:

$$
G(t) = \frac{1}{N} \left(\frac{1-y}{\left(1 + t / \xi_{free}\right)} + \frac{y}{\left(1 + t / \xi_{bound}\right)} \right) \tag{8.3a}
$$

where N is the number of fluorescent particles in the detection volume, y is the fraction of MS2-GFP molecules bound to the mRNA–ribosome complex, and ξ_{free} and ξ_{bound} are the diffusion times of free and bound MS2-GFP, respectively. The parameter y is a proxy to infer the RNA concentration in the single cell. Finally, Equation (8.3b), developed by Rigler and colleagues [23], accounts for the increase in brightness due the two *ms2*-binding sites that are present in each RNA transcript (each binding is an MS2-GFP homodimer).

$$
G(t) = \frac{1}{N} \frac{1}{\left[1+y\right]^2} \left(\frac{1-y}{\left(1 + t / \xi_{free}\right)} + \frac{4y}{\left(1 + t / \xi_{bound}\right)} \right) \tag{8.3b}
$$

This model can fit any experimental autocorrelation function in living *E. coli* by adjusting only two parameters, N and y. The lower limit of detection for N was about two transcripts per volume of detection.

When the RNA transcript containing the *ms2*-binding sites encoded the DsRed fluorescent protein, the self-calibration by FCS of RNA measurements from free and bound MS2-GFP was in good agreement with the simultaneous measurement of DsRed protein concentration. It is important to note that the genetic design used in this approach guarantees that only the freely diffusing mRNA transcripts not physically associated with an RNA polymerase or plasmid DNA are measured. To this end, the *ms2*-binding sites were located after the *DsRed* stop codon. With this orientation, the binding sites are transcribed only after *DsRed* is fully transcribed. The mRNA then becomes visible when the MS2-GFP proteins bind to the mRNA transcript with fully transcribed *ms2*-binding sites.

8.2.3   Advances in FCS for Single-Cell Measurements

Current FCS techniques require that the RNA molecules are free to diffuse. However, a recent study from Jacobs-Wagner and colleagues [24] used in situ hybridization to show that in *Caulobacter crescentus* and *E. coli*, chromosomally expressed mRNAs tend to localize near their site of transcription during their lifetime. This observation implies that the diffusion coefficients of chromosomally expressed mRNA are two orders of magnitude lower than mRNA molecules expressed from plasmids. When RNA transcripts are localized, imaging becomes necessary to quantify concentration [21,25]. To overcome this obstacle, wide-field FCS combines standard FCS measurements with imaging. This new experimental approach makes it possible, in principle, to quantify both freely diffusing and localized RNA molecules. The extension to wide-field FCS has become possible due to the development of new cameras that use an electron multiplying CCD technology (EMCCD). EMCCD cameras are ideal detectors because they combine speed and sensitivity with high quantum efficiency. These EMCCD cameras can be used as photon detectors, in kinetics mode, with only one- or two-point excitation volumes [26,27]. The reading time of the pixels from the chip of the camera is far slower than the typical dead time of avalanche photon-diodes used for standard FCS single-point detectors that have a dead time of about 70 ns. However, this lower temporal resolution should not be an issue for measurements in living cells because the typical timescales involved are of the order of a millisecond. Thus, the EMCCD cameras could be used to perform FCS measurements with a time resolution of 20 μs, which should suffice to determine the relevant timescales associated with molecular diffusion in cells [26,27]. To improve the time resolution, Heuvelman and colleagues [28] aligned a line illumination of the sample with one line of adjacent pixels in the EMCCD chip. The physical configuration of the camera in the kinetics mode simultaneously reads adjacent pixels from the same line and clears them within about 0.3 μs. This parallel multichannel acquisition of the fluorescence signal with an EMCCD camera improved the time resolution down to 14 μs. While FCS measurements were performed only along one dimension, this approach should also work with several contiguous lines to extend FCS measurements along the two dimensions with a submillisecond resolution. Wide-field FCS will most likely be immediately applicable to small rod-shaped bacteria, such as *E. coli* or *Salmonella*. For thicker eukaryotic tissue and cells, it is also necessary to scan along the z-axis. To expand FCS technique to higher throughput measurements along the z-axis, Needleman and colleagues [29] developed a pinhole array correlation imaging technique based on a stationary Nipkow disk and an EMCCD. Although this technique has not been tested on live cells, it has the potential power to perform hundreds of FCS measurements within cells thicker than bacteria, with high temporal resolution.

8.3 Spectral Analysis of Molecular Activity Fluctuations to Infer Chemical Rates

8.3.1 A Practical Aspect of the Langevin Equation

A phenomenological stochastic equation called the Langevin equation describes in a general way the fluctuations of a mass-spring system in a viscous fluid, with spring constant k_{spring}, damping constant γ, and fluctuating random force $f(t)$. When the Reynolds number is low, there is no acceleration, and friction (γ) dominates the dynamics of the particle's position $x(t)$:

$$\gamma \dot{x} = -k_{spring}x + f(t) \tag{8.4}$$

This model has been found to be very useful to interpret, in the regime of linear approximation, the behavior of a large range of dynamical systems. For example, this model can be extended to the study of biochemical reactions taking place in cells.

Taking the Fourier transform of the Langevin equation allows us to calculate the power spectrum of the fluctuations of the position δx at equilibrium

$$\dot{\delta x}^2(\omega) = \frac{2KT}{\gamma(\omega_c^2 + \omega^2)} \tag{8.5}$$

In the classic Langevin equation, the amplitude of the spontaneous fluctuations (D) defined with $\langle f(t)f(t')\rangle = D\delta(t-t')$ is obtained using the equipartition theorem that couples the thermal energy KT, and the energy stored in the spring given by

$$KT = k_{spring} \left\langle \delta x^2 \right\rangle_{time} \tag{8.6}$$

$\langle f(t)f(t')\rangle = D\delta(t-t') = 2KT\gamma\delta(t-t')$, where T is the temperature, and K is the Boltzmann constant. Equation (8.5) has several practical applications. For example, it is particularly useful to evaluate the sensitivity range of a mechanical transducer used in single-molecule experiments. In such experiments, the mechanical transducer can be the cantilever of a force microscope, a glass fiber, or the trap formed by an optical tweezers that senses the motion of a small bead submerged in solution. To determine whether to use a stiff or soft cantilever for a given experiment, the power spectrum of the fluctuations of the cantilever's position can be plotted. Spontaneous fluctuations of the mechanical transducer induced by the thermal noise are not distributed uniformly over all frequencies. The fluctuations die out for frequencies larger than the typical corner frequency $\omega_c = \frac{k_{spring}}{\gamma}$. If the transducer becomes stiffer, the spontaneous fluctuations will spread over higher frequencies because the corner frequency $\omega_c = k/\gamma$ becomes larger. Most importantly, the thermal energy stored in the system remains constant $\left\langle \delta x^2 \right\rangle_{time} = \frac{KT}{k_{spring}}$, which represents the total area defined by the power spectrum $\left\langle \delta x^2 \right\rangle = \int_0^{+\infty} \frac{2KT}{\gamma(\omega_c^2 + \omega^2)} d\omega$. In other words, this constraint implies that the amplitude of the spontaneous fluctuations of a stiff spring will be smaller at lower frequencies than those of a softer spring (Figure 8.1). If the signal we plan to measure is in a lower frequency range, then a stiffer mechanical transducer will exhibit a lower background noise in this frequency range, but the sensitivity of the transducer will be reduced by the same amount [30]. Thus, reducing the stiffness of the transducer does not improve the signal-to-noise ratio. By contrast, reducing the dimension of the transducer will reduce the friction coefficient γ, which in turn will increase the corner frequency, $\omega_c = k/\gamma$. So at equal stiffness, the background noise can be reduced by using transducers with smaller dimensions. Using this analysis, the force-extension measurements of a single DNA molecule were found to be less noisy using an optical trap [31,32] with submicron latex beads than using microglass fibers of several micron long [33].

Chapter 8

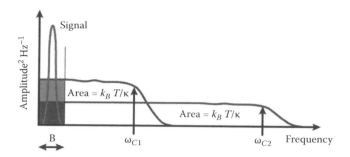

FIGURE 8.1   Thermal spontaneous fluctuations of a mechanical transducer. Changing the corner frequency, $\omega_c = k_{spring}/\gamma$, is achieved by changing the stiffness or the dimension of the mechanical transducer. (From Bustamante et al. [30]. With permission.)

8.3.2   Measuring Noise to Infer the Sensitivity of a Chemical System

For kinetic reactions, the positional coordinates $x(t)$ of the elastic spring corresponds to the concentration. Using the equation describing the fluctuations in a mass-spring system, Bialek and Setayeshgar [34] introduced a model to describe the dynamics associated with the binding and unbinding of a ligand to a receptor at equilibrium in a thermal bath. This case study is thought to be general because the fluctuations of receptor occupancy are governed solely by thermal noise. This assumption allows the authors to use the fluctuation–dissipation theorem (FDT) as a framework to relate the amplitude of the fluctuations to the macroscopic behavior of the receptor described by a kinetic equation. In the Bialek–Setayeshgar model, fluctuations in receptor activity reflect the fluctuations in the rates k^+ and k^- associated with the binding and unbinding of the ligand to the receptor. This approach is similar to that of a spring that is submerged in a thermal bath (Section 8.3.1). The only difference lies in the constraints that define the amplitude of the fluctuation D in the Langevin equation. The fluctuations of the rates k^+ and k^- are related to the fluctuations of the associated free energy that define the two receptor states. Using this noise analysis, they determined the sensitivity range of sample receptor. Their analysis is analogous to that of the mechanical transducers in Section 8.3.1.

Similarly, the fluctuations in chemical reactions can yield information about the typical chemical rates governing the reactions. In the simplest example of an isomerization reaction, a molecule switches back and forth between an active and inactive state:

$$A \underset{k^-}{\overset{k^+}{\rightleftharpoons}} A^* \tag{8.7}$$

where A and A* are the concentration of the inactive and active forms, respectively, and k^+ and k^- are the respective reaction rates from the inactive to active and the active to inactive form. The concentration of A* changes such that

$$\dot{A}^*(t) = k^+ A(t) - k^- A^*(t) \tag{8.8}$$

We linearize about the steady state using

$$A*(t) = \overline{A^*} + \delta A*(t)$$

$$A(t) = \overline{A} + \delta A(t)$$

$$k^+(t) = k^+ + \delta k^+(t)$$

$$k^-(t) = k^- + \delta k^-(t)$$

and Equation (8.8) becomes

$$\delta \dot{A}^* = \delta k^+ \overline{A}^* - \delta k^- \overline{A} + (k^+ + k^-)\delta A^*$$

The rates are related by the difference of the free energy ΔF between the active and inactive states, such that $\dfrac{k^+}{k^-} = e^{\frac{\Delta F}{KT}}$. Using this relationship, the fluctuations of the rates are given by

$$\frac{\delta k^+}{k^+} - \frac{\delta k^-}{k^-} = \frac{\delta F}{KT}$$

and Equation (8.8) takes the usual Langevin form:

$$\delta \dot{A}^*(t) = \left(k^+ + k^-\right)\delta A^*(t) + \frac{k^+ \overline{A}}{KT}\delta F(t) \tag{8.9}$$

The Langevin equation (8.9) is analogous to the equation describing fluctuations of a mass-spring system (Equation (8.4)). To apply the FDT, it is convenient to Fourier transform Equation (8.9):

$$\frac{KT}{k^+ \overline{A}}\left[i\omega - (k^+ + k^-)\right]\delta \tilde{A}^*(\omega) = \delta \tilde{F}(\omega)$$

The Fourier transform is defined with $\delta \tilde{A}^*(\omega) = \displaystyle\int_0^\infty \delta A^*(t)e^{i\omega t}\,dt$, and the response function $\chi(t)$ determines the linear response ΔA^* to a small external perturbation δF such that

$$\Delta A^*(t) = \int_0^\infty \chi(t')\delta F(t-t')\,dt'$$

$$\Delta \tilde{A}^*(\omega) = \tilde{\chi}(\omega)\delta \tilde{F}(\omega)$$

the response function becomes

$$\tilde{\chi}(\omega) = \frac{\delta \tilde{A}^*(\omega)}{\delta \tilde{F}(\omega)} = \frac{k^+ \bar{A}^*}{KT\left[i\omega - k^+ + k^-\right]} \tag{8.10}$$

Applying the FDT

$$P_{\delta A^*}(\omega) = \frac{2KT}{\omega} \text{Im}\left[\tilde{\chi}(\omega)\right] \tag{8.11}$$

Equation (8.10) becomes

$$P_{\delta A^*}(\omega) = \frac{2k^+ \bar{A}^*}{\left[\omega^2 + (k^+ + k^-)^2\right]} \tag{8.12}$$

Plotting the power spectrum is a convenient way to study the output fluctuations of a system. The corner frequency of the power spectrum in a log-log plot gives a measure of the sum of the rates, and its integral from 0 to infinity yields the total variance of the noise of this isomerization reaction (Equation (8.7)):

$$\left\langle (\delta A^*)^2 \right\rangle = \int_0^{+\infty} P_{\delta A^*}(\omega) \frac{d\omega}{2\pi} = \frac{2k^+ \bar{A}^*}{2\pi} \int_0^{+\infty} \frac{\tau^2 d\omega}{1 + (\tau\omega)^2}$$

Therefore, the total variance is

$$\left\langle (\delta A^*)^2 \right\rangle = \frac{k^+ \bar{A}^*}{(k^+ + k^-)} \tag{8.13}$$

In this approach [34], there is no assumption about the underlying statistics that govern switching between A and A*. Instead, it relies on thermodynamic equilibrium and the use of the FDT. Although this approach is powerful to describe *in vitro* systems where thermodynamic equilibrium can be well-defined, we cannot directly extend it to living cells because they are open systems and are far from thermodynamic equilibrium. Section 8.4 discusses novel experimental and theoretical approaches used to describe the linear response of living cells to a small stimulus based on an extension of the FDT.

8.3.3　Modeling Cell-to-Cell Variability from the Noise of Single Gene Expression in Living Cells—The Static Case

It also is now common to use the Langevin equation [35] to characterize the expression noise of a single gene [8,9]. While the Langevin model yields results equivalent to that of Monte Carlo simulations, its power lies in its straightforward physical interpretation that is often obscured in numerical simulations. One standard way to deal with the stochastic aspect of chemical reactions is to assume that each coordinate of a given chemical system independently obeys Poisson statistics for which the variance and average are equal, the molecular system is well mixed, and concentrations are continuous variables. For example, Ozbudak and colleagues [8] characterize the transcription of a single gene within a cell using a simple model:

$$\frac{d}{dt}RNA = k_{txn} - \gamma RNA + \eta_{txn}$$

where the number of mRNA molecules, RNA, is a continuous quantity, γ is the degradation rate of RNA, k_{txn} is the transcription rate per DNA, and η_{txn} is a random function that models the noise associated with transcription. This Langevin equation assumes that a steady state exists and that fluctuations about the steady concentration δRNA reflect the response of the system to a Gaussian white noise source, $\langle \eta_{txn}(t) \rangle = 0$ and $\langle \eta_{txn}(t)\eta_{txn}(t+\tau) \rangle = D\delta(\tau)$, where backets represent population averages, and δ represents the Dirac δ-function. Expanding around this steady state, $\langle RNA \rangle = \frac{k_{txn}}{\gamma}$, by setting $RNA = \langle RNA \rangle + \delta RNA$ gives the Langevin equation for δRNA:

$$\frac{d}{dt}\delta RNA + \gamma \delta RNA = \eta_{txn}(t) \tag{8.14}$$

Fourier-transforming this equation gives

$$\frac{\delta RNA(\omega)}{\eta_{txn}(\omega)} = \frac{1}{\gamma + i\omega}, \left\langle \left| \eta_{tex}(\omega) \right|^2 \right\rangle = D$$

so that the variance of the fluctuations is given by

$$\left\langle \delta RNA^2 \right\rangle = \int \frac{d\omega}{2\pi} \frac{1}{\gamma^2 + \omega^2} D = \frac{D}{2\gamma}$$

Although this system is not at thermodymic equilibrium and the equipartition theorem cannot be used, the value of D can be determined by assuming Poisson statistics for the fluctuations δRNA. Therefore, setting $\left\langle \delta RNA^2 \right\rangle = \langle RNA \rangle$ yields the amplitude of the input white noise $D = 2k_{txn}$[8]. Unlike in the Bialek-Setayeshgar model (Section 8.3.2), the FDT cannot be invoked because this system is not at thermodynamic equilibrium. In this static approach, there is usually no need to define the response function because, in general, the goal of these studies is to evaluate the variance associated with the transcription or translation of a specific gene. In other words, they aim to measure the cell-to-cell variability within a clonal population of cells in a steady state regime [9,36].

8.3.4 The Power of Time Series to Characterize Signaling in Single Cells

Because it is experimentally simpler, most experimental studies typically describe the noise of biological systems as cell-to-cell variability [8,9] (see Section 8.3.3). Such studies assume that biological systems are ergodic; that is, the temporal average at the single cell level is equal to the ensemble average across many cells at a fixed time point. Experimentally, this assumption implies that a snapshot of a population of cells at steady state can reveal the statistics of the system. However, this assumption is not always sufficient because the snapshot approach does not allow us to characterize the temporal correlations in protein fluctuations taking place within the individual cell. Thus, due to the difficulty of such experiments, only a few experiments have used time-series analysis to determine noise from a single gene [10,25,37–38]. For example, to distinguish which genes within a transcriptional network have an active regulatory connection, Dunlop and colleagues characterized the cross-correlation of the spontaneous fluctuations in the activity of two promoters, *galS* and *galE*, in the galactose metabolism system in *E. coli* [10]. A copy of each promoter controlling the expression of either CFP or YFP was integrated into the chromosome. It is not obvious that the cross-correlation would reveal interactions between the two genes because global noise would also simultaneously affect the activity of both promoters. However, there exists a time lag in the profile of the cross-correlation function between promoters that have active and inactive transcriptional interactions—here a feed-forward loop [39]. Therefore, time lags in cross-correlation functions can distinguish the uniform effect of global extrinsic noise from the specific cases of active regulatory interactions. While this lag time is measurable by performing the cross-correlation of the activity of two promoters, it also requires that the difference between the maturation times of CFP and YFP is small enough compared with the lag induced by the regulatory interactions.

Beyond gene expression, spectral analysis has been used to analyze the nature of the temporal fluctuations throughout the signaling cascade of the chemotaxis system in *E. coli*. This system governs the locomotion of bacteria and allows the cells to move toward the source of chemical attractants, such as amino acids [40]. In *E. coli*, chemotaxis has become a canonical system for the study of signal transduction networks because this network involves few components, and it is amenable to tractable quantitative analyses (see Chapter 13). It exhibits, however, complex behaviors, such as sensitivity to stimulus and adaptation to environmental changes. The activity of the chemotaxis kinase CheA directly reflects environmental changes in the neighborhood of the bacterium. For example, following a sudden increase in aspartate concentration in the environment, aspartate will bind to Tar receptors, which will induce a drop of activity of the kinase CheA. For the sake of simplicity, we will assume that CheA, can be either in an active (A) or inactive state (A*) (Figure 8.2). When the kinase is active, it transfers a phosphate group to the diffusible signaling response regulator CheY. Active CheYp binds to the basal part of the rotary motor that powers the rotation of a long flagellar filament and increases the probability of the motor to rotate in a clockwise (CW) direction. In swimming bacteria, CW rotation induces tumbles, and counterclockwise (CCW) rotation induces smooth runs. In the absence of a stimulus, the bacterial motor switches randomly between CCW and CW rotations, whose frequency reflects the steady state activity of the kinase and randomizes the trajectory of a swimming cell. To summarize, a drop in kinase activity, caused by a sudden stimulus of aspartate, induces CCW rotation of the motor (smooth run) and biases the random trajectory of the cell toward gradient of attractant. After its initial drop, the kinase activity adapts back to its prestimulus level. This adaptation mechanism is

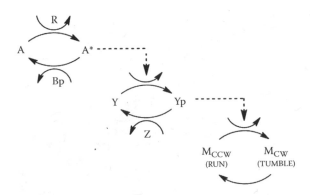

FIGURE 8.2 Receptor-kinase signaling cascade in the chemotaxis system. When ligands bind to transmembrane receptors, the receptors control the kinase CheA (*A*) activity. In its active form (A*) CheA phosphorylates the signaling molecule CheY (*Y*) into the active form CheY-P (*Y*p). CheY-P diffuses throughout the cell and interacts with the flagellar motors to induce clockwise rotation (tumble) (MCW) from counterclockwise rotation (run) (MCCW). The phosphatase CheZ (*Z*) dephosphorylates CheY-P. A sudden increase in ligand-binding causes a decrease in kinase activity. Two antagonistic enzymes regulate the activity of the kinase-receptor complexes. The methyltransferase CheR (*R*) catalyzes the autophosphorylation of CheA by methylating the receptors. The active kinase A* phosphorylates the methylesterase CheB in CheB-P (*B*p). CheB-P removes methyl groups from active receptor complexes, which catalyze kinase deactivation.

governed by two antagonistic enzymes, CheR and CheBp, which regulate the activity of the kinase-receptor complexes (Figure 8.2) [41]. The standard assumption is that nonstimulated cells exposed in a steady environment would exhibit a steady behavior. A single-cell analysis revealed that the slow fluctuations in CheA activity were reflected in the switching behavior of a single motor [42]. Power spectra were used to analyze the fluctuations of the stochastic switching events of individual motors between two states.

$$M_{ccw} \underset{k^-}{\overset{k^+}{\rightleftharpoons}} M_{cw} \tag{8.15}$$

According to this equation, the common expectation for rates k^+ and k^- is that the power spectrum associated with the fluctuations δM_{CW} should exhibit a Lorentzian profile. This power spectrum, identical to that of Section 8.3.2 (Equation (8.12)), would exhibit at long timescale a flat profile in a log-log plot:

$$P_{M_{CW}}(\omega) \sim \frac{1}{[\omega^2 + (k^+ + k^-)^2]} \tag{8.16}$$

Surprisingly, for bacteria in a steady environment, such as motility medium that does not support growth, the spectrum exhibits a corner frequency at short timescale (~1 sec), followed by a growing profile that ranges from ~10 sec to ~15 min. The slope of the power spectrum at long timescale indicates that the CCW and CW intervals from the motors are not exponentially distributed. Interestingly, the fluctuations of CheA activity are directly reflected in the distributions of runs and tumbles of an individual bacterium. Therefore, the switching events are not governed solely by the Poisson statistics of the

Chapter 8

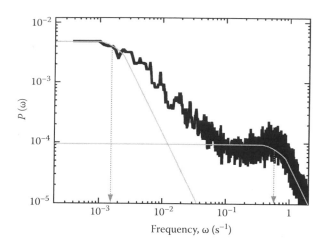

FIGURE 8.3   Noise in the chemotaxis network. (Black curve) Power spectrum of the network output from a single nonstimulated wild-type cell can be described by two superimposed Lorentzian curves (gray). The knee frequency (dashed arrows) of each Lorentzian curve comes from distinct parts of the signaling network (Figure 8.1). The higher frequency is the typical time for the motor switching. The lower frequency is the typical time for the CheA kinase fluctuations. (From Korobkova et al. [42]. With permission.)

motor but also by an additional process that takes place in the signaling cascade and that occurs at much longer timescale. The complex profile (Figure 8.3) can be understood as the superimposition of two Lorentzians, one representing the motor switching with a knee frequency at short timescale and another one with a knee frequency at a much longer timescale due to the fluctuations of the CheA kinase activity governed by the slow antagonistic action of CheR and CheBp (Figure 8.2). As a consequence, different relative expressions of CheR or CheBp concentrations would yield different amplitudes of the fluctuations of CheA activity at long timescales. The CheR/CheBp ratio controls the amplitude of an adjustable source of behavioral variability (42). By contrast, population measurements found that the distribution CCW and CW were exponentially distributed, which would yield the power spectrum described by Equation (8.16) without long timescale fluctuations observed at the single cell level [43]. This discrepancy exemplifies the non-ergodic nature of some biological systems for which the analysis of long time series in individual cells is essential.

8.4   Relationship between Fluctuations and Response, Extending the FDT to Living Cells

None of the previous examples has addressed the dynamical aspect of the FDT that relates spontaneous fluctuations with the linear response of a system to small external perturbation. Instead, this theorem has been used to evaluate only the variance associated with cell-to-cell variability within isogenic populations at a given time point. The next section explores whether the spontaneous fluctuations in biological systems is related to the response to a small external perturbation as predicted by the FDT.

8.4.1 Violation of the FDT Due to Energy-Consuming Mechanisms

A classic example of experimental violation of the fluctuation-dissipation theorem uses mechanosensitive bundles of inner hair cells (hair bundles) from bullfrogs [12].

$$\chi(t) = -\beta \frac{d}{dt} C(t) \tag{8.17}$$

While the theorem relates the response function to the derivative of the autocorrelation function, it is often more convenient to use its formulation in the frequency space:

$$\tilde{C}(\omega) = \frac{2}{\beta\omega} \text{Im} \tilde{\chi}(\omega) \tag{8.18}$$

where $\text{Im}\tilde{\chi}(\omega)$ is the imaginary part of the Fourier transform of the response function, which represents the dissipation of the system, and the function $\tilde{C}(\omega)$ is the power spectrum of the spontaneous fluctuations. In physics, β^{-1} would represent the equilibrium temperature of the environment for the system. When the system departs from the equilibrium due to a mechanism that consumes energy, the theorem is violated. A standard measure of the degree of violation of the theorem is provided by the ratio between the temperature of the system and an effective temperature, $T_{eff}(\omega)$

$$\frac{T_{eff}(\omega)}{T} = \frac{\omega\beta}{2} \cdot \frac{\tilde{C}(\omega)}{\text{Im}\tilde{\chi}(\omega)} \tag{8.19}$$

If this ratio is equal to unity, the theorem is satisfied. When the relationship is less than unity, the system contains an active mechanism that consumes energy, and the theorem is violated. Using Equation (8.19), Martin et al. compared the spontaneous fluctuations of a hair bundle from bullfrog to its response to a sinusoidal mechanical stimulation. To demonstrate that the theorem was violated [12] (Figure 8.4), they showed how far the ratio $\frac{T_{eff}(\omega)}{T}$ deviated from the unity. They concluded that the random fluctuations of hair cells are not solely due to the effect of thermal energy but also to an active mechanism within the hair cells. Such mechanism could govern active filtering and sharp frequency selectivity in this hearing system. In living systems, where many energy-consuming mechanisms exist, it is expected that the FDT is violated.

8.4.2 The Prost–Joanny–Pandaro Fluctuation–Response Theorem and Its Application to Molecular Motors

A recent theoretical analysis of the FDT demonstrated that if a system has a well-defined steady state with Markovian dynamics, the FDT is extensible to a fluctuation-response theorem even in the absence of thermodynamic equilibrium [11]. This result is extremely relevant to living cells that are not at thermodynamic equilibrium. Many biological systems display a well-defined steady state and exhibit fluctuations that are caused by underlying Markovian dynamics. Here, the term *Markovian* refers to dynamic processes

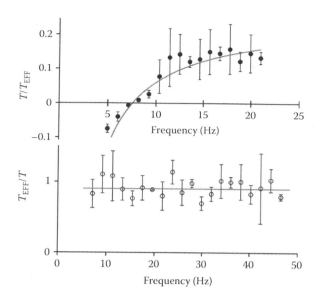

FIGURE 8.4 The effective temperature from a bundle of hair cells. (Top) An active mechanism dissipates energy and violates the FDT. (Bottom) When the active mechanism in the hair bundle is disrupted, the ratio $\frac{t_{eff}(\omega)}{T}$ is near unity, and FDT is satisfied. (From Martin et al. [12]. With permission.)

that have short *memory* so that they can relax fast enough to a well-defined steady state. In the Prost-Joanny-Parrando (PJP) fluctuation-response model [11], the steady state exists when there is a well-defined probability distribution function $\rho_{ss}(c; \lambda)$ of the variables c controlled by a set of parameters λ_α. The probability distributions are associated with the potentials $\phi(c; \lambda) = -\log[\rho_{ss}(c; \lambda)]$. The PJP fluctuation-response theorem relates the response function $\chi_{\alpha\gamma}(t-t')$ to the correlation function:

$$\chi_{\alpha\gamma}(t-t') = \frac{d}{dt}C_{\alpha\gamma}(t-t') = \frac{d}{dt}\left\langle \frac{\partial\phi(c(t); \lambda^{ss})}{\partial\lambda_\alpha}\frac{\partial\phi(c(t'); \lambda^{ss})}{\partial\lambda_\gamma}\right\rangle_{ss} \qquad (8.20)$$

This theorem is directly applicable to *in vitro* biological systems and, in particular, to the bullfrog hair cells (Section 8.4.1). As we know, the standard FDT is violated in this system. However, when the coordinates derived from the potentials ϕ are used instead of using the conventional set of variables, such as the position coordinates, the extended PJP theorem is satisfied. This theorem becomes particularly practical to identify all the slow variables of a system, which contribute to the observed fluctuations and the relaxation to a small external stimulus.

Similarly, this theorem applies to processive molecular motors [44,45] that have nonlinear dynamics and do not obey energy conservation because ATP is used as a source of energy. When an optical tweezers traps a single motor, it acts like an elastic spring opposing the motor motion. As demonstrated in the PJP paper [11], the Langevin equation is a good model for describing the position of the motor $x(t)$ subject to stochastic fluctuations $\eta(t)$

$$\dot{x}(t) = -k[x(t) - x_{ss}] + \eta(t)$$

where $x(t)$ is the position from the center of the trap of a small latex bead that is covalently attached to molecular motor. The random function $\eta(t)$ encapsulates both the complex stochastic behavior of the motor and the effect of thermal noise

$$\langle \eta(t) \rangle = 0, \ \langle \eta(t)\eta(t') \rangle = 2D\delta(t - t'),$$

Here k^{-1} is the relaxation time of this system, and $\sigma^2 = D/k$. In the case of a motor trapped in an optical tweezers, the potential $\phi(c; \lambda)$ has the following analytical expression:

$$\phi(x) = -\log\rho_{ss} = \frac{(x - x_s)^2}{2\sigma^2} + \frac{1}{2}\log(2\pi\sigma^2)$$

Using this potential, one can easily verify if the PJP fluctuation–response relationship (Equation (8.20)) is satisfied.

8.4.3 Conjecture: Fluctuation–Response Relationship and Its Implication on Single-Cell Behavior

While the formulation of the PJP fluctuation-response theorem is convenient for *in vitro* systems whose potentials can be expressed analytically, it is less useful for living cells whose potentials are not directly accessible. However, as long as the underlying dynamic processes can be approximated by Markovian statistics and the variables under study, such as the concentrations of signaling proteins or the expression level of a specific gene, have a well-defined steady state, the PJP fluctuation-response relationship should also be, in principle, valid for living organisms.

Applying the PJP fluctuation-response theorem to living cells means that we can formally write down a general relationship between the spontaneous fluctuations and the response function of the system to a small external perturbation. To illustrate the potential significance of the PJP fluctuation-response theorem in living cells, we will use bacterial chemotaxis (see Section 8.3.4). In *E. coli*, chemotaxis is one of the few biological systems in which both spontaneous fluctuations and response to a small external perturbation can be measured with high precision from individual living cells (see Chapter 13). Such accuracy is possible because the fluctuations of switching events between CW and CCW rotation of the bacterial motor reflect the fluctuations in the concentration of the signaling protein CheYp. Therefore, the output signal of the chemotaxis network is directly measurable. Moreover, this output signal can be experimentally monitored over a long timescale with high temporal precision, and its associated variance of the noise σ^2 and correlation time τ_{cor} can be characterized by computing the autocorrelation function $C(t)$. In this system, the exposure to a small step of attractant defines the external perturbation, and the linear response can be characterized by the response function $\chi(t)$ and its associated response time, τ_{res}. In the regime of linear approximation, it is expected that the response time and the correlation time from the spontaneous fluctuations before stimulus are similar, but in practice we can only expect these two times to be proportional $\tau_{cor} \propto \tau_{res}$. This relationship, which is a consequence of the linear approximation alone, yields a first practical prediction that the response time can be inferred from the spontaneous fluctuations before stimulus.

If the chemotaxis system were at thermodynamic equilibrium, the standard FDT would be satisfied:

$$\chi(t) = -\beta \frac{d}{dt} C(t)$$

Additionally, the response function and the autocorrelation functions would be directly accessible from the measurements of the time series produced from the switching behavior between CW and CCW rotational states from an individual motor. However, as in the hair bundle example [12], the chemotaxis system likely violates the FDT because of the presence of energy-consuming mechanisms. Instead, the PJP theorem can be used to assume that there exists, in principle, a relationship between fluctuations and the response of the system to a small external stimulus. Alternatively, we can formally use a Langevin equation that couples the characteristics of the spontaneous fluctuations and cellular response to a small stimulus. The fluctuations of the switching behavior of individual motors are directly related to the fluctuations, δA, of the kinase activity CheA. Therefore, the behavior of the chemotaxis system can be coarse-grained by the Langevin equation

$$\dot{\delta A^*} = -\frac{1}{\tau_{cor}} \delta A + \sqrt{D} \delta \eta(t) \qquad (8.21)$$

where $\sqrt{D} \delta \eta(t)$ is an input random function that has the characteristics of white noise and an amplitude D, and τ_{cor} is the measured correlation time in the output of the signaling system. In this phenomenological model, there exists a strict relationship between the noise amplitude of the kinase activity σ^2_{CheA} and the correlation time of the system τ_{cor} such that $\sigma^2_{CheA} = \frac{D}{2} \tau_{cor}$. Importantly, because $\tau_{cor} \propto \tau_{res}$, the noise σ^2_{CheA} that governs the behavioral variability is coupled to the cellular response τ_{res} through the coefficient D. The coupling coefficient D can depend on the cellular state in a very complex way. Unfortunately, we do not have access to all variables of this living system, and it is not realistic to predict theoretically how D depends on the cellular state. However, if the general PJP fluctuation-response theorem is valid, it implies, *in principle*, the existence of a coefficient D that governs the coupling between noise and response time for each cellular state. In chemotaxis, the noise of the kinase CheA is reflected in the behavioral variability of the motor behavior of a single bacterium, and the coefficient D should govern the coupling between the cellular response and the behavioral variability of bacteria as stated by the Langevin equation.

While it is difficult to evaluate how the coefficient D varies with different cellular states, it is straightforward to compute it for different concentrations of the chemotaxis proteins using a standard kinetic model of chemotaxis. For example, in a simple stochastic model, D was modeled using the fact that the adaptation mechanism is governed by a futile cycle [41]. Futile cycles are ubiquitous in signaling cascades and usually consist of two catalysts acting antagonistically to regulate the activity of a kinase protein.

In brief, the two catalyst enzymes act as a sort of push–pull mechanism to control the activity of the kinase CheA [46]. In chemotaxis, the coefficient D represents the strength of the spontaneous fluctuations taking place within a futile cycle associated with the methylation and demethylation reactions of receptors and was explicitly calculated in Emonet and Cluzel [41]:

$$D \sim b \cdot A^* + r \cdot A$$

where $b = \dfrac{k_b \varepsilon_{bp}}{K_b + A^*}$ and $r = \dfrac{k_r \varepsilon_r}{K_r + A}$, and where A^* is the concentration of free active kinase, ε_{bp} is the concentration of CheB-P, K_b is the Michaelis-Menten constant, k_b is the catalytic rate associated with demethylation and A is the concentration of free inactive kinase, ε_r is the concentration of CheR, K_r is the Michaelis-Menten constant, and k_r is the catalytic rate associated with methylation. Importantly, the coefficient D has strictly the same form as in [47], where D was independently calculated for a vertebrate photo-transduction cascade. This similarity results from the fact that the same futile cycle is used as an adaptive mechanism in both signaling cascades. A wide range of other signaling pathways use similar cycles, such as MAP-kinase pathways, so it is tempting to hypothesize that for most of these pathways the coupling between noise and response time should exhibit similar properties as that predicted in the chemotaxis signaling pathway. For example, in bacterial chemotaxis, we theoretically found that D varies weakly with the CheA kinase activity. As a result, we can predict that the cellular response scales linearly with behavioral variability measured before stimulus. In other words, this prediction implies that cells with the largest noise would also exhibit the longest response to a small external stimulus. It would certainly be interesting to investigate this prediction experimentally in a range of biological systems to determine when and how spontaneous fluctuations are coupled to the cellular response.

Although the discussion of this last section is highly speculative, it highlights the possibility that beyond the chemotaxis network, the PJP fluctuation-response theorem and the Langevin equation could provide a useful and general framework to characterize how noise and cellular response could be coupled in living organisms.

Acknowledgments

This chapter reflects discussions and ongoing work with Thierry Emonet, Heungwon Park, and John Marko on the conjecture relating fluctuation and response in single cells (Section 8.4.3), Jeff Moffitt and Calin Guet on technical advances of FCS (section 8.2.3), and Kevin Wood and Arvind Subramaniam on the Langevin equation (Section 8.3.1). Wendy Grus provided editorial assistance.

References

1. Brown, R. 1866. A brief account of microscopical observations made in the months of June, July and August, 1827, on the particles contained in the pollen of plants; and on the general existence of active molecules in organic and inorganic bodies. In *The miscellaneous botanical works of Robert Brown.* J. Bennett, editor. R. Hardwicke, London.

Chapter 8

2. Einstein, A. 1905. Über die von der molekularkinetischen Theorie der Wärme geforderte Bewegung von in ruhenden Flüssigkeiten suspendierten Teilchen. *Annalen der Physik* 322:549–560.

3. Elson, E. L., and D. Magde. 1974. Fluorescence correlation spectroscopy .1. conceptual basis and theory. *Biopolymers* 13:1–27.

4. Magde, D., E. L. Elson, and W. W. Webb. 1974. Fluorescence correlation spectroscopy .2. experimental realization. *Biopolymers* 13:29–61.

5. Magde, D., W. W. Webb, and E. Elson. 1972. Thermodynamic fluctuations in a reacting system—measurement by fluorescence correlation spectroscopy. *Physical Review Letters* 29:705–708.

6. Guet, C. C., L. Bruneaux, T. L. Min, D. Siegal-Gaskins, I. Figueroa, T. Emonet, and P. Cluzel. 2008. Minimally invasive determination of mRNA concentration in single living bacteria. *Nucleic Acids Research* 36–73.

7. Le, T. T., S. Harlepp, C. C. Guet, K. Dittmar, T. Emonet, T. Pan, and P. Cluzel. 2005. Real-time RNA profiling within a single bacterium. *Proceedings of the National Academy of Sciences of the United States of America* 102:9160–9164.

8. Ozbudak, E. M., M. Thattai, I. Kurtser, A. D. Grossman, and A. van Oudenaarden. 2002. Regulation of noise in the expression of a single gene. *Nature Genet* 31:69–73.

9. Elowitz, M. B., A. J. Levine, E. D. Siggia, and P. S. Swain. 2002. Stochastic gene expression in a single cell. *Science* 297:1183–1186.

10. Dunlop, M. J., R. S. Cox, J. H. Levine, R. M. Murray, and M. B. Elowitz. 2008. Regulatory activity revealed by dynamic correlations in gene expression noise. *Nature Genet* 40:1493–1498.

11. Prost, J., J. F. Joanny, and J. M. Parrondo. 2009. Generalized fluctuation–dissipation theorem for steady-state systems. *Physical Review Letters* 103:090601.

12. Martin, P., A. J. Hudspeth, and F. Julicher. 2001. Comparison of a hair bundle's spontaneous oscillations with its response to mechanical stimulation reveals the underlying active process. *Proceedings of the National Academy of Sciences of the United States of America* 98:14380–14385.

13. Krichevsky, O., and G. Bonnet. 2002. Fluorescence correlation spectroscopy: the technique and its applications. *Reports on Progress in Physics* 65:251–297.

14. Meacci, G., J. Ries, E. Fischer-Friedrich, N. Kahya, P. Schwille, and K. Kruse. 2006. Mobility of Min-proteins in *Escherichia coli* measured by fluorescence correlation spectroscopy. *Physical Biology* 3:255–263.

15. Bacia, K., S. A. Kim, and P. Schwille. 2006. Fluorescence cross-correlation spectroscopy in living cells. *Nature Methods* 3:83–89.

16. Cluzel, P., M. Surette, and S. Leibler. 2000. An ultrasensitive bacterial motor revealed by monitoring signaling proteins in single cells. *Science* 287:1652–1655.

17. Politz, J. C., E. S. Browne, D. E. Wolf, and T. Pederson. 1998. Intranuclear diffusion and hybridization state of oligonucleotides measured by fluorescence correlation spectroscopy in living cells. *Proceedings of the National Academy of Sciences of the United States of America* 95:6043–6048.

18. Alon, U., L. Camarena, M. G. Surette, B. A. Y. Arcas, Y. Liu, S. Leibler, and J. B. Stock. 1998. Response regulator output in bacterial chemotaxis. *EMBO Journal* 17:4238–4248.

19. Kuo, S. C., and D. E. Koshland. 1989. Multiple kinetic states for the flagellar motor switch. *Journal of Bacteriology* 171:6279–6287.

20. Scharf, B. E., K. A. Fahrner, L. Turner, and H. C. Berg. 1998. Control of direction of flagellar rotation in bacterial chemotaxis. *Proceedings of the National Academy of Sciences of the United States of America* 95:201–206.

21. Bertrand, E., P. Chartrand, M. Schaefer, S. M. Shenoy, R. H. Singer, and R. M. Long. 1998. Localization of ASH1 mRNA particles in living yeast. *Molecular Cell* 2:437–445.

22. Golding, I., and E. C. Cox. 2004. RNA dynamics in live *Escherichia coli* cells. *Proceedings of the National Academy of Sciences of the United States of America* 101:11310–11315.

23. Rauer, B., E. Neumann, J. Widengren, and R. Rigler. 1996. Fluorescence correlation spectrometry of the interaction kinetics of tetramethylrhodamin alpha-bungarotoxin with *Torpedo californica* acetylcholine receptor. *Biophysical Chemistry* 58:3–12.

24. Llopis, P. M., A. F. Jackson, O. Sliusarenko, I. Surovtsev, J. Heinritz, T. Emonet, and C. Jacobs-Wagner. 2010. Spatial organization of the flow of genetic information in bacteria. *Nature* 466: U77–U90.

25. Golding, I., J. Paulsson, S. M. Zawilski, and E. C. Cox. 2005. Real-time kinetics of gene activity in individual bacteria. *Cell* 123:1025–1036.

26. Burkhardt, M., and P. Schwille. 2006. Electron multiplying CCD based detection for spatially resolved fluorescence correlation spectroscopy. *Optics Express* 14:5013–5020.

27. Kannan, B., J. Y. Har, P. Liu, I. Maruyama, J. L. Ding, and T. Wohland. 2006. Electron multiplying charge-coupled device camera based fluorescence correlation spectroscopy. *Analytical Chemistry* 78:3444–3451.

28. Heuvelman, G., F. Erdel, M. Wachsmuth, and K. Rippe. 2009. Analysis of protein mobilities and interactions in living cells by multifocal fluorescence fluctuation microscopy. *European Biophysics Journal with Biophysics Letters* 38:813–828.

29. Needleman, D. J., Y. Q. Xu, and T. J. Mitchison. 2009. Pin-hole array correlation imaging: highly parallel fluorescence correlation spectroscopy. *Biophysical Journal* 96:5050–5059.

30. Bustamante, C., J. C. Macosko, and G. J. L. Wuite. 2000. Grabbing the cat by the tail: Manipulating molecules one by one. *Natural Reviews Molecular Cell Biology* 1:130–136.

31. Wang, M. D., H. Yin, R. Landick, J. Gelles, and S. M. Block. 1997. Stretching DNA with optical tweezers. *Biophysical Journal* 72:1335–1346.

32. Perkins, T. T., D. E. Smith, R. G. Larson, and S. Chu. 1995. Stretching of a single tethered polymer in a uniform-flow. *Science* 268:83–87.

33. Cluzel, P., A. Lebrun, C. Heller, R. Lavery, J. L. Viovy, D. Chatenay, and F. Caron. 1996. DNA: An extensible molecule. *Science* 271:792–794.

34. Bialek, W., and S. Setayeshgar. 2005. Physical limits to biochemical signaling. *Proceedings of the National Academy of Sciences of the United States of America* 102:10040–10045.

35. Langevin, P. 1908. The theory of brownian movement. *Comptes Rendus Hebdomadaires Des Seances De L Academie Des Sciences* 146:530–533.

36. Swain, P. S., M. B. Elowitz, and E. D. Siggia. 2002. Intrinsic and extrinsic contributions to stochasticity in gene expression. *Proceedings of the National Academy of Sciences of the United States of America* 99:12795–12800.

37. Rosenfeld, N., J. W. Young, U. Alon, P. S. Swain, and M. B. Elowitz. 2005. Gene regulation at the single-cell level. *Science* 307:1962–1965.

38. Austin, D. W., M. S. Allen, J. M. McCollum, R. D. Dar, J. R. Wilgus, G. S. Sayler, N. F. Samatova, C. D. Cox, and M. L. Simpson. 2006. Gene network shaping of inherent noise spectra. *Nature* 439:608–611.

39. Mangan, S., and U. Alon. 2003. Structure and function of the feed-forward loop network motif. *Proceedings of the National Academy of Sciences of the United States of America* 100:11980–11985.

40. Adler, J. 1966. Chemotaxis in bacteria. *Science* 153:708–716.

41. Emonet, T., and P. Cluzel. 2008. Relationship between cellular response and behavioral variability in bacterial chemotaxism. *Proceedings of the National Academy of Sciences of the United States of America* 105:3304–3309.

42. Korobkova, E., T. Emonet, J. M. G. Vilar, T. S. Shimizu, and P. Cluzel. 2004. From molecular noise to behavioural variability in a single bacterium. *Nature* 428:574–578.

43. Block, S. M., J. E. Segall, and H. C. Berg. 1983. Adaptation kinetics in bacterial chemotaxis. *Journal of Bacteriology* 154:312–323.

44. Svoboda, K., C. F. Schmidt, B. J. Schnapp, and S. M. Block. 1993. Direct observation of kinesin stepping by optical trapping interferometry. *Nature* 365:721–727.

45. Visscher, K., M. J. Schnitzer, and S. M. Block. 1999. Single kinesin molecules studied with a molecular force clamp. *Nature* 400:184–189.

46. Goldbeter, A., and D. E. Koshland. 1981. An amplified sensitivity arising from covalent modification in biological systems. *Proceedings of the National Academy of Sciences of the United States of America* 78:6840–6844.

47. Detwiler, P. B., S. Ramanathan, A. Sengupta, and B. I. Shraiman. 2000. Engineering aspects of enzymatic signal transduction: photoreceptors in the retina. *Biophysical Journal* 79:2801–2817.

Chapter 8

9. Modeling Genetic Parts for Synthetic Biology

Ying-Ja Chen
Kevin Clancy
Christopher A. Voigt

Chapter 9

Quantitative Biology: From Molecular to Cellular Systems edited by Michael E. Wall © 2012 CRC Press / Taylor & Francis Group, LLC. ISBN: 978-1-4398-2722-2

9.1   Introduction to Genetic Parts

Genetic parts are the most fundamental units of DNA in synthetic biology. They are defined by their sequence, which can be inserted into a specific location upstream, downstream, or within a gene to achieve controllable expression. By changing the DNA sequence of a part, the expression properties of its target gene such as expression level, specificity to a regulator, or response time can be altered. The precise control of these properties with parts is vital for connecting and rewiring genes into synthetic genetic circuits [1,2].

Synthetic genetic circuits are biological programs that perform desired function by connecting genes with defined functions in a way different from their natural pathway [1]. For example, by linking the activity of a membrane receptor that senses a small molecule in the environment to a reporter gene that emits light as an output, cells can be rewired to emit light when the small molecule is present. In this circuit, the connection between sensor activity and reporter expression must be controlled precisely. This is controlled by genetic parts, such as the promoter of the reporter activated by the sensor protein, whose specificity and strength determine when the reporter will be expressed and the magnitude of expression. In this circuit, the expression of the reporter must not be so strong that even its basal expression level emits detectable light, nor should it be so weak that upon full activation the emitted light is still undetectable. This dynamic range and the variability of gene expression are controlled by genetic parts. In addition, genetic parts can control the response time of a downstream gene triggered by an upstream signal. For example, by selecting a part that targets mRNA degradation of a gene, the gene will be downregulated faster than the same gene repressed through transcription by a different part. The time delay for an output to be activated upon an input signal and the duration of output can be controlled by a combination of different genetic parts.

Several properties are desirable for genetic parts to be useful for the control of expression in circuits [2]. First, the effect of the target gene expression controlled by the part must be reliable and reproducible. Second, parts must be tunable to conform to different desired outcomes when used to control different circuits. Third, when building large genetic programs, many parts of similar function, but different sequences, are required. Therefore, it is desirable to design libraries of parts that are modular, such that similar parts can be used in different modules of a more complex design. Fourth, when the same basic function is used multiple times, it is ideal to use parts that are orthogonal; in other words, they can be used simultaneously in a cell without cross-reacting. For example, orthogonal ribosomes are designed to recognize specific ribosome binding sites to selectively turn on target genes [3–8]. Finally, independence from the host cell is required so that the parts do not interfere with the host machinery.

To better design genetic parts with these properties, theory is needed to relate the sequence of a part to its function. Biophysical models are useful for this purpose. These models apply thermodynamics, kinetics, statistical mechanics, or non-linear dynamics to predict the quantitative behavior of a part. While bioinformatic algorithms analyze sequences of large protein–DNA or protein–protein interaction networks statistically, biophysical models can predict the mechanistic function of one part in its context [9–12]. These biophysical models can generate parameters necessary for models that characterize the dynamics of small gene regulatory networks. This hierarchical approach provides the simulations necessary for parts to be assembled into larger genetic networks.

With predictable function and behavior, parts can be selected and incorporated into larger *in silico* genetic circuit models on a platform provided by computer-aided design tools. Using these tools, the behavior of genetic circuits contributed by different parts can be simulated for designs of diverse functions [13,14]. The pathway from parts to programs is relevant for many applications, including the production of biofuels, biomaterials, diagnostic sensors, and therapeutics [15].

This chapter focuses on biophysical models to predict the functions of genetic parts. The models reviewed are mostly constructed in bacteria, although in vitro studies performed from eukaryotes that are relevant to prokaryotic systems are included. Figure 9.1 illustrates the genetic parts that will be introduced in this chapter. Since genes are transcribed into mRNA and translated into proteins, they can be regulated at each step. Promoters and terminators control gene expression at the transcriptional level by regulating initiation and termination, respectively. mRNA stabilizing hairpins provide regulation at the post-transcriptional level by directing cleavage, degradation, or stabilization of the target mRNA. The ribosome binding site (RBS) provides regulation at the level of translation by controlling the initiation rate. The activity of a protein also depends on post-translational regulations, such as protein folding, degradation, and post-translational modifications, which are not covered here. The following sections describe (1) the basic structural and molecular mechanism of how the part affects gene expression, (2) biophysical models developed to predict the behavior of the part, and (3) a synopsis of the current state and future prospect for the precise control of that part.

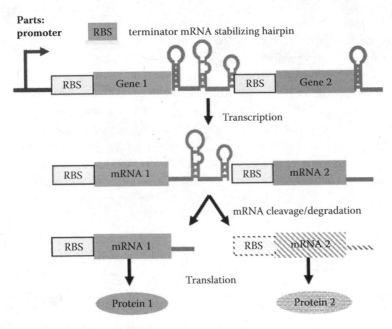

FIGURE 9.1   Genetic parts control gene expression at different steps. Genetic parts that will be introduced in this chapter are listed on the top. The expression of one gene or operon can be regulated by several genetic parts. Promoters and terminators control the initiation and termination of transcription. mRNA stabilizing hairpins regulate expression posttranscriptionally by inducing mRNA degradation, which results in decreased mRNA amount as shown for mRNA 2. Ribosome binding sites (RBS) regulate translation initiation and the protein production rate.

9.2 Promoters

9.2.1 Promoter Structure and Function

Promoters are DNA sequences upstream of genes where RNA polymerase binds and initiates transcription, which is a key step to activate gene expression. Transcription initiation involves three major steps (Figure 9.2). First, RNA polymerase binds to the promoter. After binding to the promoter, RNA polymerase melts the double-stranded DNA to form an open complex and enables ribonucleotides to polymerize. Subsequently, RNA polymerase leaves the promoter region and moves downstream to continue with transcription. Transcription enters elongation phase after these three steps are completed. Each of these steps can be regulated by transcription factors to affect promoter activity. Many transcription factors regulate transcription by binding to operator sequences in the promoter region to block or enhance RNA polymerase binding.

Promoter sequence determines its binding affinity with RNA polymerase. RNA polymerase is a multi-subunit protein containing subunits $\alpha_2\beta\beta'\omega$ as its core and a dissociable σ subunit that recognizes the promoter. With the exception of σ^{54}, most σ factors are in the σ^{70} family, which can be categorized into four groups according to their domain structure and sequence conservation, each recognizing different promoter sequences [16,17]. All bacteria contain a group 1 σ factor, which regulates housekeeping genes, for instance, σ^{70} in *Escherichia coli*. In addition, many alternative σ factors belong to groups 2–4. These σ factors regulate cellular responses to different developmental or environmental signals through their recognition of specific promoter sequences. In response to the signals, each σ factor activates selective promoters where its specific recognition sequence is present [18,19]. In general, σ factors recognize two regions, the –35 and the –10 motifs, but some of them also recognize extended regions of the promoter. Here, –35 and –10 refer to their positions upstream of the transcriptional start site (+1). A motif is a short DNA or protein sequence with a specific function, in this case, binding ability to RNA polymerase. Take σ^{70} as an example: the consensus sequence in the –10 motif is TATAAT, and the consensus sequence in the –35 motif is TTGACA. These two motifs are usually spaced 17 bp apart. The σ factor may also interact with a 4 bp extended –10 motif 1 bp upstream of the –10 motif (Figure 9.3a). In addition, RNA polymerases interact with an UP-element upstream of the –35 motif that is rich in adenine and thymine nucleotides through their

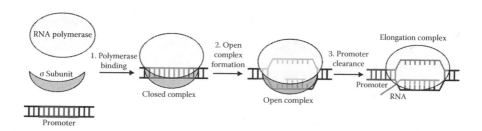

FIGURE 9.2 (See color insert.) Transcription initiation consists of three major steps. (1) In polymerase binding, RNA polymerase and the σ subunit assemble onto the promoter. (2) In open complex formation, the promoter DNA is melted by RNA polymerase in preparation for transcribing single-stranded mRNA. (3) In promoter clearance, RNA polymerase slides downstream from the promoter, releases σ, and continues to transcribe as an elongation complex.

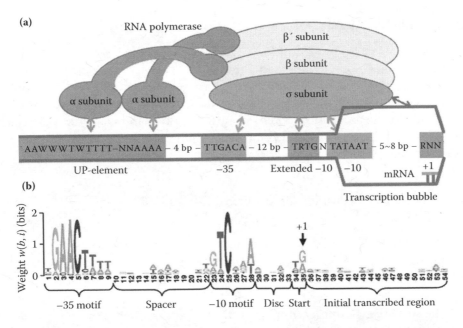

FIGURE 9.3 (See color insert.) The consensus sequence of a promoter can be represented by a position weight matrix (PWM). (a) The interaction between RNA polymerase and promoter is illustrated. The consensus sequence of the σ^{70} promoter includes a −35, a −10, and sometimes an extended −10 region that interacts with the σ subunit of RNA polymerase. Another AT-rich region called the UP-element interacts with the α subunit. (b) An example sequence for the σ^{E} promoter is represented by a position weight matrix in sequence logo form. Bases that occur more frequently in promoters are represented by larger sequence logos. The unit, bits, is adapted from information theory as a measure of information content. (From Rhodius and Mutalik [44]. With permission.)

α subunits [20]. Most σ^{70} promoters are only partially conserved with the consensus sequence because many of them are regulated by transcription factors, which may alter the binding of RNA polymerase to promoters to activate or repress gene expression.

The next key step in transcription initiation is the opening of the DNA double helix induced by RNA polymerase. This process, known as promoter melting or open complex formation, opens double-stranded DNA into single-stranded form from the −11 to the +2 position to form a transcription bubble. The mechanism for the open complex formation has not been completely determined; however, it is thought to be influenced by the untwisting of the DNA double helix and stabilized through polymerase interactions with the non-template strand [21,22].

The final step for transcription initiation is promoter clearance, also called promoter escape, referring to RNA polymerase escaping from the promoter region and entering into the elongation phase. In this step, RNA polymerase iteratively synthesizes and releases transcripts of 2–12 nucleotides without leaving the promoter in a process called abortive initiation [21]. As each nucleotide is incorporated, the polymerase scrunches on DNA by pulling downstream DNA into itself without releasing the upstream DNA. As DNA unwinds in this process, stress accumulates and may cause the initial transcripts to be ejected from the complex leading to abortive initiation. Eventually, stress provides a driving force for RNA polymerase to release the σ factor and the promoter resulting in promoter clearance and transcription elongation [23–25].

Chapter 9

9.2.2 Promoter Models

Promoters are modeled using different methods to decipher sequence and strength. Bioinformatic and thermodynamic methods have been used to locate promoter positions in the genome. Position weight matrices are simple methods used to estimate the binding strength between RNA polymerase and promoters based on sequence homology. Kinetic models predict promoter strength by taking into account the mechanism of transcription initiation including binding of RNA polymerase to the promoter, open complex formation, and/or promoter escape.

9.2.3 Predicting Promoter Locations in the Genome

Promoter locations in the genome are difficult to predict for several reasons. First, transcription start sites do not possess obvious features similar to the translation start codon, which mostly begins at ATG. Second, promoter motifs are of variable distance upstream of the transcription start site from a few to several hundred bases. Third, up to half of the positions in the promoter motifs deviate from the consensus promoter sequence. Most programs use machine learning algorithms that can infer promoter locations after determining the start of a gene using characteristics of ribosome binding sites [26–30]. In these methods, a set of known promoter sequences is used to train computers to find sequence homology, mononucleotide distribution, or other sequence features common to this training data set. Subsequently, these features are used by the computer to predict new promoter locations in the genome. Although these methods achieve a high sensitivity for detecting known promoters, the specificity is often low; in other words, most of the predicted promoters are false positives and not real promoters. Although not all of them have been verified by experimental data, these predictions are compiled in databases and regarded as the transcriptional start sites [31,32].

Besides sequence properties, DNA physical properties can also be used to predict promoter location [33–38]. While the conserved promoter sequence determines the binding energy of a promoter with the σ factor, other properties such as DNA duplex stability, B to Z conformation transition, and helix flexibility may affect the formation of RNA polymerase open complex [33]. For example, the thermal stability of DNA regions upstream of transcription start sites is lower than the remainder of the genome [34]. This property can potentially predict promoter locations that are more probable to form a transcriptional open complex.

Promoter locations may be predicted with higher accuracy by combining sequence information with DNA physical properties. One way to combine the two forms of information is to evaluate DNA duplex stability by dinucleotide thermodynamic parameters and feed this score into a machine learning model to incorporate sequence information. This method can improve both the sensitivity and specificity for promoter prediction [35].

9.2.4 Promoter Strength Predictions

Promoter strengths can be modeled by accounting for each of the three steps in transcription initiation: promoter binding, open complex formation, and promoter clearance. Examples of models are given for each of these processes.

RNA polymerase binding to the promoter has been the focus of most studies because it determines not only promoter strength but also specificity of whether a DNA sequence is a promoter or not. RNA polymerase binding affinity can be evaluated by statistical analysis of known promoter sequences compared to random sequences. Sequences that promote polymerase binding may result in strong promoter strength. The promoter sequence can be represented by a position weight matrix (PWM), as shown in Figure 9.3b. This is a matrix of the four nucleotides by the number of positions in the promoter region, where each element represents the log-probability $w(b,i)$ of observing a particular nucleotide at a particular position normalized to the genome content [39–42],

$$w(b,i) = -\ln \frac{f_{b,i}}{f_{b,genome}} \tag{9.1}$$

Here, $f_{b,i}$ is the frequency of base b observed in position i over all promoters, and $f_{b,genome}$ is the frequency of base b in the genome, where i is the position of interest in the promoter. $w(b,i)$ has units in bits, which is a unit for information content in information theory. Using a PWM, the binding energy ΔG can be calculated by taking the sum of the weights from each base, assuming that each position contributes independently to the promoter activity.

$$\Delta G = \sum_{b,i} w(b,i) \cdot s(b,i) = W \cdot S \tag{9.2}$$

Here, $s(b,i) = 1$ if the base at position i is base b and 0 otherwise. Written in vector format, the binding energy is the Frobenius inner product of the weight matrix W and the nucleotide sequence matrix S. This is useful because many algorithms are developed to calculate W, but the predictabilities of these algorithms are evaluated by calculating the ΔG of promoters (Figure 9.4). However, this binding energy ΔG is in units of bits, which is a different unit from energies measured biochemically, which are usually in kcal/mol. Binding energy ΔG determines promoter strength when binding between RNA polymerase and the promoter is the limiting factor of transcription initiation, which is often the case if open complex formation and promoter clearance are fast.

The PWM method (hereafter simply referred to as PWM) is a simple and flexible method for finding potential promoters and predicting promoter strength. Potential promoter sequences can be determined by finding sequences with ΔG above a threshold derived from the ΔG of known promoters [41]. This binding energy can be correlated to promoter strength for promoter strength prediction. In PWM, the spacing between the −10 region and the −35 region can also be written as an additional term with weights corresponding to the spacing. PWM offers the flexibility to evaluate the binding energies of different promoter motifs separately and to combine the results for prediction. Figure 9.4 shows one example of using such method to evaluate the promoter strength of an alternative σ, $σ^E$ [43,44]. Intrinsic to the method, PWM allows the contribution of each base position and spacing parameter to promoter strength to be decomposed, analyzed individually, and combined for an overall prediction. It provides a prediction of promoter strength without considering the detailed mechanisms of open complex formation and promoter clearance, which may also affect promoter activity.

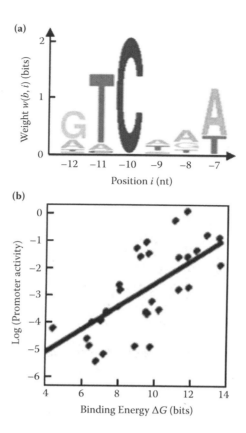

FIGURE 9.4 A position weight matrix (PWM) can be used to predict promoter strength. (a) The PWM of a –10 motif for the σ^E promoter. (b) PWM was used to predict promoter strength of σ^E promoters and compared with the promoter activity measured from a single round of in vitro transcription assay. A 10-fold cross-validation fit shows an optimized promoter model can predict the promoter strength with a correlation coefficient of 0.73. (From Rhodius and Mutalik [44]. With permission.)

9.2.5 Kinetic Models

The previous section introduced how promoter sequences can be represented by PWM to determine the binding energy of RNA polymerase to a promoter for promoter strength prediction. However, both sequence information and biophysical properties affect promoter activity. This section shows how these two properties can be incorporated into a comprehensive model. This can be achieved by using a kinetic model that describes the mechanism of open complex formation, where the DNA is opened and separated by RNA polymerase upon binding [45]. Consider the following scheme:

$$[RNAP]+[P] \underset{}{\overset{K_D}{\longleftrightarrow}} [RNAP-P]_c \overset{k_f}{\longrightarrow} [RNAP-P]_o \overset{k_c}{\longrightarrow} [RNAP]_e +[P] \tag{9.3}$$

Here, $[RNAP]$, $[P]$, $[RNA\text{-}P]_c$, $[RNAP\text{-}P]_o$, and $[RNAP]_e$ are the concentrations of free RNA polymerase, promoter, RNA polymerase–promoter complex in the closed form, the open complex, and escaped polymerase, respectively. RNA polymerase binds to the promoter with a dissociation constant K_D. Then, the open complex forms at rate k_f. After

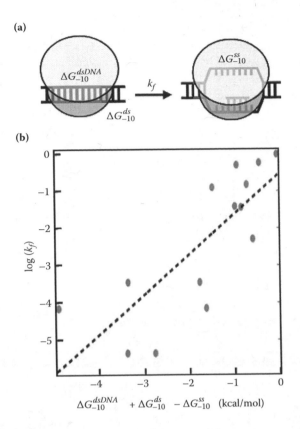

FIGURE 9.5 Kinetic models can depict the transcription initiation process to predict promoter strength. (a) A kinetic model depicting the open complex formation process can be used to predict promoter strengths when the start of transcription is rate-limited by this step. The open complex formation rate is related to free energies of interactions involved in the process. (b) The effective energy ($\Delta G_{-10}^{dsDNA} + \Delta G_{-10}^{ds} - \Delta G_{-10}^{ss}$) was calculated from the kinetic model. It is correlated to the logarithm of the open complex formation rate measured by an in vitro transcription kinetic assay with heparin. (From Djordjevic and Bundschuh [45]. With permission.)

transcription begins, RNA polymerase leaves the promoter with a promoter clearance rate k_c (Figure 9.5a).

Following Michaelis-Menten kinetics with the assumption that promoter clearance is an irreversible process, that open complex formation is the rate-limiting step ($k_f \ll k_c$), and that the promoters are not saturated by RNA polymerase binding ($K_D \gg$ [RNAP]), the transcription initiation rate r can be written as

$$r = \frac{k_f}{K_D} [\text{RNAP}] \tag{9.4}$$

K_D is related to the binding energy of RNA polymerase to the promoter

$$K_D = k_0 \exp\left(\frac{\Delta G_{-35}^{ds} + \Delta G_{spacing} + \Delta G_{-10}^{ds}}{k_B T} \right) \tag{9.5}$$

Chapter 9

where k_B is the Boltzmann constant, T is temperature, k_0 is a proportional constant, ΔG_{-35}^{ds}, ΔG_{-10}^{ds}, and $\Delta G_{spacing}$ are the binding energies of the -35 and -10 region, and the effective binding energy exerted by their spacing, respectively, which can be determined by PWMs. k_f is determined by the thermal melting of DNA and the interaction energies of RNA polymerase with double-stranded and single-stranded DNA in the -10 promoter region.

$$k_f = k_0 \exp\left(\frac{\Delta G_{-10}^{dsDNA} + \Delta G_{-10}^{ds} - \Delta G_{-10}^{ss}}{k_B T} \right),$$

(9.6)

where ΔG_{-10}^{dsDNA}, ΔG_{-10}^{ds}, and ΔG_{-10}^{ss} are the free energy of melting the DNA in the promoter, RNA polymerase interaction with the double-stranded DNA in the closed complex, and RNA polymerase interaction with the single-stranded DNA in the -10 promoter region in its open conformation, respectively. Here, the ΔG_{-10}^{ss} is subtracted because it corresponds to the stabilization of the open complex by RNA polymerase, whereas the other two terms stabilize the closed complex. By combining Equations 9.4, 9.5, and 9.6, the transcription initiation rate r is determined as

$$r = k_0 \exp\left(\frac{-\Delta G_{-35}^{ds} - \Delta G_{spacing} - \Delta G_{-12}^{ds} + \Delta G_{-10}^{dsDNA} - \Delta G_{-10}^{ss}}{k_B T} \right)$$

(9.7)

When combining the two terms, the interaction of the polymerase and the double-stranded -10 region (-12 to -7) partially cancel one another, and only the effect of the -12 position, ΔG_{-12}^{ds}, remains because the transcription bubble begins at the -11 position in the -10 region.

Using this expression, the transcription initiation rate can be predicted based on sequence information. The melting energy, ΔG_{-10}^{dsDNA}, can be calculated from the nearest-neighbor thermodynamic parameters for nucleic acids. The binding energy of the polymerase with the open complex, ΔG_{-10}^{ss}, can be calculated from biochemical measurements of mutants in the -10 region. The other binding energy terms, ΔG_{-35}^{ds}, $\Delta G_{spacing}$, and ΔG_{-12}^{ds}, can be calculated by PWMs. This analysis suggests that the rate-limiting step in open complex formation is the thermal melting of DNA and its stabilization by RNA polymerase, and these properties are captured with a mathematical model (Figure 9.5b). This provides a comprehensive framework to integrate other promoter prediction models using sequence information or physical properties.

A similar approach can be taken to describe the mechanism of promoter clearance or abortive initiation. During this process, the initial nucleotides polymerizes in the transcriptional open complex. However, RNA polymerase scrunches on the promoter by remaining bound to the promoter region while it opens downstream DNA template for transcription. The stress from polymerase scrunching causes initial short transcripts to be ejected from the complex as RNA polymerase returns to an initial state without proceeding in transcription. Eventually, polymerase releases the promoter and continues with transcription downstream. This process can be written as several competing reaction pathways after the first nucleotides are incorporated (Figure 9.6a). It involves

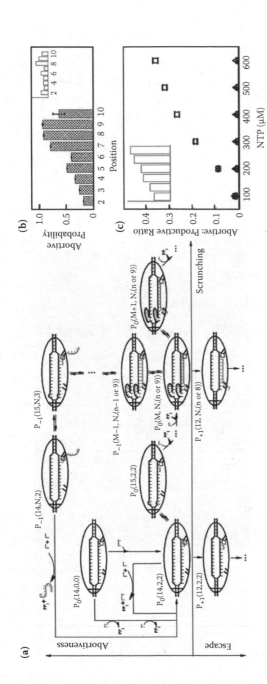

FIGURE 9.6 A kinetic model can depict several phenomena observed during abortive initiation. (a) The abortive, scrunching, and escape pathways are illustrated with the corresponding transcription initiation complex conformations. (b) The predicted abortive probability for each position for the promoter T5N25 is shown. The inset shows the experimentally measured abortive probability for each position using in vitro transcription assays followed by gel electrophoresis. (c) Assuming there is nucleotide-assisted release of nascent nucleotides in the abortive pathway (squares), the abortive to productive initiation ratio increases linearly as observed experimentally (inset). This is not observed without the nucleotide-assistance release assumption (triangles). (From Xue et al. [46]. With permission.)

the transcription initiation complex moving between different bubble conformations, which can be described by using a notation to specify different bubble conformations, $P_m(M, N, n)$. Here P_m denotes the reaction pathway that will be described below, M is the length of the bubble, N is the length of the nascent RNA, and n is the number of paired bases in the RNA–DNA hybrid. First, the initial nucleotide is incorporated. Then, RNA polymerase may enter the scrunching pathway (P_0). In this pathway, nucleotides are continuously incorporated and downstream DNA is opened one base at a time; however, RNA polymerase scrunches on the promoter without releasing it to double-stranded form. RNA polymerase may enter the escape pathway (P_{+1}) to escape from the promoter region and continues in elongation. Alternatively, RNA polymerase may undergo the abortive pathway (P_{-1}), where nascent RNA transcripts are released. In the last step of the abortive pathway, the nascent RNA is released from the complex with the assistance from the incorporation of the first two ribonucleotides, resulting in an initial complex ready to restart the initiation cycle.

By considering every conformation of the transcription initiation complex, the free energy of each state and the kinetic rates for the complex to move between each state can be calculated for a complete description of abortive initiation. The energy for each of the transcription initiation complex states can be calculated from the standard free energy of formation for each of the components:

$$\Delta G^m_{M,N,n} = \Delta G^{dsDNA}_{M,N,n} + \Delta G^{RNA:DNA}_{M,N,n} + \Delta G^{RNAP}_{M,N,n} \tag{9.8}$$

where $\Delta G^{dsDNA}_{M,N,n}$, $\Delta G^{RNA:DNA}_{M,N,n}$, and $\Delta G^{RNAP-DNA}_{M,N,n}$ are the free energy to open the DNA bubble, the RNA:DNA hybrid, and the binding energy between RNA polymerase and the promoter, respectively. The reaction rates of the scrunching and abortive pathways with respect to the nascent RNA length (k^N_m, m is pathway index) can be calculated following Michaelis-Menten kinetics with competitive inhibition.

$$k^N_m = \frac{k_{pol}[XTP]}{[XTP] + K^N_{d,m}} \tag{9.9}$$

where $[XTP]$ is the concentration of the next nucleotide to be incorporated, k_{pol} is the maximum reaction rate for RNA polymerization, and $K^N_{d,0}$ and $K^N_{d,-1}$ are the effective equilibrium dissociation constants for the scrunching and abortive pathways, respectively. $K^N_{d,0}$ and $K^N_{d,-1}$ can be calculated from the change in free energy between the current bubble conformation and all other possible bubble conformations to which it can move.

$$K^N_{d,m} = K_{d,m}\left[1 + \sum_i \exp\left(-\frac{\Delta G_i}{k_B T}\right)\right] \tag{9.10}$$

where ΔG_i represents the change of free energy between the current bubble conformation being considered and the i-th conformation to which it can move. The abortive pathway may include extra terms in its effective equilibrium constant because it involves the polymerization of the first and second nucleotide to remove the initial transcripts. Although for simplicity the exact expression is not shown here, it can be determined from Equations (9.9) and (9.10) with additional parameters including the concentration of the first two nucleotides and the dissociation constants of the first two nucleotides to the transcription initiation complex. The kinetic rate constant for the escape pathway can be calculated by Arrhenius kinetics using the free energy of the first escaped state, ΔG^{+1}, over the free energy of all other competing states (ΔG_j) as the activation energy and a constant A.

$$k_{+1}^N = \frac{A \cdot \exp\left(-\dfrac{\Delta G^{+1}}{k_B T}\right)}{\displaystyle\sum_j \exp\left(-\dfrac{\Delta G_j}{k_B T}\right)} \tag{9.11}$$

In an escaped state, the transcription bubble is 12 bp, and 9 bp of nascent RNA is hybridized to the DNA unless the transcript is shorter. Each of the aforementioned constants can be determined by experimental measured values, theoretical studies, or estimation.

This model explicitly expresses the reaction rate constants of the three competing pathways in abortive initiation by considering the free energy of associated transcriptional initiation complex conformations. This model can provide insights into how promoter sequence, position, and nucleotide concentration affect the abortive initiation process (Figures 9.6b and 9.6c) [46]. By understanding the parameters that govern abortive initiation or promoter escape, more efficient promoters may be designed that proceed through elongation at higher rates and may increase transcription rates or promoter strength.

9.2.6 Conclusion

When designing genetic circuits, promoters are the most widely used part because they determine when and how much a gene is transcribed. Promoter activity, which determines the transcription initiation rate, is controlled by the binding energy between RNA polymerase and the promoter, open complex formation, and promoter escape. Promoter strength can be predicted by methods accounting for each of these components. Position weight matrices are the simplest and most widespread models to predict promoter strength, and are based on sequence homology with known promoters. However, PWMs account for only the binding energy. On the other hand, thermodynamic models account for only the thermal melting of promoters. These two types of information can be combined by integrating thermodynamic parameters into a machine learning model to predict promoter locations within the genome, but this method has not been shown to predict promoter strength. Kinetic models account for each step in transcription initiation and can combine PWM, thermodynamic data, and *in vitro* biochemical measurements to predict promoter strength. However, this method is limited by the availability of relevant biochemical data that can accurately account for the interaction energy between RNA polymerase and promoter sequence. With the

Chapter 9

advent of experimental techniques that can characterize large quantities of sequences, these models can be trained and improved with new measurements of promoter activity. In addition, as more knowledge regarding the mechanism of open complex formation and promoter escape is elucidated, the models can be adopted to incorporate the full mechanism of transcription initiation to precisely predict promoter strength.

9.3 Transcriptional Terminators

9.3.1 Transcriptional Terminator Structure

Transcription ends when RNA polymerase reaches a DNA sequence encoding a terminator. Terminators are essential parts that isolate synthetic genetic constructs so that the expressions of neighboring genes do not interfere with each other. Bacterial terminators fall in two major categories. Intrinsic terminators are specific RNA sequences that signal for termination. ρ-dependent terminators involve a protein ρ that binds to the transcriptional elongation complex to regulate termination. Other mechanisms involve additional regulatory factors, such as viral proteins [47]. Terminators function by causing the elongating RNA polymerase to pause and destabilize through formation of a hairpin or the ρ protein. This results in its dissociation from the transcription bubble and the nascent RNA transcript is released [48,49].

The canonical structure of terminators is shown in Figure 9.7a. Intrinsic terminators usually contain a short GC-rich stem-loop followed by a stretch of U-rich sequence. The U-rich sequence forms a less stable rU·dA hybrid with the DNA template within the transcription bubble, causing RNA polymerase to pause. This allows the stem-loop in the nascent RNA to form. Stem-loop formation further destabilizes the transcriptional elongation complex, causing RNA polymerase to dissociate and transcription terminates [47,50,51]. The base composition in the stem must be GC-rich, especially toward the bottom of the stem. The stem-loops are usually short in length and may contain more A or T nucleotides if they are longer [52,53]. In addition, termination efficiency also depends on the sequence upstream of the stem-loop and downstream of the U-rich region, but their roles may involve interactions with proteins and are not well understood [54].

ρ-dependent terminators generally stretch over 200 nucleotides and are characterized by two less conserved regions. The first region is composed of a stretch of 40–85 C-rich nucleotides and is free of RNA secondary structures. This region is bound by ρ monomers to form a hexameric ring structure around the RNA. The second region is a transcriptional pause site located downstream. When RNA polymerase pauses, the ρ protein loads and unwinds RNA from the RNA–DNA hybrid through ATP hydrolysis, thereby releasing the nascent RNA [47].

9.3.2 Terminator Models

Current terminator models focus on predicting intrinsic terminators because they do not involve complex predictions of protein interactions and provide potential for designing functional terminators from DNA sequence. Intrinsic terminator loci in prokaryotes can be predicted by searching for potential hairpin structures that are more stable than

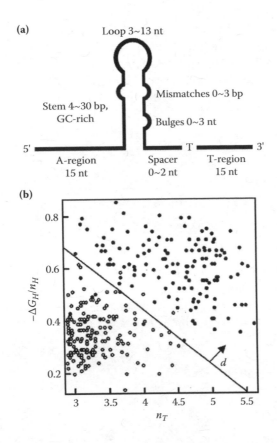

FIGURE 9.7 Terminators can be modeled by scoring a hairpin followed by a T-rich sequence. (a) The terminator structure consists of a hairpin with a short GC-rich stem and a small loop followed by the T-rich region. Short mismatches and bulges may be present. There can be a short spacer between the stem-loop and the T-region. For bidirectional terminators, the 5′-end should be an A-rich region. (b) Terminator (solid circles) and control (open circles) sequences are plotted on this plot by their hairpin scores ($-\Delta G_H/n_H$) versus T-scores (n_T). A decision line can be drawn to classify terminators from control. However, the termination efficiency cannot be determined easily from this plot. (From d'Aubenton Carafa et al. [52]. With permission.)

a threshold free energy [55–59]. These hairpins must be followed by T-rich sequences. Several conditions must be set to find hairpins that are strong enough to destabilize the elongation complex, can fold quickly before transcription proceeds beyond the terminator, and are in proximity to the U-rich region. A specific set of criteria is shown in Figure 9.7a and includes the following:

1. Hairpin structures followed by a stretch of T-rich sequences with at least 4 G·C base pairs in a 4–30 bp stem
2. Less than four mismatches and bulges allowed in the stem
3. A loop length of 3–13 nt
4. Less than 3 nt spacing between the hairpin and the first T
5. A minimum of 4 T's in the downstream region
6. A maximal total length of 75 nt
7. A minimum of 4 A's in the upstream regions for bidirectional terminators

Chapter 9

RNA secondary structure prediction programs can be used to compute the optimal hairpin structure in a given genome region by calculating the thermal stability of the potential hairpin structures [60,61]. The U-rich sequence can be scored using the following method. The term U-rich sequence is used interchangeably with T-region because U in the RNA corresponds to T in the DNA sequence. The T-region score is calculated as

$$n_T = \sum_{i=0}^{14} x_i$$

$$x_i = \begin{cases} x_{i-1} \cdot 0.9 & \text{if the i-th nucleotide is T} \\ x_{i-1} \cdot 0.6 & \text{if the i-th nucleotide is not T} \end{cases} \tag{9.12}$$

where n_T is the score for the T-region, and $x_0 = 0.9$ [52]. This scoring method assigns higher penalties to non-T nucleotides in the 5' position. To combine the hairpin score and the T-region score for terminator prediction, the statistical analysis of *E. coli* terminators were used to generate a decision rule:

$$d = 96.59 \left(\frac{-\Delta G_H}{n_H} \right) + 18.16 n_T - 116.87 \tag{9.13}$$

where ΔG_H is the Gibbs free energy of the hairpin, and n_H is the number of nucleotides in the hairpin [52]. Figure 9.7b graphically shows the *d* score on the plot of $-\Delta G_H/n_H$ versus n_T. This *d* score is positively correlated with termination efficiency and generates a simple decision rule (*d* > 0) to predict whether a given sequence is a terminator.

Another method to predict terminators is to score potential terminator sequences according to thermodynamic free energy. Here, the free energy of the T-region is subtracted from the free energy of the stem-loop (Figure 9.8). This is because the ability for the hairpin formation, which is crucial for the destabilization and dissociation of the transcription elongation complex, is dependent on the disruption of the RNA:DNA duplex within the T-region. In the T-region, nucleotides in proximity to the hairpin have more profound effects than those more downstream. Therefore, the T-region is divided into proximal, distal, and extra regions, each consisting of 3 nt.

$$\Delta G_{total} = \Delta G_{stemloop} - [\Delta G_{spacer} + \Delta G_{proximal}] - 0.5 \Delta G_{distal} - 0.01 \Delta G_{extra} \tag{9.14}$$

Here, ΔG_{total}, $\Delta G_{stemloop}$, $\Delta G_{proximal}$, ΔG_{distal}, and ΔG_{extra} are the total free energy score, and free energies of the stem-loop, proximal, distal, and extra T-regions, respectively. The free energy in the T-regions are calculated as the free energy for RNA:DNA duplex formation in these regions. ΔG_{spacer} is a free-energy score that penalizes any nucleotide that is not T between the stem-loop and the T-region. For the distal and extra T-regions to have reduced effects, they are weighted by 0.5 and 0.01, respectively [56].

Kinetic models have been constructed to explain the mechanism of transcriptional pausing [62,63]. Pausing is relevant to termination because it often occurs before the

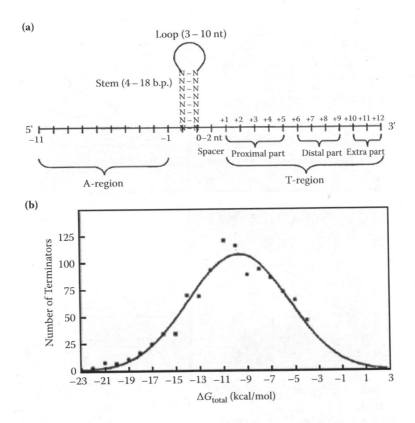

FIGURE 9.8 Thermodynamics can be used to predict terminators. (a) The terminator structure can be divided into the stem-loop and the T-region. The T-region can be further divided into proximal, distal, and extra regions according to their distance from the stem-loop. The spacer accounts for the distance between the stem-loop and the first T-nucleotide in the T-region. (b) Terminators are scored by combining the thermodynamic free energy of each of the components. This scoring method was applied to all 3'-ends of the *E. coli* genes. The number of terminators with each score is plotted here and is distributed as a Gaussian distribution with a threshold free energy score of −4 kcal/mol. (From Lesnik et al. [56]. With permission.)

transcriptional elongation complex dissociates at terminators. To describe the mechanism of pausing, let us first describe transcription elongation, which can be dissected into three steps: (1) nucleotide binding to the elongation complex; (2) nucleotide incorporation with the elongating RNA; and (3) translocation of the elongation complex to the next position for transcription (Figure 9.9a):

$$(N,1) + NTP \xleftrightarrow{K_d} (N,1) \cdot NTP \xrightarrow{k_{inc}} (N+1,0) \xleftrightarrow{K_{tr}} (N+1,1) \qquad (9.15)$$

Here, the state of the transcription elongation complex is represented by (N, m), where N is the transcript length, m is the translocation position, 0 is the pre-translocation position after nucleotide incorporation, and 1 is the mode when the previous nucleotide is fully incorporated, translocated, and ready to accept the next nucleotide. K_d is the dissociation constant of the nucleotide binding, K_{tr} is the equilibrium constant between the two translocation positions, and k_{inc} is the incorporation rate constant for the irreversible nucleotide incorporation including pyrophosphate release. K_d is determined by the

Chapter 9

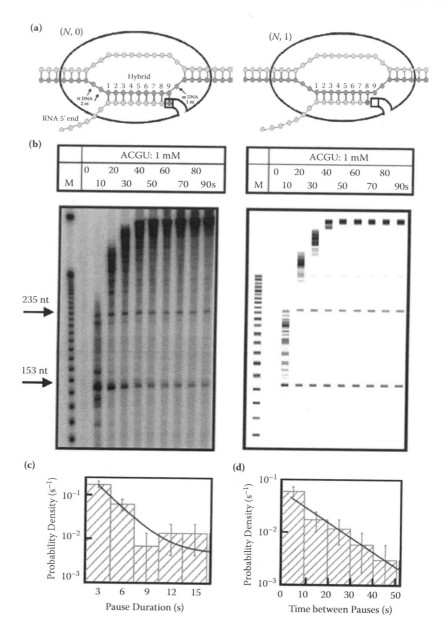

FIGURE 9.9   A kinetic model accounting for every transcription bubble conformation can be used to predict transcription pausing. (a) Transcription elongation complex is shown as an oval with the nucleotide incorporation site shown as a square. The template DNA, nontemplate DNA, and nascent RNA nucleotides are shown as top, middle, and bottom circles, respectively. Transcription elongation is a two-step process. First, a new nucleotide (right-most circle on the bottom strand) is incorporated to form a nascent RNA of length N. This is state $(N, 0)$. Next, the transcription elongation complex translocates to the next position, at state $(N, 1)$, ready for incorporation of the next nucleotide. (b) The kinetic model can be used to simulate transcription pausing visualized by gel electrophoresis. The left panel is a photograph from a bulk transcription assay of the *E. coli rpoB* gene. The arrows show the two pause sites at 153 and 235 nt, respectively. The right panel is the simulated transcription gel for the same sequence. A similar pattern including the two major pause sites can be seen. (c–d) The model can predict the probability of pause duration (c) and time between pauses (d) in single-molecule experiments. The bars are experimental data, and the line is the predicted pause duration or time between pauses. The two correlate well. (From Bai et al. [62]. With permission.)

free energy for the two transcriptional elongation complex conformations. Accounting for the translocation reaction that occurs right before the next nucleotide incorporation, the effective dissociation constant from a pretranslocated state $(N,0)$ to the pretrans-located state after the next nucleotide is incorporated $(N + 1,0)$ can be determined by Michaelis-Menten kinetics, and the elongation rate k_e is

$$k_e = \frac{k_{inc}[NTP]}{K_d\left[1+\exp\left(\dfrac{\Delta G_{N,1}-\Delta G_{N,0}}{k_B T}\right)\right]+[NTP]} \tag{9.16}$$

Here, $\Delta G_{N,m}$ is the free energy of the transcription elongation complex in state (N, m) and can be calculated from the free energy to open the transcription bubble ($\Delta G_{N,m}^{dsDNA}$), the RNA–DNA hybrid ($\Delta G_{N,m}^{RNA:DNA}$), and the binding energy between RNA polymerase and the transcription bubble ($\Delta G_{N,m}^{RNAP}$):

$$\Delta G_{N,m} = \Delta G_{N,m}^{dsDNA} + \Delta G_{N,m}^{RNA:DNA} + \Delta G_{N,m}^{RNAP} \tag{9.17}$$

During transcription elongation, RNA polymerase may backtrack along the DNA, causing transcription to pause. Instead of moving from $(N,0)$ to $(N,1)$, the transcription elongation complex moves from $(N,0)$ to $(N,-1)$, and may keep backtracking until the kinetic energy barrier to backtrack is high. The kinetics of backtracking from (N, m) to $(N, m-1)$ can be described by difference between the energy barrier $\Delta G_{N,m-1}^{\dagger}$ and the free energy of the initial mode:

$$k_{N,m\rightarrow m-1} = k_0 \exp\left(-\frac{\Delta G_{N,m-1}^{\dagger}-\Delta G_{N,m}}{k_B T}\right) \tag{9.18}$$

where $k_{N,m\rightarrow m\rightarrow 1}$ is the kinetic rate constant for backtracking from (N, m) to $(N, m - 1)$, and k_0 is a constant that accounts for the attack frequency for this energy barrier. $\Delta G_{N,m-1}^{\dagger}$ can be estimated and tuned according to a few assumptions. Backtracking energy barrier can be assumed to be constant because the transcript length remains the same and the transcription bubble conformation does not subject to significant change. On the other hand, a forward-tracking energy barrier may increase linearly because it includes the energy required to shorten the RNA–DNA hybrid in the transcription bubble. Although these parameters cannot be measured directly, reasonable estimates can be determined from transcription assays. The rate constants k_e and $k_{N,m\rightarrow m\rightarrow 1}$ determine the energy barrier, and therefore the probability, for the transcription elongation complex to undergo elongation or backtracking. Using these equations, the effective elongation rate at each nucleotide can be calculated for a given DNA sequence. Pause sites are nucleotide positions where the elongation rates are substantially slower than the average. Figures 9.9b, 9.9c, and 9.9d show how this model can be used to simulate pausing characteristics observed experimentally. This model takes into account each possible transcription bubble conformation and backtracking positions to describe pausing. It predicts the elongation kinetics during transcription, the pause positions, and pause durations at those positions [62,63]. Because transcriptional termination often occurs after transcription pauses, this model may be extended to predict terminators. This model provides a method that accounts for every possible transcription bubble conformation during transcription elongation and estimates the kinetic rate constant for the elongation complex to move between each conformation. It introduces a way to calculate the energy barrier for elongation, which can be used to calculate the kinetic rate of the elongation reaction competing with termination. This model may be extended to depict the mechanism for transcriptional termination and predict the termination efficiency.

9.3.3 Conclusion

It is desirable to predict and design intrinsic transcriptional terminators for incorporation into genetic circuits to prevent transcription from proceeding into downstream genes on synthetic constructs. These terminators, which do not depend on the protein factor ρ for termination, can be characterized by a short stem-loop structure followed by a stretch of U-rich sequence. The detailed sequence properties of known terminators have been analyzed statistically, and can predict the location of intrinsic terminators. While these methods cannot predict termination efficiency, a score correlated to termination efficiency can be generated. In some bacterial species, intrinsic terminators do not require the U-rich region, and additional effort is required to accurately predict such terminators [57,64]. Kinetic models have been used to predict transcription pause sites. A similar approach may be able to predict terminator positions and strength. Moreover, it would be useful to develop programs that design transcriptional terminators to attain the desired termination efficiency for use in larger genetic circuits.

9.4 Messenger RNA Stability

9.4.1 Mechanisms to Control mRNA Stability

RNA stability regulates the amount and the response time of gene expression. With low mRNA stability, the protein expression can be turned off by degrading all available mRNA. However, a highly transcribed gene with low mRNA stability results in a protein expression with a high turnover rate. Conversely, stable mRNA can maintain a sustaining basal protein expression. Several mechanisms can be used to design mRNA stability parts that regulate gene expression from this aspect.

The stability of mRNA is controlled by ribonucleases (RNases), small RNAs (sRNAs), and riboswitches. RNases are enzymes that control the degradation of mRNA by cleaving it internally or from one end in a single-stranded form or at double-stranded hairpins (Figure 9.10a). RNase E is the major endoribonuclease that cleaves mRNA internally from 5' to 3' and induces further degradation by other RNases. There is no known consensus cleavage site for RNase E; however, preference for cleavage at AU-rich regions has been observed. In addition, hairpin structures in the 5' untranslated region of mRNAs can protect against cleavage by RNase E [65].

Small noncoding RNAs (sRNAs) can regulate gene expression through base pairing with their specific mRNA targets (Figure 9.10b). These sRNAs are 50–200 nt RNA transcripts either cis-encoded on the antisense strand of the target gene, which transcribes

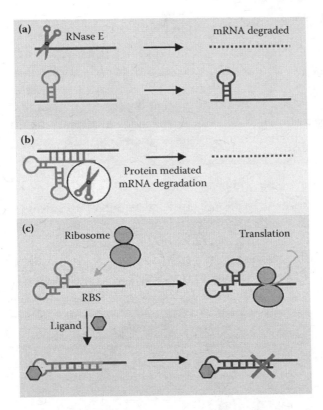

FIGURE 9.10   RNA stability is regulated by several small RNAs. The regulatory RNA parts are shown in violet. (a) A hairpin structure inserted to the 5'-untranslated region of an mRNA can protect the mRNA from being degraded by RNase E (illustrated by scissors). (b) sRNAs can base pair with their target mRNAs and induce degradation mediated by proteins (illustrated by a circle with scissors inside). (c) A riboswitch changes conformation upon binding to a ligand (octagon), which blocks the ribosome binding site and inhibits translation.

into a perfectly complementary RNA to its cognate mRNA, or trans-encoded at a distant location, which transcribes into a partially complementary RNA. They often regulate translation by base pairing with their target mRNA at their 5'-untranslated region close to the ribosome binding site. This base pairing may be mediated by the Hfq protein and can act in conjunction with RNases to degrade or cleave the target mRNA. Small RNAs can also regulate translation through blocking or unblocking ribosome binding by base pairing at the ribosome binding site or a secondary structure that inhibits ribosome binding. The base pairing regions are only 6–12 bp, but this can yield specific sRNA–mRNA interactions [66–71].

Riboswitches are RNAs that regulate expression by changing secondary structures upon binding to small molecule ligands (Figure 9.10c). They are often encoded on the 5'-end on the same RNA strand as their target mRNA or on a separate locus. They control gene expression by changing their base-pairing conformations in the presence of a small molecule ligand. In the absence of ligand, their most stable secondary structures may include part of a transcriptional terminator hairpin or the ribosome binding site. In the presence of ligand, the secondary structure changes to expose the terminator or ribosome binding site, resulting in transcriptional termination or translational initiation.

Chapter 9

9.4.2   RNA Stability and Regulation Models

Three types of RNA stability parts are introduced in this section. RNA hairpins or small regulatory RNAs can be designed to control mRNA stability through mathematical models that calculate the thermal stability of RNA hairpins designed to protect against RNase E cleavage. Mathematical models that describe the dynamics of sRNA and target mRNA transcription can explain the behavior of sRNA regulation. For the design of riboswitches, a kinetic model can relate their RNA sequence to their quantitative function.

9.4.2.1   RNase binding and protection sites

RNA hairpin structures in the 5'-untranslated region of an mRNA can protect against cleavage by the major endoribonuclease in *E. coli*, RNase E. Because RNase E cleavage triggers downstream RNA degradation mechanisms, this protective hairpin can increase mRNA stability. The thermal stability of a hairpin calculated by RNA secondary structure prediction programs may be used to predict the stability of an mRNA (Figure 9.11) [72]. However, the correlation is weak, and more information regarding the

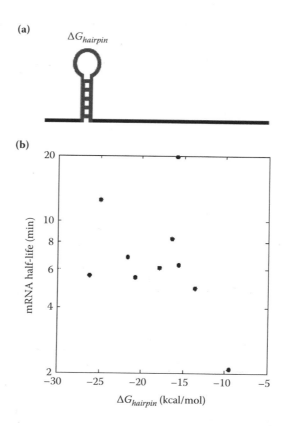

FIGURE 9.11   mRNA stability can be controlled by 5'-hairpin structures. (a) Different hairpin structures with different thermal stabilities calculated by their $\Delta G_{hairpin}$ can be added to the 5'-untranslated region to protect the mRNA from being cleaved by RNase E. (b) The effect of 5'-hairpins on mRNA half-life was measured from constructs containing synthetic hairpins in the 5'-untranslated region of a reporter protein using Northern blot analysis. The correlation between the free energies of the hairpins and the measured mRNA half-lives is plotted. (From Carrier and Keasling [72]. With permission.)

cleavage properties of RNase E is necessary for the accurate prediction of mRNA stability. By considering the correlation between thermal stability of a protective hairpin to the stability of the mRNA, mathematical models may facilitate the design of a hairpin part inserted at the 5'-end of mRNAs to control their stability.

9.4.2.2 sRNA models

sRNAs can target specific mRNAs and regulate their expression. In the genome, sRNAs are identified by searching for sequence conservation across species in intergenic regions associated with putative promoters and intrinsic terminators [73,74]. Other properties that can be used to find sRNAs include conserved secondary structure and high GC content, where stable RNA secondary structures are more likely to form. Another method is to search in evolutionarily conserved sequences for compensatory mutations. Potential sRNAs are conserved regions where one mutation can be compensated by another mutation that base pairs with that position in the RNA structure [75,76]. Once an sRNA is identified, its target RNA can be predicted by searching for mRNA sequences that can base-pair with the sRNA in their translation initiation regions [77].

The quantitative characteristics of sRNA regulation can be described by a mathematical model (Figure 9.12a) [78]. The concentration of the sRNA (s) and its target mRNA (m) can be written as two mass-action equations:

$$\frac{dm}{dt} = r_m - \beta_m m - kms$$

$$\frac{ds}{dt} = r_s - \beta_s s - kms \tag{9.19}$$

where r_m and r_s are the transcription rates of mRNA and sRNA, respectively, β_m and β_s are the turnover rates of the mRNA and sRNA, respectively, and k is the second-order kinetic constant that describes the coupled degradation between the sRNA and its target mRNA. By solving the equation under steady state, the steady state mRNA level m is

$$m = \frac{1}{2\beta_m}[(r_m - r_s - \lambda) + \sqrt{(r_m - r_s - \lambda)^2 + 4r_m\lambda}] \tag{9.20}$$

Here $\lambda = \beta_m\beta_s/k$, which characterizes the rate of mRNA turnover that is not due to the sRNA. When λ is reasonably small, the expression level of the target mRNA increases linearly with its own transcription rate beyond a threshold determined by the transcription rate of the sRNA (Figure 9.12b). When the mRNA transcription rate is lower than the sRNA transcription rate, all mRNAs are bound by sRNA and cannot be translated. As the mRNA transcription rate exceeds the sRNA transcription rate, the free mRNAs can translate into proteins with translation level linearly related to the concentration of the mRNA. This characteristic provides a guideline for controlling the expression level of a target protein by controlling the transcription rates of the regulatory sRNA with respect to mRNA [78]. However, this method is applicable only to known sRNA–mRNA pairs.

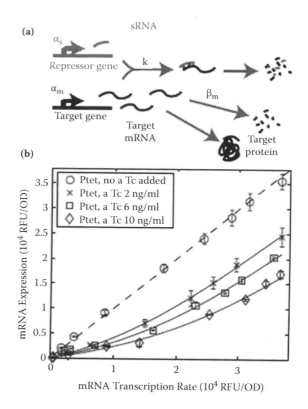

FIGURE 9.12 (See color insert.) The quantitative behavior of sRNAs can be modeled using a kinetic model. (a) The sRNA and the target mRNA genes are shown in red and black, respectively. Only the target mRNAs that are not bound by sRNAs can translate into the target protein. (b) The expressions of several sRNA–mRNA pairs show the predicted behavior. The expression of the target gene, *rhyB*, increases linearly with its transcription rate after a threshold. This threshold is determined by the transcription rate of the sRNA, which is regulated by the small molecule aTc for the P_{tet} promoter that drives the sRNA expression. (From Levine et al. [78]. With permission.)

9.4.2.3 Riboswitch models

The protein expression of a target mRNA regulated by a riboswitch can be captured using a kinetic model. Consider a riboswitch that alternates between an active and an inactive form. A small molecule ligand binds and stabilizes the riboswitch in the inactive conformation. In the active form, the target mRNA expression is repressed by the riboswitch.

$$A \xleftrightarrow{K_{part}} I \xleftrightarrow{K_{bind}} I:L \tag{9.21}$$

where A and I are the active and inactive conformations of the riboswitch, respectively, L is the ligand, K_{part} is the partitioning constant of the riboswitch, and K_{bind} is the binding constant of the inactive riboswitch with the ligand. The efficacy e of the riboswitch is the relative target gene expression with respect to ligand concentration, and is nonlinearly related to the fraction of active riboswitch.

$$e = 1 - e_A \cdot \left[\frac{1}{1 + K_{part}(1 + K_{bind} \cdot L)} \right]^h \tag{9.22}$$

where, e_A is the efficiency of the riboswitch in its active form, and h is the Hill coefficient to account for nonlinearity between the concentration of active riboswitch and target gene expression. This equation describes the transfer function of the system, which is the relationship between input and output. In this case, the input is the ligand concentration and the output is the relative target mRNA expression.

In (9.22), the equilibrium constant between the active and inactive forms of the riboswitch K_{part} is related to the change in free energy between the two forms, ΔG, which can be calculated from the thermal stability of the two base-pairing conformations. To use ΔG to predict K_{part}, the basal expression of the riboswitch or target mRNA can be fit to the following equation that takes the same form as Equation (9.22) when L is zero.

$$e_{fit} = 1 - C_1 \left[C_2 + \exp\left(-\frac{\Delta G}{k_B T} \right) \right]^{-C_3} \tag{9.23}$$

where T is temperature, k_B is the Boltzmann constant, C_1, C_2, and C_3 are fit constants, and e_{fit} is the measured basal expression of the target gene for the fit curve. After C_{1-3} are determined, K_{part} can be expressed in terms of ΔG and C_{1-3} by equating e_{fit} in Equation (9.23) to e in Equation (9.22) when L is zero. By plugging K_{part} back into Equation (9.22), the predicted efficacy of the riboswitch can be determined as

$$e_{predicted} = 1 - e_A \cdot \left[1 + \left[\sqrt[h]{\frac{e_A}{C_1} \left[C_2 + \exp\left(-\frac{\Delta G}{k_B T} \right) \right]^{C_3}} - 1 \right] (1 + K_{bind} \cdot L) \right]^{-h} \tag{9.24}$$

Here, C_{1-3}, K_{bind}, e_A, and h are empirically determined values. With this model, the efficacy of a riboswitch can be designed by tuning the stability of the two base-pairing conformations to modify ΔG (Figure 9.13). This model can also guide the design of additional riboswitches optimized for particular transfer functions [79,80].

9.4.3 Conclusion

RNase binding and protection sites, sRNAs and riboswitches are genetic parts that can be used to control mRNA stability. mRNA stability can be altered by adding hairpin structures to the 5'-end to prevent cleavage from RNase E. Both sRNA and riboswitches rely on the base pairing of the regulatory RNA with its target mRNA. The base-pairing strength can be determined from thermodynamic models. By combining the thermodynamic parameters of RNA base pairing, mathematical models can predict the behavior of a riboswitch or sRNA sequence. For genes that are known to be regulated by sRNA or riboswitches, these methods can be used to tune RNA regulators to desired efficiency without interfering with other genes. However, the design of RNA stability parts for any

Chapter 9

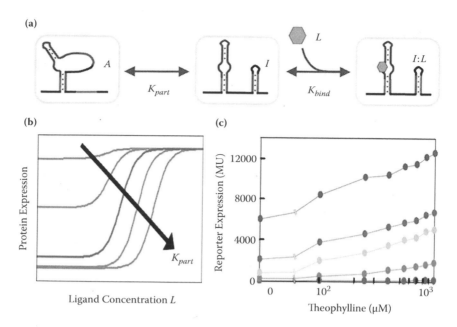

FIGURE 9.13 (See color insert.) A mathematical model guides the design of riboswitches. (a) A riboswitch can fold into two secondary structures that are either active (A) or inactive (I) for the downstream regulation. A ligand (L) binds to the inactive form and stabilizes the riboswitch in the inactive form. (b) The transfer function of the riboswitch regulation changes with the equilibrium constant K_{part} between the active and inactive forms as determined by their secondary structure stability. (c) The protein expression of the regulated gene with respect to ligand (theophylline) responds differently according to the different secondary structure stabilities as predicted by the model. The colors red, orange, yellow, green, and blue denote riboswitch designs that each stabilizes the riboswitch in the inactive form more than the previous riboswitch, leading to elevated basal expression as well as dynamic range in the target protein expression as predicted by the model. Reporter expression is measured by a β-galactosidase assay in Miller units. (From Beisel and Smolke [80]. With permission.)

given gene will require more detailed understanding regarding the mechanism sRNA regulation and structure than what is known today. These parts can provide regulation of gene expression at the post-transcriptional level, which offers faster response time than transcriptional control and also control of protein turnover rate.

9.5 Ribosome Binding Sites

9.5.1 Ribosome Binding Sequences

Protein expression level depends on both transcription and translation. Translation initiation depends on the assembly of the 30S ribosomal subunit onto the mRNA transcript. This occurs at the ribosome binding site (RBS) in the 5'-untranslated region of the mRNA (Figure 9.14a). This process requires the RBS of the mRNA to be unfolded so that it can be recognized by the 30S ribosomal subunit for binding. After the assembly of the 30S ribosomal subunit and the initiator tRNA, the 50S ribosomal subunit assembles and forms the complete 70S ribosome, which is fully functional to translate mRNA into protein.

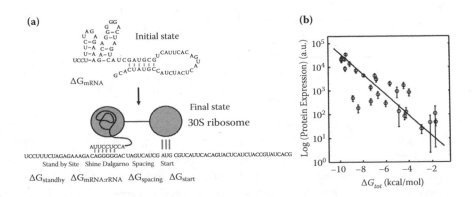

FIGURE 9.14 (See color insert.)   The strength of a ribosome binding site can be predicted by a thermodynamic model. (a) The ribosome binding site interacts with the 30S ribosomal subunit to initiate translation. To interact with the 16S rRNA in the 30S ribosomal subunit, several regions in the RBS must be unfolded, including the standby site, the Shine-Dalgarno sequence, the start codon, with a defined spacing between the Shine-Dalgarno sequence and the start codon. (b) Using a thermodynamic model, the protein expression level of a given RBS sequence can be predicted within two folds from the total free energy accounting for each of the sequence contributions. The expression level of reporter fluorescence proteins is measured by flow cytometry. (From Salis et al. [101]. With permission.)

The RBS consists of the start codon, a Shine-Dalgarno (SD) sequence, and a standby site. The start codon is recognized by the initiating tRNA through base-pairing with the anticodon. The SD sequence is upstream of the start codon and binds to the 3'-end of the 16S rRNA in the 30S ribosomal subunit. The standby site is upstream of the SD sequence where the 30S ribosomal subunit binds nonspecifically before the SD sequence unfolds and become available for binding [81]. The start codon is usually AUG, but UUG and GUG are also present in bacteria. The consensus SD sequence in *E. coli* is UAAGGAGGU, although only a portion of this sequence is found in the RBS of each gene.

The effects of different RBS components on translation initiation rate have been determined through thermodynamic models, structural studies, and mutagenesis studies. The translation initiation rate is reduced when a hairpin structure blocks the ribosome from assembling onto the mRNA [82,83]. The SD sequence does not have to be entirely present, but must be able to bind to the 16S rRNA [84–86]. In addition, the –3 position from the start codon and the second codon may affect the translation initiation rate greatly [84,87,88]. Furthermore, an optimal spacing of 12 nt between the SD sequence and the start codon exists, which relates to the spatial structure of the ribosome between the two binding sites [84,88–90]. The role of the standby site is less clear. It has been proposed that the single-stranded mRNA region 40–100 nt upstream of the start codon may function as a docking site for the nonspecific binding of the ribosome while waiting for the RBS to unfold [81,91]. Each of these components contribute to the strength of an RBS, which determines the translation initiation rate of a protein.

9.5.2   RBS Models

RBS models were first constructed to predict the start positions of genes in the genome [92–95]. These models are inferred from 16S rRNA sequences, obtained by alignment of sequences upstream of start codons, or trained using machine learning methods to distinguish RBS sequences from background [30, 93–99]. By locating strong RBS sequences within a defined distance from the start codon, promoter locations and translation initiation sites are predicted [26–30]. Several modules further identify RBSs through post-processing after the genes are identified using multiple sequence alignment or probabilistic models [92,96,100]. In addition to identifying RBS sequences, it is desired to predict RBS strengths from these sequences.

The expression level of proteins in genetic circuits can be predicted by thermodynamic models that can predict the translation initiation rate based on the strength of the RBS calculated from the mRNA sequence [101]. Consider translation initiation as a reversible transition between two states: the initial state is the folded mRNA and free 30S ribosomal complexes, and the final state is the 30S ribosome assembled on the RBS of the mRNA (Figure 9.14a). The translation initiation rate r_{tl} is proportional to the concentration of 30S ribosomes bound to mRNA and can be written as [101]

$$r_{tl} = k_0 \exp\left(-\frac{\Delta G_{tot}}{k_B T}\right) \tag{9.25}$$

where T is temperature, k_B is the Boltzmann constant, k_0 is a proportional constant that includes the total 30S ribosome concentration and the mRNA concentration, and ΔG_{tot} is the change in free energy between the initial state of a folded mRNA and the final state of the mRNA transcript loaded with the 30S ribosome preinitiation complex. ΔG_{tot} can be calculated from the contributions of five individual terms:

$$\Delta G_{tot} = \Delta G_{mRNA:rRNA} + \Delta G_{start} + \Delta G_{spacing} - \Delta G_{standby} - \Delta G_{mRNA} \tag{9.26}$$

where $\Delta G_{mRNA:rRNA}$ is the energy released when the last nine nucleotides of the 16S rRNA hybridize to the mRNA. ΔG_{start} is the energy released when the start codon hybridizes to the initiating tRNA anticodon. $\Delta G_{spacing}$ is the free energy penalty caused by nonoptimal spacing between the SD sequence and the start codon relative to the 30S ribosomal complex. $\Delta G_{standby}$ is the work required to unfold the secondary structures sequestering the standby site defined by the 4 nucleotides upstream of the 16S rRNA binding site. ΔG_{mRNA} is free energy of the secondary structure in the mRNA. The value of each of these terms, except for the $\Delta G_{spacing}$ term, can be calculated from RNA secondary structure prediction programs. These terms depend only on the thermodynamic parameters related to the sequence of the mRNA. The $\Delta G_{spacing}$ term is empirically determined.

This thermodynamic model can be used to design RBS sequences to achieve a desired protein expression level [101]. This is achieved by combining the above model with a simulated annealing optimization algorithm. Starting from a randomly generated initial RBS sequence, the RBS sequence is mutated by one nucleotide at a time and evaluated by the model previously described. The new sequence is accepted or rejected at a predetermined rate and mutated again for the next cycle. The process continues iteratively until the target translation initiation rate is reached (Figure 9.14b).

Kinetic models can also predict RBS strength as well as describe the mechanism of translation initiation [88,102]. Consider translation initiation as the kinetic process of

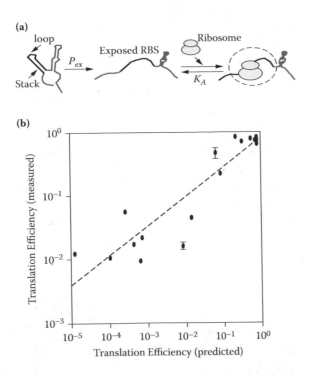

FIGURE 9.15   A kinetic model can be used to predict the translation efficiency from the mRNA sequence. (a) Translation initiates after the ribosome binds to the exposed RBS, which depends on the exposure probability P_{ex} and the binding constant K_A. (b) The model can be used to design RBS sequences with the desired translation efficiency with a correlation coefficient between the predicted and measured translation efficiency to be 0.87. The translation efficiency is measured by β-galactosidase assay. (From Na et al. [102]. With permission.)

free ribosomes binding to the RBS of exposed mRNAs to become ribosome-bound mRNA (Figure 9.15a). The association constant for the ribosome to bind to mRNA is

$$K_A = \frac{m_R}{m_E \cdot \dfrac{R_F}{n_R}} = \exp\left(-\frac{\Delta G_{mRNA:rRNA}}{k_B T}\right) \tag{9.27}$$

where R_F, m_E, m_R, and n_R are the number of free ribosomes, RBS-exposed mRNAs, ribosome-bound mRNA, and ribosomes in a transcript, respectively. K_A can be determined by the free energy of binding by the SD sequence in the mRNA to the 16S rRNA in the ribosome, $\Delta G_{mRNA:rRNA}$. The number of mRNAs with an exposed RBS (m_E) depends on the exposure probability p_{ex}.

$$m_E = (m_T - m_R)p_{ex} \tag{9.28}$$

where m_T is the total number of mRNAs. p_{ex} depends on the secondary structure in the RBS. The RBS is exposed if the SD sequence is in a linear region or the loop of a hairpin. It cannot bind to the ribosome if it is part of a base-paired stack structure. p_{ex} can be calculated as the sum of probabilities for each base-pairing conformation S_i, weighted

Chapter 9

by the probability of the mRNA to fold into that conformation, which is determined by its free energy. Assuming that each nucleotide in the RBS contributes to the overall exposure of the RBS independently, the exposure probability can be decomposed into the contribution of each nucleotide in the RBS.

$$
p_{ex} = \sum_i \frac{\exp\left(\dfrac{\Delta G(S_i)}{k_B T}\right)}{\sum_l \exp\left(\dfrac{\Delta G(S_l)}{k_B T}\right)} \prod_j \theta_{i,j}
\tag{9.29}
$$

where S_i is the i-th base-pairing conformation of the mRNA, and $\theta_{i,j}$ is the exposure probability of base j in conformation S_i, which is determined as

$$
\theta_{i,j} = \begin{cases} 1 & \text{if } j \text{ in loop} \\[2ex] \sqrt[L]{\dfrac{1}{1+\exp\left(\dfrac{\Delta G_{i,j}}{k_B T}\right)}} & \text{if } j \text{ in stack} \end{cases}
\tag{9.30}
$$

When the base j in a conformation is part of a loop, it is assumed to be exposed and available for ribosome binding; when base j is part of a stack of length L, the exposure probability depends on the thermal stability of that stack (Figure 9.15a).

By combining K_A and p_{ex} determined by Equations 9.27–9.30 with the ordinary differential equations that govern the change in m_E, m_R, and R_F at steady state, the probability of translation p_{tr} can be determined as

$$
p_{tr} = \frac{m_R}{m_T} = \frac{1}{2m_T K_R p_{ex}} + \frac{R_T}{2m_T n_R} + \frac{1}{2} - \sqrt{\left(\frac{1}{2m_T K_R p_{ex}} + \frac{R_T}{2m_T n_R} + \frac{1}{2}\right)^2 - \frac{R_T}{m_T n_R}}
\tag{9.31}
$$

This probability determines protein expression level and can be viewed as the relative translation initiation rate. This relative translation initiation rate here represents a probability, which is on a scale from 0 to 1. It is different from that in the thermodynamic model introduced previously, which is on a scale from 0 to 100,000 determined from the scale of the data for model correlation. In theory, these two scales both correlate to the translation initiation rate and can be mapped to each other. However, the RBS strength determined from these models are on a relative scale, which means that the RBS strength of one gene can be improved using the model prediction and design, but the translation initiation rate of different genes cannot be compared on the same scale. Similar to the thermodynamic model, this kinetic model can be used to design synthetic RBS sequences by using a genetic algorithm that iteratively evaluates and mutates candidate sequences until the target translation initiation rate is reached (Figure 9.15b) [102]. The thermodynamic model captures the key factors of translation initiation and can predict the relative translation initiation rate using a simple summation of free energy terms. On the other hand, the kinetic model determines the kinetics during the ribosome binding step and can be used to determine details in the process, such as how each mRNA structure affects translation and how many mRNAs are bound by ribosomes. Both models can be used to predict the translation efficiency of a sequence and design ribosome binding sites with desired efficiency, although the relative translation initiation rates are on different scales in the two models.

9.5.3   Conclusion

Translation initiation depends mainly on the ribosome binding to the RBS of an mRNA. This binding event depends on the unfolding of the mRNA to expose the RBS and the binding of the 16S rRNA to the exposed Shine-Dalgarno sequence. Both thermodynamic and kinetic models capturing these factors can predict the strength of an RBS. In the thermodynamic model, the ribosome binding event is captured by summing several free energy terms that describe the thermal stability from the mRNA sequence. In the kinetic model, the RBS exposure probability and the binding constant between the mRNA and the ribosome can be calculated from thermal stability of mRNA conformations to derive the translation probability. Both models can be used to design RBS sequences to a desired protein expression level. Without affecting the protein coding sequence and promoter, the RBS can be designed to easily tune protein expression at the translation level, which is useful for matching upstream inputs to downstream outputs in genetic circuits.

9.6   Conclusion

In this chapter, we introduced four genetic parts: promoters, terminators, mRNA stability tags, and ribosome binding sites. These parts are identified by their specific sequence structures and control gene expression at a different time in transcription or translation. The function of the parts can be predicted using biophysical models. These models are constructed hierarchically with different levels of details. Statistical analysis of part sequences can reveal sequence properties that contribute to part function. The function of a part can be determined by accounting for the kinetic rate constants for every mechanistic step in the process. These kinetic rate constants can be calculated from free energies of each thermodynamic state of the part. For example, the open complex formation rate for RNA polymerase bound to the promoter can be calculated from the change in free energy between the closed complex and the open complex. The free energy can be calculated using thermodynamic principles. For open complex formation, the change in free energy includes the thermal melting of double-stranded DNA and the interaction energy between the RNA polymerase and the single-stranded DNA and the downstream double-stranded DNA. Thermodynamic nearest-neighbor models and RNA secondary structure prediction programs can calculate free energy based on DNA or RNA sequence. Therefore, sequence information can be integrated with kinetics to predict part function. Different models utilizing this approach to predict the function of different parts were reviewed in each section. It is important to be able to predict the function of parts in order to assemble these parts into large genetic circuits. The design of genetic circuits requires many modular parts to connect a complex circuit. For the circuit to function correctly, the expression of input and output genes must be controlled precisely by parts. Biophysical models can predict the quantitative behavior of parts from their sequences. They can also be used to design parts, which can be incorporated into genetic circuits. The characterization and prediction of parts is the first step toward the construction of large genetic circuits for a wide range of applications.

References

1. Voigt, C. A. 2006. Genetic parts to program bacteria. *Curr. Opin. Biotechnol.* 17:548–557.
2. Andrianantoandro, E., S. Basu, D. K. Karig, and R. Weiss. 2006. Synthetic biology: new engineering rules for an emerging discipline. *Mol Syst Biol* 2:2006.0028.
3. Mukherji, S., and A. van Oudenaarden. 2009. Synthetic biology: understanding biological design from synthetic circuits. *Nat. Rev. Genet.* 10:859–871.
4. Lucks, J. B., L. Qi, W. R. Whitaker, and A. P. Arkin. 2008. Toward scalable parts families for predictable design of biological circuits. *Curr. Opin. Microbiol.* 11:567–573.
5. An, W., and J. W. Chin. 2009. Synthesis of orthogonal transcription–translation networks. *Proc. Natl. Acad. Sci. USA* 106:1–6.
6. Barrett, O. P. T., and J. W. Chin. 2010. Evolved orthogonal ribosome purification for in vitro characterization. *Nucleic Acids Res.* 38:2682–2691.
7. Chubiz, L. M., and C. V. Rao. 2008. Computational design of orthogonal ribosomes. *Nucleic Acids Res.* 36:4038–4046.
8. Rackham, O., and J. W. Chin. 2005. A network of orthogonal ribosome x mRNA pairs. *Nat. Chem. Biol.* 1:159–166.
9. Beyer, A., S. Bandyopadhyay, and T. Ideker. 2007. Integrating physical and genetic maps: from genomes to interaction networks. *Nat. Rev. Genet.* 8:699–710.
10. De Smet, R., and K. Marchal. 2010. Advantages and limitations of current network inference methods. *Nat. Rev. Microbiol.* 8:717–729.
11. Lee, N. H. 2005. Genomic approaches for reconstructing gene networks. *Pharmacogenomics* 6:245–258.
12. Veiga, D. F., B. Dutta, and G. Balazsi. 2009. Network inference and network response identification: moving genome-scale data to the next level of biological discovery. *Mol. Biosyst.* 6:469–480.
13. Kaznessis, Y. N. 2007. Models for synthetic biology. *BMC Systems Biology* 1:47.
14. Marchisio, M. a., and J. Stelling. 2009. Computational design tools for synthetic biology. *Curr. Opin. Biotechnol.* 20:479–485.
15. Khalil, A. S., and J. J. Collins. 2010. Synthetic biology: applications come of age. *Nat. Rev. Genet.* 11:367–379.
16. Gruber, T. M., and C. A. Gross. 2003. Multiple sigma subunits and the partitioning of bacterial transcription space. *Annu. Rev. Microbiol.* 57:441–466.
17. Lonetto, M., M. Gribskov, and C. A. Gross. 1992. The σ70 family: sequence conservation and evolutionary relationships. *J. Bacteriol.* 174:3843–3849.
18. Paget, M. S., and J. D. Helmann. 2003. The σ70 family of sigma factors. *Genome Biol.* 4:203.
19. Typas, A., G. Becker, and R. Hengge. 2007. The molecular basis of selective promoter activation by the σS subunit of RNA polymerase. *Mol. Microbiol.* 63:1296–1306.
20. Ross, W., K. K. Gosink, J. Salomon, K. Igarashi, C. Zou, A. Ishihama, K. Severinov, and R. L. Gourse. 1993. A third rocognition element in bacterial promoters DNA binding by the alpha subunit of RNA polymearse. *Science* 262:1407–1413.
21. Young, B. A., T. M. Gruber, C. A. Gross, and S. Francisco. 2002. Views of transcription initiation. *Cell* 109:417–420.
22. DeHaseth, P. L., and J. D. Helmann. 1995. Open complex formation by *Escherichia coli* RNA polymerase: the mechanism of poiymerase-Induced strand separation of double helical DNA. *Mol. Microbiol.* 16:817–824.
23. Kapanidis, A. N., E. Margeat, S. O. Ho, E. Kortkhonjia, S. Weiss, and R. H. Ebright. 2006. Initial transcription by RNA polymerase proceeds through a DNA-scrunching mechanism. *Science* 314:1144–1147.

24. Revyakin, A., C. Liu, R. H. Ebright, and T. R. Strick. 2006. Abortive initiation and productive initiation by RNA polymerase involve DNA scrunching. *Science* 314:1139–1143.

25. Hsu, L. M. 2002. Promoter clearance and escape in prokaryotes. *Biochim Biophys Acta* 1577:191–207.

26. Towsey, M., P. Timms, J. Hogan, and S. A. Mathews. 2008. The cross-species prediction of bacterial promoters using a support vector machine. *Comput. Biol. Chem.* 32:359–366.

27. Gordon, J. J., M. W. Towsey, J. M. Hogan, S. A. Mathews, and P. Timms. 2006. Improved prediction of bacterial transcription start sites. *Bioinformatics* 22:142–148.

28. Gordon, L., A. Y. Chervonenkis, A. J. Gammerman, I. A. Shahmuradov, and V. V. Solovyev. 2003. Sequence alignment kernel for recognition of promoter regions. *Bioinformatics* 19:1964–1971.

29. Li, Q.-Z., and H. Lin. 2006. The recognition and prediction of σ70 promoters in *Escherichia coli* K-12. *J. Theor. Biol.* 242:135–141.

30. Makita, Y., M. J. L. de Hoon, and A. Danchin. 2007. Hon-yaku: a biology-driven Bayesian methodology for identifying translation initiation sites in prokaryotes. *BMC Bioinformatics* 8:47.

31. Gama-Castro, S., V. Jiménez-Jacinto, M. Peralta-Gil, A. Santos-Zavaleta, M. I. Peñaloza-Spinola, B. Contreras-Moreira, J. Segura-Salazar, L. Muñiz-Rascado, I. Martínez-Flores, H. Salgado, C. Bonavides-Martínez, C. Abreu-Goodger, C. Rodríguez-Penagos, J. Miranda-Ríos, E. Morett, E. Merino, A. M. Huerta, L. Treviño-Quintanilla, and J. Collado-Vides. 2008. RegulonDB (version 6.0): gene regulation model of *Escherichia coli* K-12 beyond transcription, active (experimental) annotated promoters and Textpresso navigation. *Nucleic Acids Res.* 36:D120–124.

32. Mendoza-Vargas, A., L. Olvera, M. Olvera, R. Grande, L. Vega-Alvarado, B. Taboada, V. Jimenez-Jacinto, H. Salgado, K. Juárez, B. Contreras-Moreira, A. M. Huerta, J. Collado-Vides, and E. Morett. 2009. Genome-wide identification of transcription start sites, promoters and transcription factor binding sites in *E. coli*. *PLoS One* 4:e7526.

33. Lisser, S., and H. Margalit. 1994. Determination of common structural features in *Escherichia coli* promoters by computer analysis. *Eur. J. Biochem.* 223:823–830.

34. Kanhere, A., and M. Bansal. 2005. A novel method for prokaryotic promoter prediction based on DNA stability. *BMC Bioinformatics* 6:1.

35. Askary, A., A. Masoudi-nejad, R. Sharafi, A. Mizbani, S. N. Parizi, and M. Purmasjedi. 2009. N4: A precise and highly sensitive promoter predictor using neural network fed by nearest neighbors. *Genes Genet. Syst.* 84:425–430.

36. Wang, H., and C. J. Benham. 2006. Promoter prediction and annotation of microbial genomes based on DNA sequence and structural responses to superhelical stress. *BMC Bioinformatics* 7:248.

37. Alexandrov, B. S., V. Gelev, S. W. Yoo, A. R. Bishop, K. O. Rasmussen, and A. Usheva. 2009. Toward a detailed description of the thermally induced dynamics of the core promoter. *PLoS Comp. Biol.* 5:e1000313.

38. Alexandrov, B. S., V. Gelev, S. W. Yoo, L. B. Alexandrov, Y. Fukuyo, A. R. Bishop, K. Ø. Rasmussen, and A. Usheva. 2010. DNA dynamics play a role as a basal transcription factor in the positioning and regulation of gene transcription initiation. *Nucleic Acids Res.* 38:1790–1795.

39. Hawley, D. K., W. R. Mcclure, and I. R. L. P. Limited. 1983. Compilation and analysis of *Escherichia coli* promoter DNA sequences. *Nucleic Acids Res.* 11:2237–2255.

40. Lisser, S., and H. Margalit. 1993. Compilation of *E. coli* mRNA promoter sequences. *Nucleic Acids Res.* 21:1507–1516.

41. Mulligan, M. E., D. K. Hawley, R. Entriken, and W. R. Mcclure. 1984. *Escherichia coli* promoter sequences predict in vitro RNA polymerase selectivity. *Nucleic Acids Res.* 12:789–800.

42. Staden, R. 1989. Methods for calculating the probabilities of finding patterns in sequences. *Bioinformatics* 5:89–96.

43. Weindl, J., P. Hanus, Z. Dawy, J. Zech, J. Hagenauer, and J. C. Mueller. 2007. Modeling DNA-binding of *Escherichia coli* σ70 exhibits a characteristic energy landscape around strong promoters. *Nucleic Acids Res.* 35:7003–7010.

44. Rhodius, V. A., and V. K. Mutalik. 2010. Predicting strength and function for promoters of the *Escherichia coli* alternative sigma factor, σE. *Proc. Natl. Acad. Sci. USA* 107:2854–2859.

45. Djordjevic, M., and R. Bundschuh. 2008. Formation of the open complex by bacterial RNA polymerase—a quantitative model. *Biophys. J.* 94:4233–4248.

46. Xue, X. C., F. Liu, and Z. C. Ou-Yang. 2008. A kinetic model of transcription initiation by RNA polymerase. *J. Mol. Biol.* 378:520–529.

47. Nudler, E., and M. E. Gottesman. 2002. Transcription termination and anti-termination in *E. coli*. *Genes Cells* 7:755–768.

Chapter 9

48. von Hippel, P. H. 1998. An integrated model of the transcription complex in elongation, termination, and editing. *Science* 281:660–665.

49. Gusarov, I., and E. Nudler. 1999. The mechanism of intrinsic transcription termination. *Mol. Cell* 3:495–504.

50. Christie, G. E., P. J. Farnham, and T. Plattt. 1981. Synthetic sites for transcription termination and a functional comparison with tryptophan operon termination sites in vitro. *Proc. Natl. Acad. Sci. USA* 78:4180–4184.

51. Larson, M. H., W. J. Greenleaf, R. Landick, and S. M. Block. 2008. Applied force reveals mechanistic and energetic details of transcription termination. *Cell* 132:971–982.

52. d'Aubenton Carafa, Y., E. Brody, and C. Thermest. 1990. Prediction of rho-independent *Escherichia coli* transcription terminators a statistical analysis of their RNA stem-loop structures. *J. Mol. Biol.* 216:835–858.

53. Wilson, K. S. 1995. Transcription termination at intrinsic terminators: The role of the RNA hairpin. *Proc. Natl. Acad. Sci. USA* 92:8793–8797.

54. Reynolds, R., and M. J. Chamberlin. 1992. Parameters affecting transcription termination *Escherichia coli* RNA II: construction and analysis of hybrid terminators. *J. Mol. Biol.* 224:53–63.

55. Argaman, L., R. Hershberg, J. Vogel, G. Bejerano, E. G. H. Wagner, H. Margalit, and S. Altuvia. 2001. Novel small RNA-encoding genes in the intergenic regions of *Escherichia coli*. *Curr. Biol.* 11:941–950.

56. Lesnik, E. A., R. Sampath, H. B. Levene, T. J. Henderson, J. A. Mcneil, and D. J. Ecker. 2001. Prediction of rho-independent transcriptional terminators in *Escherichia coli*. *Nucleic Acids Res.* 29:3583–3594.

57. Unniraman, S., R. Prakash, and V. Nagaraja. 2002. Conserved economics of transcription termination in eubacteria. *Nucleic Acids Res.* 30:675–684.

58. Kingsford, C. L., K. Ayanbule, and S. L. Salzberg. 2007. Rapid, accurate, computational discovery of Rho-independent transcription terminators illuminates their relationship to DNA uptake. *Genome Biol.* 8:R22.

59. de Hoon, M. J. L., Y. Makita, K. Nakai, and S. Miyano. 2005. Prediction of transcriptional terminators in *Bacillus subtilis* and related species. *PLoS Comp. Biol.* 1:e25.

60. Mathews, D. H., J. Sabina, M. Zuker, and D. H. Turner. 1999. Expanded sequence dependence of thermodynamic parameters improves prediction of RNA secondary structure. *J. Mol. Biol.* 288:911–940.

61. Zuker, M. 2003. Mfold web server for nucleic acid folding and hybridization prediction. *Bioinformatics* 31:3406–3415.

62. Bai, L., A. Shundrovsky, and M. D. Wang. 2004. Sequence-dependent kinetic model for transcription elongation by RNA polymerase. *J. Mol. Biol.* 344:335–349.

63. Tadigotla, V. R., A. M. Sengupta, V. Epshtein, R. H. Ebright, E. Nudler, and A. E. Ruckenstein. 2006. Thermodynamic and kinetic modeling of transcriptional pausing. *Proc. Natl. Acad. Sci. USA* 103:4439–4444.

64. Mitra, A., K. Angamuthu, H. V. Jayashree, and V. Nagaraja. 2009. Occurrence, divergence and evolution of intrinsic terminators across eubacteria. *Genomics* 94:110–116.

65. Condon, C. 2007. Maturation and degradation of RNA in bacteria. *Curr. Opin. Microbiol.* 10:271–278.

66. Waters, L. S., and G. Storz. 2009. Regulatory RNAs in bacteria. *Cell* 136:615–628.

67. Masse, E. 2003. Regulatory roles for small RNAs in bacteria. *Curr. Opin. Microbiol.* 6:120–124.

68. Guillier, M., and S. Gottesman. 2008. The 5' end of two redundant sRNAs is involved in the regulation of multiple targets, including their own regulator. *Nucleic Acids Res.* 36:6781–6794.

69. Kawamoto, H., Y. Koide, T. Morita, and H. Aiba. 2006. Base-pairing requirement for RNA silencing by a bacterial small RNA and acceleration of duplex formation by Hfq. *Mol. Microbiol.* 61:1013–1022.

70. Pfeiffer, V., K. Papenfort, S. Lucchini, J. C. D. Hinton, and J. Vogel. 2009. Coding sequence targeting by MicC RNA reveals bacterial mRNA silencing downstream of translational initiation. *Nat. Struct. Mol. Biol.* 16:840–846.

71. Udekwu, K. I., F. Darfeuille, J. Vogel, J. Reimegård, E. Holmqvist, and E. G. H. Wagner. 2005. Hfq-dependent regulation of OmpA synthesis is mediated by an antisense RNA. *Genes Dev.* 19:2355–2366.

72. Carrier, T. A., and J. D. Keasling. 1999. Library of synthetic 5' secondary structures to manipulate mRNA stability in *Escherichia coli*. *Biotechnol. Prog.* 15:58–64.

73. Livny, J., and M. K. Waldor. 2007. Identification of small RNAs in diverse bacterial species. *Curr. Opin. Microbiol.* 10:96–101.

74. Busch, A., A. S. Richter, and R. Backofen. 2008. IntaRNA: efficient prediction of bacterial sRNA targets. *Bioinformatics* 24:2849–2856.

75. Rivas, E., and S. R. Eddy. 2001. Noncoding RNA gene detection using comparative sequence analysis. *BMC Bioinformatics* 19:1–19.

76. Rivas, E., R. J. Klein, T. A. Jones, and S. R. Eddy. 2001. Computational identification of noncoding RNAs in *E. coli* by comparative genomics. *Curr. Biol.* 11:1369–1373.

77. Vogel, J., and E. G. H. Wagner. 2007. Target identification of small noncoding RNAs in bacteria. *Curr. Opin. Microbiol.* 10:262–270.

78. Levine, E., Z. Zhang, T. Kuhlman, and T. Hwa. 2007. Quantitative characteristics of gene regulation by small RNA. *PLoS Biol.* 5:e229.

79. Beisel, C. L., T. S. Bayer, K. G. Hoff, and C. D. Smolke. 2008. Model-guided design of ligand-regulated RNAi for programmable control of gene expression. *Mol. Syst. Biol.* 4:224.

80. Beisel, C. L., and C. D. Smolke. 2009. Design principles for riboswitch function. *PLoS Comp. Biol.* 5:e1000363.

81. de Smit, M. H., and J. van Duin. 2003. Translational standby sites: how ribosomes may deal with the rapid folding kinetics of mRNA. *J. Mol. Biol.* 331:737–743.

82. de Smit, M. H., and J. van Duin. 1990. Secondary structure of the ribosome binding site determines translational efficiency: A quantitative analysis. *Proc. Natl. Acad. Sci. USA* 87:7668–7672.

83. de Smit, M. H., and J. van Duin. 1994. Control of translation by mRNA secondary structure in *Escherichia coli*. *J. Mol. Biol.* 244:144–150.

84. Lee, K., C. A. Holland-Stley, and P. R. Cunningham. 1996. Genetic analysis of the Shine-Dalgarno interaction: selection of alternative functional mRNA-rRNA combinations. *RNA* 2:1270–1285.

85. Osada, Y., R. Saito, and M. Tomita. 1999. Analysis of base-pairing potentials between 16S rRNA and 5′ UTR for translation initiation in various prokaryotes. *Bioinformatics* 15:578–581.

86. Schurr, T., E. Nadir, and H. Margalit. 1993. Identification and characterization of *E. coli* ribosomal binding sites by free energy computation. *Nucleic Acids Res.* 21:4019–4023.

87. Barrick, D., K. Villanueba, J. Childs, R. Kalil, T. D. Schneider, E. Lawrence, L. Gold, and G. D. Stormo. 1994. Quantitative analysis of ribosome binding sites in *E. coli*. *Nucleic Acids Res.* 22:1287–1295.

88. Ringquist, S., S. Shinedling, D. Barrick, L. Green, J. Binkley, G. D. Stormo, and L. Gold. 1992. Translation initiation in *Escherichia coli*: sequences within the ribosome-binding site. *Mol. Microbiol.* 6:1219–1229.

89. Chen, H., M. Bjerknes, R. Kumar, and E. Jay. 1994. Determination of the optimal aligned spacing between the Shine-Dalgarno sequence and the translation initiation codon of *Escherichia coli* mRNAs. *Nucleic Acids Res.* 22:4953–4957.

90. Vellanoweth, R. L., and J. C. Rabinowitz. 1992. The influence of ribosome-binding-site elements on translational efficiency in *Bacillus subtilis* and *Escherichia coli* in vivo. *Mol. Microbiol.* 6:1105–1114.

91. Unoson, C., and E. G. H. Wagner. 2007. Dealing with stable structures at ribosome binding sites. *RNA Biol.* 4:113–117.

92. Besemer, J., A. Lomsadze, and M. Borodovsky. 2001. GeneMarkS: a self-training method for prediction of gene starts in microbial genomes. Implications for finding sequence motifs in regulatory regions. *Nucleic Acids Res.* 29:2607–2618.

93. Hannenhalli, S. S., W. S. Hayes, A. G. Hatzigeorgiou, and J. W. Fickett. 1999. Bacterial start site prediction. *Nucleic Acids Res.* 27:3577–3582.

94. Lukashin, A. V., and M. Borodovsky. 1998. GeneMark.hmm: new solutions for gene finding. *Nucleic Acids Res.* 26:1107–1115.

95. Yada, T., Y. Totoki, T. Takagi, and K. Nakai. 2001. A novel bacterial gene-finding system with improved accuracy in locating start codons. *DNA Res.* 106:97–106.

96. Suzek, B. E., M. D. Ermolaeva, and M. Schreiber. 2001. A probabilistic method for identifying start codons in bacterial genomes. *Bioinformatics* 17:1123–1130.

97. Ou, H.-Y., F.-B. Guo, and C.-T. Zhang. 2004. GS-Finder: a program to find bacterial gene start sites with a self-training method. *Int. J. Biochem. Cell Biol.* 36:535–544.

98. Zhu, H.-q., G.-q. Hu, Z.-q. Ouyang, J. Wang, and Z.-S. She. 2004. Accuracy improvement for identifying translation initiation sites in microbial genomes. *Bioinformatics* 20:3308–3317.

99. Tech, M., N. Pfeifer, B. Morgenstern, P. Meinicke, A. Bioinformatik, and G.-a.-u. Göttingen. 2005. TICO: a tool for improving predictions of prokaryotic translation initiation sites. *Bioinformatics* 21:3568–3569.

100. Hasan, S., and M. Schreiber. 2006. Recovering motifs from biased genomes: application of signal correction. *Nucleic Acids Res.* 34:5124–5132.

101. Salis, H. M., E. A. Mirsky, and C. A. Voigt. 2009. Automated design of synthetic ribosome binding sites to control protein expression. *Nat. Biotechnol.* 27:946–950.

102. Na, D., S. Lee, and D. Lee. 2010. Mathematical modeling of translation initiation for the estimation of its efficiency to computationally design mRNA sequences with desired expression levels in prokaryotes. *BMC Systems Biology* 4:71.

Chapter 9

10. Modeling Cellular Variability

Brian Munsky

Quantitative Biology: From Molecular to Cellular Systems edited by Michael E. Wall © 2012 CRC Press / Taylor & Francis Group, LLC. ISBN: 978-1-4398-2722-2

Chapter 10

10.1　Introduction

Phenotypical diversity may arise despite clonal genetics for a number of reasons, many of which are due to fluctuations in the environment: for example, cells nearer to nutrient sources grow faster, and those subjected to heat, light, or other inputs will respond accordingly. But even cells in carefully controlled, homogenous environments can exhibit diversity, and a strong component of this diversity arises from the rare and discrete nature of genes and molecules involved in gene regulation. In particular, many genes have only one or two copies per cell and may be inactive for large portions of the cell cycle. The times at which these genes turn on or off depend upon many random or chaotic events, such as thermal motion, molecular competitions, and upstream fluctuations.

To illustrate how diversity may arise despite identical genetics and initial conditions, Figure 10.1 shows a cartoon of a simple gene regulatory system. At the initial time (Figure 10.1a), the cell could have a single copy of an important gene, such as green fluorescent protein (GFP), and a single copy of an unstable activator molecule, A. The dynamics of the process is a race—will A bind and activate *gfp* first (Figure 10.1c), or will A degrade before it has a chance to bind (Figure 10.1b)? Although real biological systems are far more complex than this toy cartoon, the principles remain the same: genes may be active or inactive simply due to a chance reaction with another molecule. Regulatory molecules may undergo many different reactions (e.g., degradation, dimerization, folding), which may impede or help them to bind to the proper gene regulatory site. Smaller

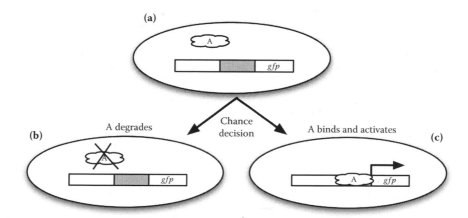

FIGURE 10.1　Cartoon depiction of stochastic gene regulation. (a) The cell begins with a single gene and a single, unstable activator protein (A). (b) Through a chance event, the activator molecule degrades, and the gene remains inactive. (c) Through a different chance event, the activator may bind and activate the gene making the cell active.

copy numbers typically result in more variable responses since a single-molecule event represents a much higher relative change. In particular, with one gene copy, switches can be all-or-nothing; with more copies, the response can be much more graded.

Once rarity and discreteness causes variability to arise in a single gene or regulatory molecule, it can affect the system's downstream elements as the products of one gene activate or repress another [1–7]. How the system survives and even exploits this variability depends upon the mechanisms of the underlying gene regulatory network. In many cases, variability is undesirable, in which case evolution will have favored mechanisms that diminish the level of variability. For example, negative feedback (especially auto-regulation) mechanisms can reduce variability for a given mean level signal [8–10], and such autoregulatory mechanisms are present in about 40% of the transcription factors in *E. coli* [11]. Section 10.5.1 considers a model of gene transcription and translation in which a protein represses the activation of its own gene. This auto-regulation enables the system to exhibit less variability for the same mean level of expression. In another context, dynamics in one part of a regulatory network can help to filter out certain fluctuation frequencies coming from other sources via low-pass or band-pass filters [12,13]. For example, the simple system examined in Section 10.5.2 acts as a low-pass filter.

In other circumstances, discrete variations may be exploited to realize different desired cellular behaviors. When passed through certain nonlinear processes, such as sequential molecular binding events or oligomerization, external signals can be amplified [14] or damped as a result of system stochasticity. Section 10.5.2 illustrates such an example, where the nonlinearity arrises as a result of the binding of gene regulatory factors. Other mechanisms use fluctuations to excite or improve the robustness of resonant behaviors [15]. An example of this behavior will be presented in Section 10.5.4.

One of the most important and obvious of stochastic phenomena is that of stochastic switching, where cells can express two or more very different phenotypes despite identical genotypes and environments [16–19]. In single-cell organisms, the ability to switch at random is an important evolutionary survival trait in competitive or uncertain environments. If a species behavior is completely uniform, then a single new mechanism may be sufficient for a competitor to overcome that species (i.e., a single antibiotic would destroy all bacteria). If a species switches too predictably between phenotypes, then competitors can evolve their own switching strategies to outperform that organism in every circumstance (i.e., a given sequence of antibiotics would destroy all bacteria). But if a given cell type switches at random among many unpredictable phenotypes, then such strategies become much more difficult to devise, and perhaps no strategy would suffice (i.e., some bacteria would survive no matter what sequence of antibiotics are applied). Even in the absence of direct competition, the ability to switch at random is also important for survival in an uncertain environment [20]. To illustrate stochastic switching behavior, Section 10.5.3 provides an example of the analysis of a genetic toggle switch that has been used to sense and record environmental conditions such as ultraviolet radiation [21,22].

10.1.1   Measurement of Single–Cell Variability

A number of well-established experimental techniques can be used to measure the phenotypic or molecular variability of single cells [23]. In many of these techniques, cells are prepared so that traits of interest (e.g., gene expression, protein-membrane localization) are made visible due to the presence or activation of fluorescent markers. For example, fluorescent dyes can be attached to antibodies, which bind to specific cellular proteins or phosphoproteins of interest. Using fluorescence in situ hybridization (FISH) techniques, fluorophores can also be attached via oligomers to DNA and RNA molecules [24–26]. By combining multiple dyes, researchers can simultaneously measure multiple different molecule types, or examine colocalization or conformational changes in molecules via fluorescence resonance energy transfer (FRET) techniques (see Chapters 5 and 7). Alternatively, genes in cell strains may be cloned to include coding for fluorescent proteins such as green, yellow, or cyan fluorescent protein (GFP, YFP, CFP, respectively) instead of, or in addition to, their naturally expressed proteins. Like FRET, introduction of split GFP [27–29] enables a similar ability to measure the colocalization of important proteins.

In addition to the many means of marking individual biological molecules in a cell, there are also many different ways of measuring these markings within a cell. A natural approach is fluorescence microscopy, with can be used to simply look at the cells and directly observe which cells are active and which are not. With confocal microscopy, it is possible to resolve individual fluorescently tagged molecules such as DNA and RNA in fixed cells [26]. For example, with single molecule FISH and automated three-dimensional image processing software, it is then possible to count mRNA molecules in hundreds to thousands of different cells—thereby obtaining precise distributions in carefully controlled experimental conditions. With time lapse microscopy and noninvasive reporters like fluorescent proteins, it is possible to track individual living cells as they change over time. In either confocal or time-lapse microscopy approach, each cell reveals a huge amount of data in the form of multiple multicolor images, which must be manually or automatically processed.

Another common technique is flow cytometry [30]. In this approach, individual cells pass through an excitation laser (or lasers) and detectors record how much light is reflected, scattered, or absorbed and reemitted at various wavelengths of light. The high throughput nature of this approach enables researchers to measure millions of cells in a minute. With autosampling techniques, researchers can test hundreds of cultures in an hour—each with different inputs or conditions. More recent technologies are currently under development to combine the strengths of microscopy with the high throughput nature of flow cytometry. In particular, new imaging flow cytometers can capture multiple fluorescent images of each cell as it passes through the excitation lasers—in some cases these images may be used to count individual fluorescently tagged molecules.

10.1.2   Using Measurements of Cellular Variability to Infer System Properties

Each of these experimental approaches enables thorough quantitative measurements of gene expression. However, the resulting data are vast and often difficult to analyze. In many computational studies, cell-to-cell variability has been viewed as a computational nuisance. Certain cellular behaviors can be understood only in the context of intrinsic variability, but including this variability in a computational model can result in an explosion of computational complexity. Researchers have made significant progress on developing methods to handle this computational challenge including kinetic Monte Carlo (MC) algorithms [31–36], linear noise approximations [37,38], moment closure [39,40] and matching [41] techniques, moment-generating functions [42], spectral methods [43], and finite state projection approaches [44–47]. While no single computational approach applies to all biological systems, the growing arsenal of tools makes it more likely that some approach may suffice for a given system of interest. Sections 10.3 and 10.4 review a couple of these approaches.

By integrating stochastic modeling approaches with experimental measurements of single-cell variability, it becomes possible to obtain a better understanding of the dynamics of biochemical networks. Such analyses provide a new tool with which to compare and contrast different possibilities for evolutionary design [20]. These analyses of cellular variability may also help to determine what mechanisms are being employed by a particular biological system [48–50]. For example, different logical structures such as AND or OR gates can be discovered in two component regulatory systems by examining the stationary transmission of the cell variability through the network [48], or correlations of different aspects of cell expression at many time points can reveal different causal relationships between genes within a network [49]. Similarly, measuring and analyzing the statistics of gene regulatory responses in certain conditions can help to identify system parameters and develop quantitative, predictive models for certain systems [22,51–53]. Section 10.6 provides such an example on the identification of a gene regulation model from single-cell flow cytometry data.

10.1.3 Chapter Focus

This chapter discusses phenomena of cell-to-cell variability in biological systems and illustrates a few computational analyses of these phenomena. Section 10.2 examines the mesoscopic scale for modeling intracellular processes as discrete state Markov processes; the chemical master equation (CME) describing such processes is derived; and a few kinetic Monte Carlo algorithms often used to simulate these processes are reviewed. Section 10.3 describes the finite state projection (FSP) approach to solving the chemical master equation, and Section 10.4 includes step-by-step examples on using the FSP approach to analyze and identify stochastic models of gene regulation. Each of these approaches is further illustrated with graphical user interface MATLAB® software, which can be downloaded online (http://cnls.lanl.gov/~munsky) or requested from the author. Section 10.5 illustrates the use of the FSP approach and software on a few examples of stochastic gene regulation, and Section 10.6 uses this software and flow cytometry data to identify a model of *lac* regulation in *E. coli*. Finally, Section 10.7 briefly summarizes the state-of-the-art in the analysis and identification of single-cell variability in gene regulatory systems.

10.2 Mesoscopic Modeling of Biomolecular Reactions

An attempt could be made to analyze biochemical reaction networks at many different scales. At the microscopic scale, molecular dynamics simulations could be used to explore how individual protein molecules move, fold, and interact with surrounding

Chapter 10

molecules. At the macroscopic scale, large-volume chemical processes are treated with continuous-valued concentrations that evolve according to deterministic ordinary differential equations. However, single-cell data of the types discussed in Section 10.1.2 require an intermediate approach, typically referred to as the *mesoscopic* scale. At this scale, each chemical species, $\{S_1,...,S_N\}$, is described with an integer population; that is, the population of S_k is denoted by the integer $\xi_k \geq 0$. At any point the state of the system is then given by the integer population vector $[\xi_1,...,\xi_N]$, and reactions correspond to transitions from one such state to another. Typically, the process is assumed to evolve according to Markovian dynamics, meaning that reaction rates depend only upon the current state of the system and not upon how that state has been reached. For this assumption to hold, the system must be well mixed in some sense as follows.

Gillespie's 1992 paper [54] provides a derivation of Markovian chemical kinetics based upon a literal sense of a well-mixed chemical solution. To understand this argument, consider two spherical molecules, s_1 and s_2, in a volume of Ω. A reaction occurs when the two molecule centers come within a certain distance, r, of one another. During an infinitesimal time period dt, the molecule s_1 moves with an average speed u, covers a distance udt, and sweeps a region $d\Omega \approx \pi r^2 udt$ relative to the center of molecule s_2. Assuming that the system is physically well mixed, the probability that the two molecules react is $\pi r^2 u\Omega^{-1}dt$. If there were ξ_1 molecules of s_1 and ξ_2 molecules of s_2, then the probability that *any* such reaction will occur is given by $w(\xi_1,\xi_2)dt = \xi_1\xi_2\pi r^2 u\Omega^{-1}dt$. In this formulation, the key result is that the infinitesimal probability of reaction has the form $w(\xi_1,\xi_2)dt$, which depends only upon the population $\{\xi_1,\xi_2\}$ at the current time and not upon the history of these populations. We note that according to this derivation, reaction propensities are restricted to at most second order.

At face value the Markovian description derived by Gillespie does not seem to apply to most biological systems, where molecules are spatially concentrated and nonspherical. However, the *well-mixed* concept need not be so literal—the memoryless nature of Markovian dynamics can also result from the overwhelming complexity of biochemical reactions [55]. Many biochemical reactions, such as transcription, translation, degradation, protein assembly, and folding are composed of numerous substeps, each of which adds a new opportunity for the given reaction to reverse, abort, or otherwise fail to complete. As more and more of these subreactions occur, the distribution of the corresponding substates quickly equilibrates to a quasi-steady distribution. Thus, after a short transient period, the system's transition probabilities attain a well-mixed quasi-steady equilibrium, which is defined by the current coarse state of the system. Unlike the formulation in [54], this concept of well-mixedness supports far more complicated formulation for the stochastic reaction rates, including Michaelis-Menten, Hill, and other more complicated functions.

This chapter assumes the most general Markov form for a discrete-value, continuous-time chemical process. The reaction rates are given by propensity function $w(\xi_1,...,\xi_N,t)$ dt, where w can be any nonlinear function of the species populations and the current time. For later convenience, specific Markov processes are referred to with the notation, M, and they are thought of as random walks on a discrete lattice as shown in as shown in Figure 10.2a. The next subsection presents the chemical master equation, which describes the dynamics of the probability distributions for such a process.

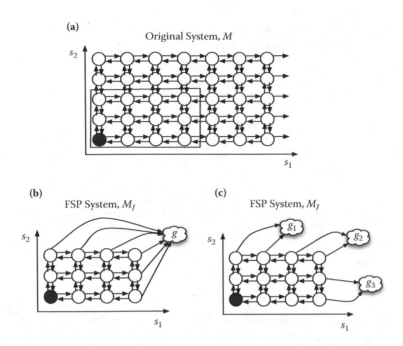

FIGURE 10.2 (a) A Markov chain for a two-species chemically reacting system, M. The process begins in the configuration shaded in gray and undergoes four reactions to increase or decrease the two different species populations. The dimension of the master equation is equal to the total number of configurations in M and is too large to solve exactly. (b) In the FSP algorithm a configuration subset $\mathbf{X}_J$ is chosen, and all remaining configurations are projected to a single absorbing point g. This results in a small dimensional Markov process, M_J. (c) By using multiple absorbing sites, it is possible to track of how the probability measure leaves the projection space. (From Munsky and Khammash [62]. With permission.)

10.2.1 The Chemical Master Equation

A chemical solution of N species, $\{S_1,\ldots,S_N\}$ is described by its state $\mathbf{x} = [\xi_1,\ldots,\xi_N]$ (see also Chapter 3). Each μ-th reaction is a transition from some state $\mathbf{x}_i$ to some other state $\mathbf{x}_j = \mathbf{x}_i + \nu_\mu$. Here, ν_μ is known as the *stoichiometric vector*, and it describes how the μ-th reaction changes the system's state. For example, the reaction $s_1 + s_2 \to s_3$ has the stoichiometric vector $\nu = [-1,-1,1]^T$. As described already, each reaction has a *propensity function*, $w_\mu(\mathbf{x},t)dt$, which is the probability that the μ-th reaction will happen in a time step of length dt. This function of the system population vector and the current time is allowed to have any arbitrary form, provided that it does not allow for reactions that lead to negative numbers.

The stoichiometry and propensity functions for each of the M possible reactions fully define the system dynamics and are sufficient to find sample trajectories with the kinetic Monte Carlo methods as discussed in Section 10.2.2. However, for many interesting gene regulatory problems, individual system trajectories are not the best description. Instead, it is desirable to analyze the dynamics in terms of probability distributions. For this it is useful to derive the chemical master equation, or master equation (ME).

Suppose the probability of all states $\mathbf{x}_i$ at time t are known; then the probability that the system will be in the state $\mathbf{x}_i$ at time, $t + dt$, is equal to the sum of (i) the probability that the system begins in the state $\mathbf{x}_i$ at t and remains there until $t + dt$, and (ii) the probability that the system is in a different state at time t and will transition to $\mathbf{x}_i$ in the considered time step, dt. This probability can be written as

$$p(\mathbf{x}_i;t+dt) = p(\mathbf{x}_i;t)\left(1 - \sum_{\mu=1}^{M} w_\mu(\mathbf{x},t)dt\right) + \sum_{\mu=1}^{M} p(\mathbf{x}_i - \mathbf{v}_\mu;t)w_\mu(\mathbf{x}_i - \mathbf{v}_\mu,t)dt \qquad (10.1)$$

If one enumerates all possible $\mathbf{x}_i$ and defines the probability distribution vector

$$\mathbf{P}(t) = [p(\mathbf{x}_1;t), p(\mathbf{x}_2;t),\ldots]^T$$

then it is relatively easy to derive the set of linear ordinary differential equations, known as the CME [37]:

$$\dot{\mathbf{P}}(t) = \mathbf{A}(t)\mathbf{P}(t) \qquad (10.2)$$

Although the master equation is linear, its dimension can be extremely large or even infinite, and it is unusually impossible to solve exactly. In many cases, the master equation can be solved only by using kinetic Monte Carlo (MC) to simulate numerous trajectories for its dynamics. Such approaches are discussed in the following subsection. In other cases, certain projection approaches make it possible to obtain approximate solutions for the master equation, as is discussed in Section 10.3.

10.2.2 Kinetic Monte Carlo Methods (Stochastic Simulation Algorithm)

The majority of analyses at the mesoscopic scale have been conducted using kinetic MC algorithms. The most widely used of these algorithms is Gillespie's stochastic simulation algorithm (SSA) [31], which is very easy to apply (see also Chapters 3 and 12). Each step of the SSA begins at a state $\mathbf{x}$ and a time t and is composed of three tasks: (1) generate the time until the next reaction; (2) determine which reaction happens at that time; and (3) update the time and state to reflect the previous two choices. For a single reaction with propensity function, $w(\mathbf{x})$, the time of the next reaction, τ, is an exponentially distributed random variable with mean $1/w(\mathbf{x})$. For M different possible reactions with propensities $\{w_\mu(\mathbf{x})\}$, τ is the minimum of M such random variables, or, equivalently an exponentially distributed random variable with mean equal to $\left(\sum_{\mu=1}^{M} w_\mu(\mathbf{x})\right)^{-1}$. To determine which of the M reactions occurs at $t + \tau$, a second random variable must be generated from the set $\mu = \{1,2,\ldots,M\}$ with the probability distribution given by $P(\mu) = w_\mu(\mathbf{x})\left(\sum_{\mu=1}^{M} w_\mu(\mathbf{x})\right)^{-1}$. Once τ and μ have been chosen, the system can be updated to $t = t + \tau$ and $\mathbf{x} = \mathbf{x} + \mathbf{v}_\mu$.

Researchers have proposed many accelerated approximations of the SSA. In the first such approximation, the system is partitioned into slow and fast portions. In [34] the system is separated into slow *primary* and fast *intermediate* species. This method uses three random variables at each step: first, the primary species' populations are held constant, and the population of the intermediate species is generated as a random variable from its quasi-steady state (QSS) distribution. The dynamics of the primary species are then found with two more random variables, similar to the SSA but with propensity functions depending upon the chosen populations of the intermediates species. The slow-scale SSA (ssSSA) [35] is very similar in that the system is again separated into sets of slow and fast species. The ssSSA differs in that it does not explicitly generate a realization for the fast species but instead uses the QSS distribution to scale the propensities of the slow reactions. Each of these QSS assumption-based approaches leads to a reduced process where all states must retain exponential waiting times. In contrast, similar reductions based upon concepts of stochastic path integrals and moment generating functions have yielded coarse-grained realizations that allow for nonexponential waiting times and thereby preserve the statistical characteristics of the original dynamics [42]. So-called hybrid methods (e.g., [56,57]) also separate the system into fast and slow reactions, but these methods do not then rely upon a QSS approximation. Instead, the fast reactions are approximated with deterministic ordinary differential equations (ODEs) or as continuous valued Markov processes using Langevin equations, and the slow reactions are treated in a manner similar to the SSA except now with time varying propensity functions.

In a second approach to accelerating the SSA, researchers frequently assume that propensity functions are constant over small time intervals. With this "τ leap assumption" one can model each of the M reaction channels as an independent Poisson random process [32]. Beginning at time t and state $\mathbf{x}(t)$, the state at the end of a time step of length τ is approximated as $\mathbf{x}(t+\tau) = \mathbf{x}(t) + \sum_{\mu=1}^{M} k_\mu \nu_\mu$, where each k_μ is a random variable chosen from the Poisson distribution $k_\mu \in P(w_\mu(X(t)), \tau)$. The accuracy of τ leaping methods depends only upon how well the τ leap assumption is satisfied. Naturally, the τ leap assumption is best satisfied when all species have sufficiently large populations and all propensity functions are relatively smooth. Otherwise small changes in populations could result in large relative changes in propensities. Ignoring these changes can easily lead to unrealistic predictions of negative populations and/or numerical stiffness. One may avoid negative populations by using a Binomial τ leap strategy [58] or by adaptively choosing the size of each τ leap [59]. One can also ameliorate the problem of numerical stiffness using implicit methods such as that in [60].

When the populations are very large, and the propensity functions are very smooth, the chemical species may be more easily modeled with continuous variables using the *chemical Langevin equation* [61]. In this solution scheme, one assumes that many reactions will occur in the *macroscopic infinitesimal* times step dt without violating the τ leap assumption. One can therefore replace the Poisson distributions with Gaussian distributions, and treat the resulting process as a stochastic differential equation driven by white noise [61].

A single simulation using kinetic Monte Carlo algorithms, such as the SSA and its modifications, describes a possible trajectory of one cell as it changes over time. These trajectories may then be compared directly to experimental data such as time lapse fluorescence microscopy studies, with which it is possible to track the dynamics of single cells. Unfortunately, because these trajectories are random, two identical cells may show very different trajectories, and this comparison can be difficult to make or even misleading. To avoid these problems, it is often useful to collect statistics from many such trajectories and try to make comparisons on the levels of these statistics rather than at the level of a single trajectory. The next section discusses an alternate approach that can directly generate these statistics for certain systems.

10.3 Analyzing Population Statistics with FSP Approaches

As discussed above, there are a number of experimental techniques to measure and quantify cell to cell variability. In particular, many of these approaches such as flow cytometry and FISH are capable of taking only images or measurements from a given cell at a single time point in its development. With these approaches, one cannot measure trajectories of a given cell, but it is very easy to establish probability distributions for a population of cells. Thus, to better compare models and data, it is useful to use modeling approaches to generate these distributions. This is equivalent to solving the master equation at certain instances in time. With the KMC approaches described above, this corresponds to running many different simulations and collecting the ensemble statistics. Alternatively, one could attempt to directly solve for the population statistics or distributions. This section discusses one such approach—FSP—to solve the master equation.

10.3.1 Notation for the FSP

To describe the FSP, some convenient notation must first be introduced in addition to what has already been presented. As before, the population of the system is composed of the integer populations of the different species, $\{\xi_1,...,\xi_N\} \in \mathbb{Z}_{\geq 0}$. The states can be enumerated, meaning that each can be assigned a unique index i such that the state $\mathbf{x}_i$ refers to the population vector, $[\xi_1^{(i)},...,\xi_N^{(i)}]$.

Let $J = \{j_1, j_2, j_3,...\}$ denote a set of indices in the following sense. If $\mathbf{X}$ is an enumerated set of states $\{\mathbf{x}_1, \mathbf{x}_2, \mathbf{x}_3,...\}$, then $\mathbf{X}_J$ denotes the subset $\{\mathbf{x}_{j1}, \mathbf{x}_{j2}, \mathbf{x}_{j3},...\}$. Let J' denote the complement of the set J. Furthermore, let $\mathbf{v}_J$ denote the subvector of $\mathbf{v}$ whose elements are chosen according to J, and let $\mathbf{A}_{IJ}$ denote the submatrix of $\mathbf{A}$ such that the rows have been chosen according to I and the columns have been chosen according to J. For example, if I and J are defined as $\{3, 1, 2\}$ and $\{1, 3\}$, respectively, then:

$$\begin{bmatrix} a & b & c \\ d & e & f \\ g & h & k \end{bmatrix}_{IJ} = \begin{bmatrix} g & k \\ a & c \\ d & f \end{bmatrix}$$

For convenience, let $\mathbf{A}_J := \mathbf{A}_{JJ}$. With this notation, the main result of the FSP approach [44,46] can be stated, which is presented as described in [62].

The infinite state Markov process, M, is defined as the random walk on the configuration set $\mathbf{X}$, as shown in Figure 10.2a. The master equation for this process is $\dot{\mathbf{P}}(t) = \mathbf{A}(t)\mathbf{P}(t)$, with initial distribution $\mathbf{P}(0)$ as described in Section 10.2. A new Markov process M_J can be defined such as that in Figure 10.2b, composed of the configurations indexed by J plus a single absorbing state. The master equation of M_J is given by

$$
\begin{bmatrix} \dot{\mathbf{P}}_J^{FSP}(t) \\ \dot{g}(t) \end{bmatrix} = \begin{bmatrix} \mathbf{A}_J & \mathbf{0} \\ -\mathbf{1}^T \mathbf{A}_J & 0 \end{bmatrix} \begin{bmatrix} \mathbf{P}_J^{FSP}(t) \\ g(t) \end{bmatrix}
\tag{10.3}
$$

with initial distribution

$$
\begin{bmatrix} \mathbf{P}_J^{FSP}(0) \\ g(0) \end{bmatrix} = \begin{bmatrix} \mathbf{P}_J(0) \\ 1 - \sum \mathbf{P}_J(0) \end{bmatrix}
$$

10.3.2 FSP Theorems and Results

The finite state process M_J has a clear relationship to the original M. First, the scalar $g(t)$ is the *exact probability* that the system has been in the set $\mathbf{X}_{J'}$ at *any* time $\tau \in [0,t]$. Second, the vector $\mathbf{P}_J^{FSP}(t)$ are the *exact joint probabilities* that the system (i) is in the corresponding states $\mathbf{X}_J$ at time t, and (ii) the system has remained in the set $\mathbf{X}_J$ *for all* $\tau \in [0,t]$. Note that $\mathbf{P}_J^{FSP}(t)$ also provides a finite dimensional approximation of the solution to the CME as follows.

First, it is guaranteed that $\mathbf{P}_J(t) \geq \mathbf{P}_J^{FSP}(t) \geq 0$ for any index set J and any initial distribution $\mathbf{P}(0)$. This is a consequence of $\mathbf{P}_J^{FSP}(t)$ being a more restrictive joint distribution than $\mathbf{P}_J(t)$. Second, the actual 1-norm distance between $\mathbf{P}(t)$ and $\mathbf{P}^{FSP}(t)$ is easily computed as

$$
\begin{aligned}
\left\| \begin{bmatrix} \mathbf{P}_J(t_f) \\ \mathbf{P}_{J'}(t_f) \end{bmatrix} - \begin{bmatrix} \mathbf{P}_J^{FSP}(t_f) \\ \mathbf{0} \end{bmatrix} \right\|_1 &= \left| \mathbf{P}_J(t_f) - \mathbf{P}_J^{FSP}(t_f) \right|_1 + \left| \mathbf{P}_{J'}(t_f) \right|_1 \\
&= \left| \mathbf{P}_J(t_f) \right|_1 - \left| \mathbf{P}_J^{FSP}(t_f) \right|_1 + \left| \mathbf{P}_{J'}(t_f) \right|_1 \\
&= 1 - \left| \mathbf{P}_J^{FSP}(t_f) \right|_1 \\
&= g(t)
\end{aligned}
\tag{10.4}
$$

10.3.3 The FSP Algorithm

Equation (10.4) suggests an FSP algorithm [44], which examines a sequence of finite projections of the ME. For each projection set, an accuracy guarantee can be obtained using Equation (10.4). If this accuracy is insufficient, more configurations can be added to the projection set, thereby monotonically improving the accuracy. The full algorithm can be stated as follows:

The Original Finite State Projection Algorithm

Inputs *Propensity functions and stoichiometry for all reactions.*
Initial probability density vector, $\mathbf{P}(0)$.
Final time of interest, t_f.

Total amount of acceptable error, ε > 0.
Step 0 *Choose an initial finite set of states,* $\mathbf{X}_{J_0}$, *for the FSP.*
Initialize a counter, i = 0.
Step 1 *Use propensity functions and stoichiometry to form* $\mathbf{A}_{J_i}$.
Compute g(t_f) by solving Equation (10.3)
Step 2 *If g(t_f) ≤ ε,* **Stop.**
$\mathbf{P}^{\mathrm{FSP}}(t)$ *approximates* $\mathbf{P}(t_f)$ *to within a total error of ε.*
Step 3 *Add more states to find* $\mathbf{X}_{J_{i+1}}$.
Increment i and return to **Step 1.**

In this FSP algorithm, there are many way to choose and expand the projections space in Steps 0 and 3, respectively. The following subsections present a couple of such approach, although others may be equally good.

10.3.3.1　Choosing the Initial Projection Space

A number of different approaches have been proposed for choosing the initial guess for the projection space. In previous work [44], the initial projection set $\mathbf{X}_{J_0}$ was an arbitrarily chosen set of configurations reachable from the initial condition. The most obvious choice is for $\mathbf{X}_{J_0}$ to contain only the initial configuration: $\mathbf{X}_{J_0} = \{\mathbf{x(0)}\}$. The problem with this approach is that the initial projection space is likely to be far too small. In [63] we proposed initializing $\mathbf{X}_{J_0}$ with a set of states determined by running a few trial SSA trajectories. If we use more SSA runs, $\mathbf{X}_{J_0}$ will likely be larger and therefore retain a larger measure of the probability distribution in the specified time interval. As one uses more SSA runs in the initialization portion of Step 0, fewer iterations of the FSP algorithm are necessary, but there is an added computation cost for running and recording the results of the SSA runs. This study and the codes provided use a mixture of the two approaches.

First, a projection space is defined by a set of nonlinear inequalities:

$$\mathbf{X}_J = \{\mathbf{x}_i\}, \text{ such that } \{f_k(\mathbf{x_i}) \le b_k\} \text{ for all constraints } k = \{1, 2, \ldots, K\} \tag{10.5}$$

where the functions $\{f_k(\mathbf{x})\}$ are fixed functions of the populations, and the bounds $\{b_k\}$ are changed to expand or contract the projection space. For example, in the two species $\{\xi_1, \xi_2\}$ systems, these projection shape functions are used:

$$f_1 = -\xi_1, f_2 = -\xi_2, f_3 = \xi_1, f_4 = \xi_2$$

$$f_5 = \max(0, \xi_1 - 4)\max(0, \xi_2 - 4)$$

$$f_6 = \max(0, \xi_1 - 4)^2 \max(0, \xi_2 - 4)$$

$$f_7 = \max(0, \xi_1 - 4)\max(0, \xi_2 - 4)^2$$

With $b_1 = b_2 = 0$, the first of these two constraints specify that the both species must have nonnegative populations. The third and fourth constraints specify the max populations of each species, and the remaining constraints specify additional upper bounds on various products of the population numbers. For all constraints, it is important that increases in the values $\{b_k\}$ correspond to relaxations of the associated constraints and increases in the projections space. In practice, these constraints functions are easily changed—the best choice of constraints remains an open problem that will differ from one system to another. Next, a single SSA simulation is run, and all of the states (ξ_1, ξ_2) visited in that simulation are recorded. Finally, the boundary values $\{b_k\}$ are increased until the inequalities in (10.5) are satisfied. Thus, we arrive at an initial guess for the projection space; the next step is to expand that projection space until the FSP error meets the specified tolerance.

10.3.3.2   Updating the Projection Space

In Step 3 of the FSP algorithm it is necessary to expand the projection space. In [44] the space was expanded to include all of the states that are reachable in one reaction from the current set. Because not all reactions are equal, this is a very inefficient approach to expanding the projection space—it can lead to expanding too far in one direction or too little in another. Here an approach is tailored similar to that in [63] to match our definition of the projection space given in Equation (10.5). For this, K absorbing points $\{g_1, \ldots, g_K\}$ are chosen, where each $g_k(t)$ corresponds to the probability that the system has left the set $\mathbf{X}_J$ in such a way as to violate the k-th boundary condition. To do this, we simply split the index set, J', into K different subsets, $\{J'_1, \ldots, J'_K\}$ where J'_k is the set of states that satisfy the first $(k - 1)$ boundary constraints, but not the k-th boundary constraint:

$$J'_k = \{i\} \text{ such that } \{f_1(\mathbf{x}_i) \le b_1, \ldots, f_{k-1}(\mathbf{x}_i) \le b_{k-1}, f_k(\mathbf{x}_i) > b_k\}$$

With these index sets, we arrive at a new projection for the master equation:

$$\frac{d}{dt}\begin{bmatrix} \mathbf{P}_J^{FSP}(t) \\ g_1(t) \\ \vdots \\ g_K(t) \end{bmatrix} = \begin{bmatrix} \mathbf{A}_J(t) & \mathbf{0} \\ -\sum \mathbf{A}_{J'_1 J}(t) & \mathbf{0} \\ \vdots & \mathbf{0} \\ -\sum \mathbf{A}_{J'_K J}(t) & \mathbf{0} \end{bmatrix} \begin{bmatrix} \mathbf{P}_J^{FSP}(t) \\ g_1(t) \\ \vdots \\ g_K(t) \end{bmatrix} \tag{10.6}$$

The solution of (10.6) at a time t_f yields all of the same information as above. In particular, the sum of the vector $\mathbf{g}(t)$ provides the exact distance between the FSP approximation and the true solution, as was observed in Equation (10.4). In addition, each element, $g_k(t)$, is the probability that the k-th boundary condition was violated and that this violation occurred before the $\{1, 2, \ldots, k - 1\}$-th boundary conditions were violated. This knowledge is easily incorporated into Step 3 of the FSP algorithm. If the k-th boundary condition is violated with high probability, $\mathbf{X}_J$ is expanded by increasing b_k to relax that boundary condition.

10.3.4   Advancements to the FSP Approach

Linear systems theory provides many tools with which the order of the chemical master equation may be reduced and the efficiency of the FSP may be improved. In most of these reductions, the goal is to approximate the vector $\mathbf{P}(t) \in \mathbb{R}^n$ (here, n may or may not be finite) as some linear transformation of a lower dimensional vector, $\mathbf{\Phi P}(t) = \mathbf{q}(t) \in \mathbb{R}^{m \leq n}$. For example, the original FSP is one such projection in which the elements of $\mathbf{q}(t)$ correspond to $\mathbf{P}_j(t)$. There are many other possible projection choices, each of which takes advantage of a different common trait of discrete state Markov processes.

In [62,64], theory concepts of controllability and observability are used to obtain a minimal basis set for the space in which the solution to the FSP evolves. This approach takes into account that not all points in the full probability distribution are necessary and that only a solution for a coarser-level behavior, such as population means, variances, or extreme value events, may be desired. In this case, the vector $\mathbf{q}$ corresponds to the observable and controllable states of the master equation and can have a very low dimension [62].

Alternatively, in [65,66], timescale separations may be used to project the system onto a spaces defined by the system's fast or slow dynamics. In this case, the projection operator, $\mathbf{\Phi}$ is spanned by the appropriately chosen sets of eigenvectors, and $\mathbf{q}(t)$ refers to the dynamics in that space. For long times, $\mathbf{\Phi}$ consists of the eigenvectors corresponding to the slow eigenvalues. Conversely, for short times, $\mathbf{\Phi}$ should consist of the eigenvectors that correspond to the fast eigenvalues. This approach is similar to the ssSSA [35], in that the existence of fast and slow species does indeed result in a separation of time scales in the ME. However, timescale-based reductions to the master equation are more general in that they may be possible even in the absence of clear separations of fast and slow species.

For a third projection-type reduction, the probability distribution is assumed to vary smoothly over some portions of the configuration space. In this case, the FSP problem is solved on a coarse grid and the distributions are interpolated from the solution on the coarse grid. This approach has shown great promise for certain problems [46], particularly for systems with high populations, where the full FSP may have an exorbitant dimension size. Furthermore, it is relatively easy to formulate an algorithm to systematically refine the interpolation grid to attain more precise solutions, which are better tailored to a given system.

For some systems, probability distributions may drift over large portions of the state space yet remain relatively tight in that they are sparsely supported during any given instant in time [45,63]. By splitting the full time interval into many small subintervals, computational effort can be reduced by considering much smaller portions of the state space during each time increment. For further improvements in efficiency, many of these multiple time interval solutions of the FSP can readily be combined with the projection-based reductions.

10.4   Description of the FSP Two-Species Software

Before studying any specific biological problem, it is useful to introduce the FSP Toolkit through a simple tutorial example.*

---

* All examples shown in this tutorial can be accessed by typing "FSP_ToolKit_Main" in the MATLAB command window and then clicking the appropriate button in the resulting graphical user interface.

10.4.1   System Initialization

The first task in analyzing any system is to specify the system mechanics, parameters, and initial conditions. This first example considers a process of gene transcription and translation [3], where the reaction mechanisms are defined as

$$R_1: \quad \phi \rightarrow R \qquad\qquad R_2: \quad R \rightarrow \phi,$$

$$R_3: \quad R \rightarrow R + P \qquad R_4: \quad P \rightarrow \phi.$$

The propensity functions of these reactions are

$$w_1 = k_R \qquad\qquad w_2 = \gamma_R x$$

$$w_3 = k_P x \qquad\qquad w_4 = \gamma_P y$$

where the rates are $\{k_R = 5, \gamma_R = 1, k_P = 5, \gamma_P = 1\}$, and the initial condition is given as five molecules of mRNA ($x(0) = 5$) and two protein molecules ($y(0) = 2$). For this example, a final time of 10 time units has been chosen. For the FSP approach, it is necessary to specify the maximum allowable 1-norm error in the solution of the master equation. For all examples presented in Sections 10.4 and 10.5, a strict accuracy requirement of $\varepsilon = 10^{-6}$ has been set. For the FSP Toolkit, these mechanisms and parameters can be entered or changed directly in the graphical interface, or they can be predefined in a user-specified file.

10.4.2   Generating Stochastic Trajectories

Once the system is specified, it can be solved with many different approaches. Perhaps the simplest such approach is to run a stochastic simulation to find a sample trajectory for the process. In the FSP Toolkit, this can be accomplished simply by pressing the button "Run SSA." Figure 10.3 illustrates three different aspects of sample trajectories obtained from this approach. Figures 10.3a and 10.3b show the populations of x and y, respectively, as functions of time, and Figure 10.3c shows the trajectories in the x–y plane. Because the process is stochastic, each trajectory achieved in this manner will be different. Although a few SSA runs do not provide a solution to the master equation, they do provide a good sense of the system's dynamics. Furthermore, the SSA runs help to choose an initial projections space for use in the FSP solution.

10.4.3   Solving the Master Equation

In the FSP Toolkit, all aspects of the FSP solution can be acquired simply by pressing the button marked "FSP–Solve It" on the graphical interface. For the casual user, this is sufficient to specify and solve the master equation for many problems. However, for the advanced user, it is important to understand how this solution is obtained. This process is described as follows, beginning with the definition of the master equation.

Chapter 10

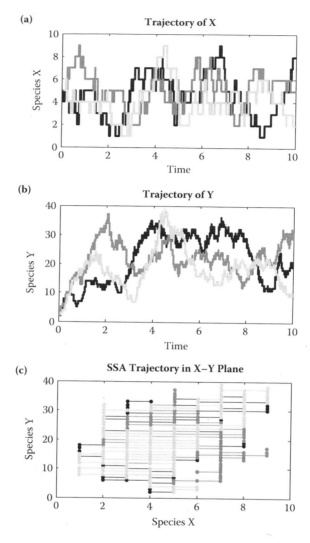

FIGURE 10.3 Three stochastic trajectories for the example system in Section 10.4. (a) Trajectories for species X versus time. (b) Trajectories for species Y versus time. (c) Trajectories on the X–Y plane.

10.4.3.1 Defining the Full Master Equation

With the definition of the mechanisms and parameters, one can also define the master equation, $\dot{\mathbf{P}}(t) = \mathbf{A}(t)\mathbf{P}(t)$. For this, the infinitesimal generator matrix, $\mathbf{A} = \{A_{i,j}\}$ is defined as

$$\mathbf{A}_{ij} = \begin{cases} -\sum_{\mu=1}^{M} w_\mu(\mathbf{x}_i) & \text{for}\,(i=j) \\ w_\mu(\mathbf{x}_j) & \text{for all } j \text{ such that}\,(\mathbf{x}_i = \mathbf{x}_j + \nu_\mu) \\ 0 & \text{Otherwise} \end{cases} \tag{10.7}$$

For the initial distribution, we use the distribution $\mathbf{P}(0) = \{P_i(0)\}$ given as

$$P_i(0) = \begin{cases} 1, & \text{if } \mathbf{x}_i = [5,2] \\ 0, & \text{otherwise} \end{cases}$$

which corresponds to a specific initial condition of five mRNAs and two proteins.

10.4.3.2 Defining the Projected Master Equation

The finite state projection space is governed by the boundary shape functions, $\{f_k\}$, as defined in Equation (10.5). To initialize the constraints, $\{b_k\}$, we use the previous SSA runs as follows. If $\mathbf{X}_{SSA}$ refers to the set of all states, $\{\mathbf{x}\}$, that were visited during the SSA runs, then the initial value for each b_k is set to

$$b_k = \max_{\mathbf{x} \in \mathbf{X}_{SSA}} f_k(\mathbf{x})$$

In turn, the index sets for the projection are defined by the functions $\{f_k\}$ and the constraints $\{b_k\}$ as follows:

$$J = \{i\} \text{ such that } \{f_1(\mathbf{x}_i) \le b_1, \ldots, f_k(\mathbf{x}_i) \le b_k\}$$

$$J_1' = \{i\} \text{ such that } \{f_1(\mathbf{x}_i) > b_1\}$$

$$J_2' = \{i\} \text{ such that } \{f_1(\mathbf{x}_i) \le b_1, f_2(\mathbf{x}_i) > b_2\}$$

$$\vdots = \vdots$$

$$J_K' = \{i\} \text{ such that } \{f_1(\mathbf{x}_i) \le b_1, \ldots, f_{K-1}(\mathbf{x}_i) \le b_{K-1}, f_K(\mathbf{x}_i) > b_k\}$$

With these index sets we can define the projections matrix $\mathbf{A}_J$ and the row vectors $\left\{ \sum \mathbf{A}_{JkJ}(t) \right\}$ for use in Equation (10.6).

10.4.3.3 Solving the Projected Master Equation

Once defined, Equation (10.6) can be solved in a number of different ways depending upon the system. For systems with time varying reaction rates (see Section 10.5.2), a more general stiff ODE solver is necessary. For systems where the matrix $\mathbf{A}$ is constant, the solution can be found with Krylov subspace methods included in Roger Sidje's expokit ([67]; http://www.expokit.org). With either solution scheme, the solution of (10.6) is solved incrementally in time from initial time t_0 to time $t = min(t_f, t_v)$, where t_v is the time at which the FSP error tolerance is first observed to be violated. This definition of t_v is necessary only for efficiency reasons—at time t_v the algorithm already knows that the current projection is insufficient and which of the K boundary conditions have been violated. By exiting early from the ODE solver, the solution need not be computed over the interval (t_v, t_f), and there is a significant computational savings.

10.4.3.4 Updating the Projection

Upon solving Equation (10.6), there are two possibilities: either the truncation error was small enough $\left(\sum_{k=1}^{K} g_k(t_f) \le \varepsilon \right)$ and the solution is acceptable; or the projection must

Chapter 10

be expanded to encompass more states. In the latter case, the values of $\{g_k(t_v)\}$ are used to increase the boundary constraint constants. For the examples shown here, the simple expansion rule is used:

If $g_k(t_v) \geq \varepsilon / K$ then $b_k + 0.05|b_k| \mapsto b_k$

Once the boundary constants have been updated in this manner, the next projection can be defined, and the FSP algorithm may continue.

10.4.3.5 Analyzing FSP Solutions

Once the FSP error tolerance has been met, there are a number of ways to represent and understand the acquired solution. Figure 10.4 shows a couple of these representations that are automatically plotted using the FSP Toolkit. Figure 10.4a shows a contour plot of the joint probability distribution for the populations of mRNA and protein molecules and Figures 10.4c and d show marginal distributions for each of these species separately. Figure 10.4b shows a plot of the projection space that was found during the FSP algorithm, where the region in white is included in the projection and the rest in black is excluded.

All of the results in Figure 10.4 correspond to the solution at the final time of $t_f = 10$ time units. However, once a projection is found to be sufficient for the final time, t_f, that projection will also be satisfactory for all times between t_0 and t_f. With this in mind, the

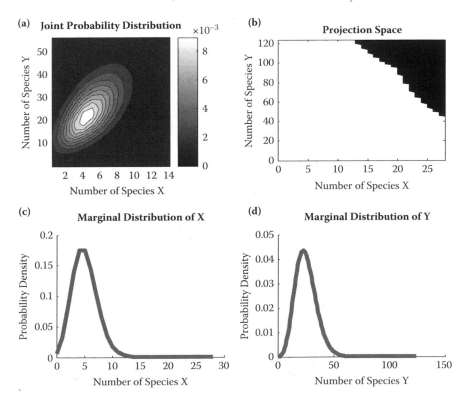

FIGURE 10.4 The FSP solution of the master equation at the final time. (a) The joint probability distribution of species X and Y. (b) The automatically chosen projection space that satisfies a stopping criteria of $\sum (g_{t_f}) \leq= 10^{-6}$. (c–d) The marginal distributions of Species X and Y, respectively.

dynamics of the means for each of the species can be computed as functions of time. For example, the trajectory for species Y is plotted in Figure 10.6a. To generate these plots, the button "Show FSP Dynamics" in "FSP Toolkit" can simply be pressed. In addition to showing trajectories of the means and standard deviation, this will also create a movie of the joint distribution as a function of time.

10.5 Examples of Stochastic Analyses

This section uses the aforementioned stochastic analysis tools to illustrate some important stochastic phenomena in biological systems. All of these examples can be implemented in MATLAB using the codes provided and with minimal user input. The reader is strongly encouraged to work through each example using this software.

10.5.1 Example 1: Using Autoregulation to Reduce Variability in Gene Expression

To illustrate the importance of feedback, the fist example considers the autoregulation of a single gene whose protein inhibits its own transcription. The simplified model is composed of four simple reactions:

$$R_1: \quad \phi \rightarrow R \qquad\qquad R_2: \quad R \rightarrow \phi$$

$$R_3: \quad R \rightarrow R + P \qquad\qquad R_4: \quad P \rightarrow \phi$$

The propensity functions of these reactions are

$$w_1 = k_R / (1 + k_f y) \qquad\qquad w_2 = \gamma_R x$$

$$w_3 = k_P x \qquad\qquad w_4 = \gamma_P y$$

where the term k_f denotes the strength of the negative feedback. For the nominal model without feedback, a parameter set of $\{k_R = 5, \gamma_R = 1, k_P = 5, \gamma_P = 1, k_f = 0\}$, in nondimensional time units, has been chosen. For the feedback model, the feedback term has been set to unity, $k_f = 1$, and the basal transcription rate has been adjusted to $k_R = 120$ to maintain the same mean levels of 5 mRNA transcripts and 25 proteins at the final time. For each parameter set, the distribution of mRNAs and proteins after $t_f = 10$ time units is plotted in Figure 10.5, where the solid lines correspond to the nominal system and the dashed lines correspond to the system with feedback.

By adding negative feedback, the variance in the mRNA levels is reduced by about 20%, whereas the variance in the protein levels is reduced by about 40% (see Figure 10.5). Figure 10.6 shows the trajectories for the mean level of proteins for the two systems with and without feedback. Although both systems eventually achieve the same mean level of protein, the actual dynamics are slightly different. In addition to having less variance, the autoregulated system has a faster response time. However, the increased speed and lower variability in the feedback mechanism come at a cost of protein overproduction, which may be valuable in terms of cellular resources.

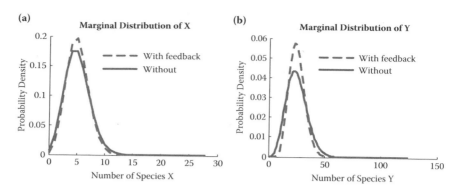

FIGURE 10.5 The effect of feedback in gene regulation. (a) The marginal distributions of Species X. (b) The marginal distributions of Species Y.

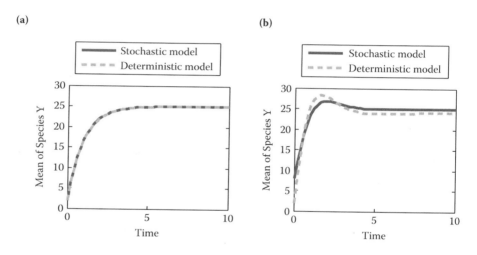

FIGURE 10.6 (See color insert.) The effect of feedback in gene regulation. (a) Trajectory for the mean level of species Y in the absence of feedback. (b) Trajectory for the mean level of species Y with autoregulatory feedback. The dashed lines correspond to the solution to the deterministic ODE model, and the solid lines correspond to the mean of the stochastic model.

10.5.2 Example 2: Using Nonlinearities and Stochasticity to Amplify or Damp External Signals

For systems with *linear* reaction rates, stochastic models will exhibit the exact same mean-level behavior as the corresponding deterministic description. This connection rests on the fact that the expected value of a linear function is equal to the same linear function applied to the mean of its argument. In other words, $E\{f(\tilde{x})\} = f(\{E\{\tilde{x}\})$, for any distribution of $\tilde{x}$, when $f(.)$ is linear. For nonlinear functions, this equivalence typically does not hold, and nonlinearities can result in big differences between the stochastic and deterministic representations of the system. Of particular interest is when $f(x)$ has a significant curvature over the support of the variable x. According to Jensen's inequality, if $f(.)$ is convex over the support of x, then $E\{f(x)\} \geq f(\{E\{x\})$. In this case, nonlinearities and stochasticities combine to amplify the signal. If the function is concave, then the inequality is reversed, and the signal is effectively damped by the nonlinear and stochastic effects. This example shows how this nonlinearity can help to amplify or damp a system's response to an external signal.

This example considers a two-species system composed of four simple reactions:

$$R_1: \quad \phi \to X \qquad\qquad R_2: \quad X \to \phi$$

$$R_3: \quad \phi \to Y \qquad\qquad R_4: \quad Y \to \phi$$

In this system it is assumed that the production of the first species, X, is modulated by an external signal according to $w_1 = k_x(1 + e^{-i(\Omega t - \pi/2)})$, and the degradation of X is given by the standard $w_2 = \gamma_x x$. Three subcases are considered for the production of species Y:

(a) Species X *activates* the production of species Y according to the *linear* propensity function $w_3^- = k_y x / 2$.

(b) Species X *activates* production of species Y according to the *concave* propensity function $w_3^\cap = k_y x / (1+x)$.

(c) Species X *represses* the production of species Y according to the *convex* propensity function $w_3^\cup = k_y / (1+x)$.

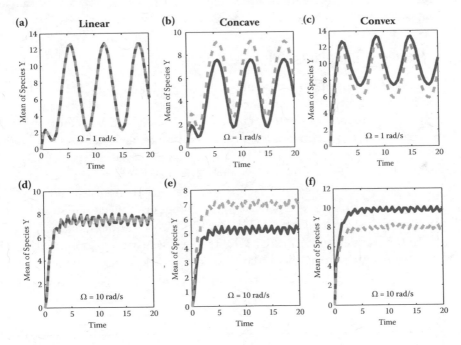

FIGURE 10.7 (See color insert.) The effects of nonlinearities and stochasticity on signal transduction. (a) Trajectory for the mean level of species Y with when species X activates Y through linear regulation. (b) Trajectory for the mean level of species Y when species X activates Y with a concave function. (c) Trajectory for the mean level of species Y when species X represses Y with a convex function. (a)–(c) correspond to the same systems as (a)–(c), but where the external signal varies with a frequency of 1 rad/s. (d)–(f) correspond to a system where the external signal varies with a frequency of 10 rad/s. In all plots, the dashed lines correspond to the solution to the deterministic ODE model, and the solid lines correspond to the mean of the stochastic model.

Chapter 10

In all three subcases, the degradation of species Y is given by $w_4 = \gamma_y y$. The parameters are kept the same in all three cases: {$k_x = 10s^{-1}$, $\gamma_x = 10s^{-1}$, $k_y = 15s^{-1}$, $\gamma_y = 1s^{-1}$, $\Omega = 1s^{-1}$}.

The FSP Toolkit can be used to solve for and generate movies of the distributions of X and Y as functions of time. The solid lines in Figure 10.7 illustrate the dynamics of the mean of X for each of the different subcases where the propensity for the production is linear, convex, or concave. For comparison, the dashed lines show the corresponding solution to the deterministic ordinary differential equation:

$$\frac{dx}{dt} = w_1(x,t) - w_2(x); \quad \frac{dy}{dt} = w_3(x) - w_4(y)$$

As expected, the mean level of the system with the linear reaction rates is exactly the same as the solution of the deterministic ODEs (Figure 10.7a). For the concave activation of Y, it is evident that the nonlinear stochastic effects dampen the response to the external signal (Figure 10.7b). Finally, for the convex repression of Y, it is clear that the nonlinear stochastic effects amplify the response to the external signal (Figure 10.7c).

As a side note, the system considered here also acts as a low-pass filter of the external signal. At a frequency of 1 rad/s, the external signal is easily passed through the system. However, if the external fluctuations are much higher in frequency, say 10 rad/s, then the fluctuations become much smaller (Figures 10.7d through 10.7f).

10.5.3 Example 3: Stochastic Toggle Switch

One of the most obvious of stochastic phenomena in biological systems is that of stochastic switching. To illustrate this phenomenon and how it could be analyzed, this example considers the toggle switch composed of genes *lacI* and λcI, which inhibit each other. This system was experimentally constructed in [68] and later used as a sensor of ultraviolet light in the environment [21]. In the switch the proteins λcI and LacI inhibit each other, as shown in Figure 10.8. The switch works as a sensor because the degradation rate of λcI is sensitive to various factors in the external environment, and the system is tuned so that its phenotype is sensitive to changes in this degradation rate. With low degradation rates, λcI will outcompete LacI—in high λcI degradation conditions, LacI will win the competition. In [21], a GFP reporter was used to quantify the expression level of LacI. While many models are capable of describing this and other toggle switches (see, e.g., [19,21,69,70]), a relatively simple model from [22] is considered here. This model is described as follows.

FIGURE 10.8 Schematic of the toggle switch model. Two proteins, λcI and LacI, inhibit each other. Environmental influences (ultraviolet radiation) increase the degradation rate of λcI and affect the trade-off between the two regulators.

It is assumed that four nonlinear production–degradation reactions can change the populations of λcI and LacI according to

$$R_1 \quad ; \quad R_2 \quad ; \quad R_3 \quad ; \quad R_4$$
$$\phi \to \lambda cI \quad ; \quad \lambda cI \to \phi \quad ; \quad \phi \to LacI \quad ; \quad LacI \to \phi \tag{10.8}$$

The rates of these reactions, $\mathbf{w}(\lambda cI, LacI, \Lambda) = [w_1(\lambda cI, LacI, \Lambda), \ldots, w_4(\lambda cI, LacI, \Lambda)]$ depend upon the populations of the proteins, λcI and LacI, and the parameters in

$$\Lambda = \left\{ k_{\lambda cI}^{(0,1)}, \alpha_{LacI}, \eta_{LacI}, k_{LacI}^{(0,1)}, \alpha_{\lambda cI}, \eta_{\lambda cI}, \delta_{LacI}, \delta_{\lambda cI}(UV) \right\} , \text{ according to}$$

$$w_1 = k_{\lambda cI}^{(0)} + \frac{k_{\lambda cI}^{(1)}}{1 + \alpha_{LacI}[LacI]^{\eta_{LacI}}} \qquad w_2 = \delta_{\lambda cI}(UV)[\lambda cI]$$

$$w_3 = k_{LacI}^{(0)} + \frac{k_{LacI}^{(1)}}{1 + \alpha_{\lambda cI}[\lambda cI]^{\eta_{\lambda cI}}} \qquad w_4 = \delta_{LacI}[LacI]$$

In the model, the λcI degradation parameter, $\delta_{\lambda cI}$, takes on different values depending upon the ultraviolet radiation level, whereas the remaining parameters are assumed to be independent of environmental conditions. As in [22], a reference parameter set has been chosen as follows:

$$
\begin{aligned}
& k_{\lambda cI}^{(0)} = 6.8 \times 10^{-5} \text{ s}^{-1} && k_{\lambda cI}^{(1)} = 1.6 \times 10^{-2} \text{ s}^{-1} && \alpha_{LacI} = 6.1 \times 10^{-3} \text{ N}^{-\eta_{LacI}} \\
& k_{LacI}^{(0)} = 2.2 \times 10^{-3} \text{ s}^{-1} && k_{LacI}^{(1)} = 1.7 \times 10^{-2} \text{ s}^{-1} && \alpha_{\lambda cI} = 2.6 \times 10^{-3} \text{ N}^{-\eta_{\lambda cI}} \\
& \eta_{LacI} = 2.1 \times 10^{-0} && \eta_{\lambda cI} = 3.0 \times 10^{-0} && \delta_{LacI} = 3.8 \times 10^{-4} \text{ N}^{-1}\text{s}^{-1}
\end{aligned}
\tag{10.9}
$$

where the notation N corresponds to the integer number of molecules of the relevant reacting species. The following is also assumed for the degradation rate of λcI:

$$\delta_{\lambda cI}(UV) = 3.8^{-4} + \frac{0.002UV^2}{1250 + UV^3}$$

which has been chosen to approximate the values of

$$\{ \delta_{\lambda cI}(0) = 0.00038\text{s}^{-1}, \delta_{\lambda cI}(6) = 0.00067\text{s}^{-1}, \delta_{\lambda cI}(12) = 0.0015\text{s}^{-1} \}$$

as used in [22].

The left side of Figure 10.9 shows numerous trajectories of the system beginning at two different initial conditions in the absence of ultraviolet radiation: one corresponding to high expression of $\lambda cI = 30$ and low expression of LacI $= 5$; and the other corresponding to high expression of LacI $= 60$ and low expression of $\lambda cI = 10$. Both conditions are quite stable over a period of 5 hours when there is no ultraviolet radiation. When ultraviolet radiation is applied, the degradation rate of λcI increases and the stability of the high λcI state is decreased, leading to switching to the high LacI state. This can be observed in the center and right columns corresponding to 6 and 12J/m² UV radiation

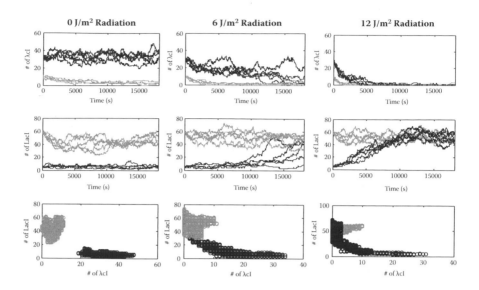

FIGURE 10.9 Trajectories of the genetic toggle switch. Two separate initial conditions are considered corresponding to (λcI,LacI) = (30,5) in black and (λcI,LacI) = (10,60) in gray. Three different ultraviolet radiation levels are considered: 0 J/m², 6 J/m², and 12 J/m² in the left, center, and right columns, respectively. Different aspects of the trajectories are shown in the rows: λcI versus time (top), LacI versus time (middle), λcI versus LacI (bottom).

levels. As the radiation level increases, the rate of switching also increases (compare center and right columns).

To quantify the probability of switching from the high to low λcI state, the master equation can be solved easily using the FSP Toolkit. Figure 10.10 shows the probability distribution for the amount of LacI after 1, 2, 4, and 8 hours for each of the three different ultraviolet levels. In all cases, it is assumed that the system began with 30 molecules of λcI and 5 molecules of LacI.

10.5.4 Example 4: Stochastic Resonance

To illustrate the phenomenon of stochastic resonance, this example turns to a very simple theoretical model of circadian rhythm. The model consists of a singe gene that can have two states, s_1 and s_2. When it is the system is in state s_1, it is active and rapidly produces a protein denoted as Y. This protein is assumed to bind to s_1 with a very high cooperativity factor transforming the gene into state s_2. State s_1 is active and allows for the production of Y, and state s_2 is assumed to be inactive. In addition to binding and forming state s_2, the product Y also stabilize s_2 state, also with a high cooperativity. Mathematically, these reactions are described by

$$R_1: \quad s_1 \rightarrow s_2 \qquad\qquad R_2: \quad s_2 \rightarrow s_1$$

$$R_3: \quad s_1 \rightarrow s_1 + Y \qquad R_4: \quad Y \rightarrow \phi$$

The propensity functions of these reactions are

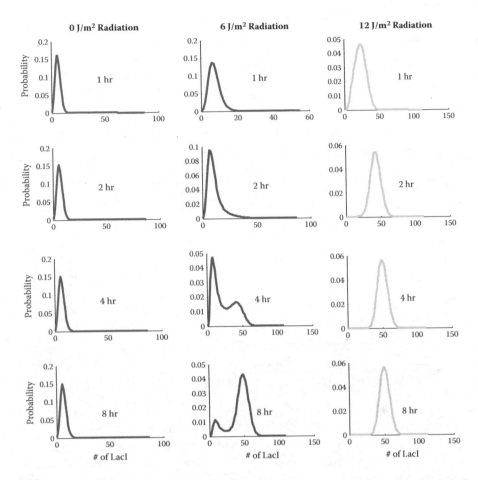

FIGURE 10.10   Distributions of LacI at different times (rows) and ultraviolet radiation levels (columns). All cells start with a high λcI and low LacI expression level (λcI = 30, LacI = 5). Without radiation (left column) the low LacI state is very stable, and very few cells switch. At high ultraviolet radiation (right column) almost all cells switch to the high expression state within about two hours. At a moderate ultraviolet level (center), all cells will eventually switch, but the time to do so is much longer (a significant low LacI population still exists after eight hours).

$$w_1 = \frac{s_1 y^{10}}{1000^{10} + y^{10}} \qquad w_2 = \frac{100 s_2}{1 + y^{10}}$$

$$w_3 = 20 s_2 \qquad w_4 = 0.02 y$$

It is assumed that the system begins at an initial condition where the gene is in s_1 state, and there are 100 molecules of Y. With these reactions and initial conditions, Figure 10.11a illustrates two stochastic trajectories of the system. From the figure, it is clear that the process maintains a strong oscillatory behavior, although the deterministic model of the same process reaches a steady state in only one oscillation (see the dashed lines in Figure 10.11b). The oscillations of the stochastic process would continue in perpetuity, although without restoring interactions, different trajectories become desynchronized after numerous cycles. To illustrate this desynching of the processes,

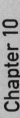

Chapter 10

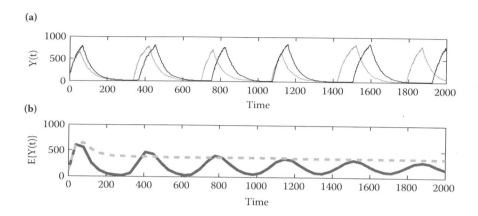

FIGURE 10.11 The effect of noise to induce and maintain oscillations. (a) Two stochastic trajectories of the theoretical circadian rhythm model. (b) The mean level of species Y as a function of time as computed with the stochastic model (solid line) or with the deterministic ODE model (dashed line).

the solid line in Figure 10.11b shows the mean level of the stochastic process over time as computed with the FSP approach. Even after five cycles, the mean level of the population is still showing significant oscillations.

10.6 Identifying Stochastic Models of Gene Regulation

As has been seen already in the previous examples, different mechanisms or parameters can cause biochemical systems to exhibit different behaviors with respect to their fluctuations and cell-to-cell variability. As a result, these fluctuations contain additional information about the underlying system, which might not be obtainable from the mean level behavior [52]. In turn, this information could enable researchers to identify mechanisms and parameters of gene regulatory constructs [22,52]. To illustrate this approach, a model of regulation of the *lac* operon in *E. coli* will be identified under the induction of isopropyl β-D-1-thiogalactopyranoside (IPTG) (Chapter 3 contains information about the regulation of *lac* by thio-methylgalactoside). For this identification, experimental data from [52] are used, but an attempt is made to fit a simpler model to these data.

The chosen model corresponds to a gene that can be in three distinct states, denoted as g_{off^2}, g_{off^1} and g_{on} corresponding to when two, one, or zero molecules of LacI are bound to the *lac* operon (Figure 10.12). The unbinding of LacI is assumed to be the same for both molecules and can be described by the reactions

$$R_1 : g_{off^2} \xrightarrow{w_1} g_{off^1}, \quad R_2 : g_{off^1} \xrightarrow{w_2} g_{on}$$

where each unbinding transition is assumed to depend upon the level of IPTG according to

$$w_1 = \left(\kappa_0 + \kappa_1 [\text{IPTG}]\right) g_{off^2}, \quad w_2 = \left(\kappa_0 + \kappa_1 [\text{IPTG}]\right) g_{off^1}$$

In this model, the total level of LacI, $[\text{LacI}]_{\text{Tot}}$, is assumed to be constant, but the effective amount of LacI free to bind is diminished through the action of IPTG according to

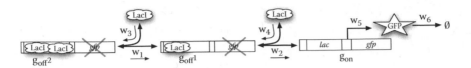

FIGURE 10.12 Schematic of the *lac* induction model with cooperative activation by IPTG.

$$[\text{LacI}]_{\text{eff}} = \frac{[\text{LacI}]_{\text{Tot}}}{\beta + [\text{IPTG}]}$$

The binding rate is then simply a constant times the effective LacI level. To capture the effect of cooperatively, this constant is allowed to depend upon whether it is the first or second LacI molecule to bind.

$$w_3 = \frac{\alpha_1}{\beta + [\text{IPTG}]} g_{\text{off1}}, \quad w_4 = \frac{\alpha_2}{\beta + [\text{IPTG}]} g_{\text{on}}$$

For the current study, the level of IPTG is assumed to be constant for all times $t > 0$.

Production of GFP occurs only when the gene is in the g_{on} state and has the rate $w_5 = k_G g_{\text{on}}$. The propensity for degradation of GFP is the standard linear degradation: $w_6 = \gamma_{\text{GFP}} y$. Because GFP is known to be a stable protein, its degradation rate is set to the dilution rate of $\gamma_{\text{GFP}} = 3.8 \times 10^{-4}$, and there remain five unknown positive real parameters for the regulatory system:

$$\Lambda = \{\kappa_0, \kappa_1, \alpha_1, \alpha_2, \beta, k_G\} \in \mathbb{R}_+^6.$$

In addition to these parameters describing the evolution of the probability distribution for the GFP population, it is also necessary to account for the background fluorescence and variability in the fluorescence of individual GFP molecules. The values for these quantities were obtained in [52]. In particular, the background fluorescence was assumed to be independent of the IPTG levels, and was measured at each instant in time [52]. The mean, $\mu_{\text{GFP}} = 220$ AU, and standard deviation, $\sigma_{\text{GFP}} = 390$ AU, in the fluorescence per GFP molecule are also taken from [52]. It is important to note that the parameters μ_{GFP} and σ_{GFP} are highly dependent upon the flow cytometer and its measurement settings, particularly the thresholds for event detection and the amplification of the fluorescence detector.

The current model is fit to the measurements of GFP fluorescence at {5, 10, 20, 40, 100} µM IPTG and at times of {0, 3, 4, 5} hours after induction. These data are in arbitrary units of fluorescence and have been collected into sixty logarithmically distributed increments between 10 and 10^5. The two separate data sets are shown with solid lines in Figure 10.13, where the columns correspond to different time points and the rows correspond to different IPTG induction levels. For the initial conditions, it was assumed that every cell begins in the g_{off2} state with zero molecules of GFP and no IPTG at a time of three hours before induction ($t = -3$hr). The fit is begun with an initial guess of unity for β and 10^{-4} for the remaining five parameters, and the search is run using numerous iterations of MATLAB's *fminsearch* and a simulated annealing algorithm. The objective for the fit is to match the fluorescence distribution as close as possible in

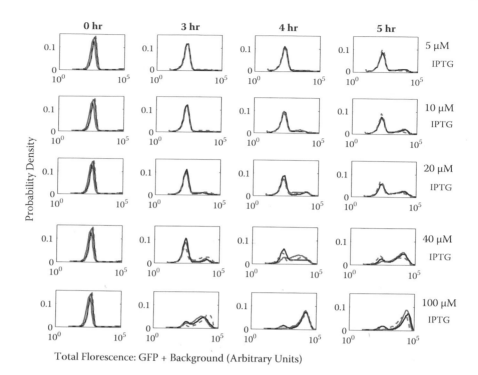

Total Florescence: GFP + Background (Arbitrary Units)

FIGURE 10.13 (See color insert.) Measured (solid lines) and computed (dashed lines) histograms of GFP under the control of the *lac* operon and induced with IPTG. The columns correspond to different measurement times {0,3,4,5} hr after induction. The rows correspond to different levels of extracellular IPTG induction {5,10,20,40,100} µM. Experimental data are reproduced from Munsky et al. [52] (with permission), but a different model is used to fit this data as described in the text.

the 1-norm sense for all times and IPTG levels. The electronic data and the codes for fitting this data can be downloaded online (http://cnls.lanl.gov/~munsky) or requested from the author.

For the chosen model, Figure 10.13 shows the distributions of the measured and fitted GFP fluorescence levels in the various experimental conditions. The figure shows that this simplified model does indeed capture the qualitative and quantitative features of the distributions at the different times and IPTG induction levels (compare dashed lines to solid lines). The final parameter values of the fit were found to be

$$\kappa_0 = 1.91 \times 10^{-5}\,\mathrm{s}^{-1} \qquad \kappa_1 = 3.21 \times 10^{-6}\,\mu\mathrm{M}^{-1}\mathrm{s}^{-1} \qquad \beta = 4.88 \times 10^2\,\mu\mathrm{M}$$
$$\alpha_1 < 1.0 \times 10^{-10}\,\mu\mathrm{Ms}^{-1} \qquad \alpha_2 = 5.36 \times 10^{-1}\,\mu\mathrm{Ms}^{-1} \qquad k_G = 8.09 \times 10^{-2}\,\mathrm{s}^{-1}$$

It is useful to note that the parameter, α_1 is many orders of magnitude smaller than the parameter α_2, which has the same units. Furthermore, the time scale of reaction three is on the order of $w_3^{-1} \approx \beta/\alpha_1 \gg 10^{10}\,\mathrm{s}^{-1}$, which is far longer than the experimental time. This suggests that this parameter is not necessary for the model to fit the data. Indeed, setting $\alpha_1 = 0$ results in no appreciable difference in the model fits for any of the cases. In other words, the state $g_{\mathrm{off}2}$ is not needed in the current model to capture the data, suggesting that this state is unobservable from the current data set.

For the fits shown in Figure 10.13, all of the data were used, and a single parameter set was found. It is also interesting to determine how well smaller subsets of the data would do to (1) constrain the model parameters and (2) enable predictions of the other conditions. To examine this, an attempt is made to identify the model from each possible combination of three or four different IPTG levels. Table 10.1 and Figure 10.14 show the parameters that have been identified with each of these data sets, and Table 10.2 shows the one norm errors in each of the different data sets, where 1-norm prediction errors are shown in bold. From the fits resulting from the various data subsets, it is possible to determine which data sets are the most predictive of the remaining conditions. In particular, when four data sets are available, the best overall fit is found when all but the 40 μM IPTG concentration is used, meaning that the information learned from that condition is redundant to the information contained in the other conditions. Leaving two data sets out, shows that the 20 μM IPTG concentration is also easily predicted from the remaining data sets. By leaving out one or two data sets, we are able to characterize the uncertainty in the parameter values. With three different IPTG concentrations, the uncertainty on the parameters can be estimated as

$$\kappa_0 = 1.62\times10^{-5} \pm 7.75\times10^{-6} \text{ s}^{-1} \qquad \kappa_1 = 3.45\times10^{-6} \pm 4.44\times10^{-7} \text{ μM}^{-1}\text{s}^{-1}$$

$$\beta = 1.78\times10^{3} \pm 3.05\times10^{3} \text{ μM} \qquad \alpha_2 = 1.74\times10^{0} \pm 2.65\times10^{0} \text{ μMs}^{-1}$$

$$k_G = 8.30\times10^{-2} \pm 1.91\times10^{-2} \text{ s}^{-1}$$

Table 10.1   Parameter Sets for the Various Data Subsets for the *lac* Regulation Model

IPTG Levels Used for Fit (μM)	One Norm Differences from Data					
	5 μM	10 μM	20 μM	40 μM	100 μM	Total
{5,10,20}μM	8.35e−01	1.02e+00	1.03e+00	**1.69e+00**	**1.93e+00**	6.51e+00
{5,10,40}μM	8.47e−01	1.02e+00	**1.07e+00**	1.64e+00	**1.82e+00**	6.40e+00
{5,10,100}μM	8.57e−01	1.01e+00	**1.05e+00**	**1.65e+00**	1.78e+00	6.35e+00
{5,20,40}μM	8.34e−01	**1.05e+00**	1.04e+00	1.65e+00	**1.81e+00**	6.38e+00
{5,20,100}μM	8.36e−01	**1.04e+00**	1.04e+00	**1.65e+00**	1.79e+00	6.36e+00
{5,40,100}μM	8.34e−01	**1.06e+00**	**1.07e+00**	1.64e+00	1.79e+00	6.39e+00
{10,20,40}μM	**8.85e−01**	1.00e+00	1.04e+00	1.65e+00	**1.84e+00**	6.42e+00
{10,20,100}μM	**8.91e−01**	1.00e+00	1.04e+00	**1.66e+00**	1.78e+00	6.37e+00
{10,40,100}μM	**8.97e−01**	1.01e+00	**1.07e+00**	1.65e+00	1.77e+00	6.40e+00
{20,40,100}μM	**9.20e−01**	**1.12e+00**	1.03e+00	1.65e+00	1.88e+00	6.60e+00
{5,10,20,40}μM	8.39e−01	1.03e+00	1.04e+00	1.66e+00	**1.81e+00**	6.37e+00
{5,10,20,100}μM	8.59e−01	1.01e+00	1.04e+00	**1.66e+00**	1.78e+00	6.35e+00
{5,10,40,100}μM	8.60e−01	1.01e+00	**1.07e+00**	1.65e+00	1.78e+00	6.36e+00
{5,20,40,100}μM	8.35e−01	**1.04e+00**	1.04e+00	1.65e+00	1.79e+00	6.36e+00
{10,20,40,100}μM	**8.80e−01**	1.00e+00	1.05e+00	1.65e+00	1.78e+00	6.36e+00
{5,10,20,40,100}μM	8.55e−01	1.02e+00	1.04e+00	1.65e+00	1.78e+00	6.35e+00

(a)

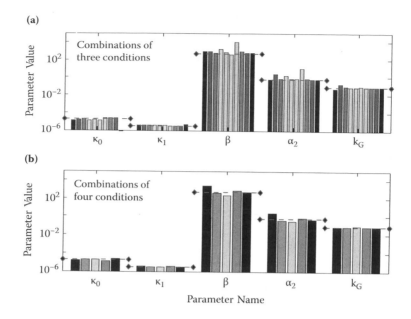

(b)

FIGURE 10.14 (See color insert.)   Identified parameters for the induction of *lac* with IPTG. (a) Parameters identified with every possible combination of three different IPTG concentration from {5,10,20,40,100} μM. (b) Parameters identified with every possible combination of four different IPTG concentrations. In each set of bars, the diamonds and the horizontal dashed lines correspond to the parameter set identified from all five IPTG levels.

where the values are listed as the mean plus or minus one standard deviation. From these results it is clear that the values of κ_0, κ_1, and k_G are well determined from just three different IPTG concentrations, but the other values are more poorly constrained. By adding a fourth IPTG concentration, the uncertainty drops considerably for all six parameters as follows:

$$\kappa_0 = 1.84 \times 10^{-5} \pm 4.18 \times 10^{-6} \text{ s}^{-1} \qquad \kappa_1 = 3.32 \times 10^{-6} \pm 2.20 \times 10^{-7} \text{ μM}^{-1}\text{s}^{-1}$$

$$\beta = 8.47 \times 10^2 \pm 9.26 \times 10^2 \text{ μM} \qquad \alpha_2 = 8.63 \times 10^0 \pm 8.43 \times 10^{-1} \text{ μMs}^{-1}$$

$$k_G = 8.01 \times 10^{-2} \pm 7.44 \times 10^{-3} \text{ s}^{-1}$$

From these values it is clear that the addition of the fourth concentration goes a long way toward helping to constrain the parameters.

The code "'FSP Fit_Tools" provides a simple graphical user interface with which these different fits can be obtained and plotted for each of the different subsets of data. This example considered only a single model for the IPTG induction of the *lac* operon, and a single parameter set was obtained with which this model does a good job of capturing the observed experimental behavior. Other models will perform better or worse than that presented here. The interested reader is encouraged to use the provided FSP Toolkit codes to propose and test alternate models for this system. Also included in the online software is an example of identifying a model of the toggle switch (Section 10.5.3) from simulated data of the marginal and full distributions at different levels of ultraviolet radiation (see also [22]).

Table 10.2   One Norm Errors in the Distributions for the Various Fits

IPTG Levels Used for Fit (µM)	Best Fit Parameter Values					
	κ_0 s^{-1}	κ_1 $\mu M^{-1}s^{-1}$	β μM	α_1 μMs^{-1}	α_2 μMs^{-1}	k_G s^{-1}
{5,10,20}µM	1.22e−05	3.94e−06	7.76e+02	<1e−10	5.52e−01	5.53e−02
{5,10,40}µM	1.74e−05	3.34e−06	7.84e+02	<1e−10	1.66e+00	1.29e−01
{5,10,100}µM	1.89e−05	3.43e−06	6.23e+02	<1e−10	7.53e−01	8.54e−02
{5,20,40}µM	1.18e−05	3.60e−06	1.72e+03	<1e−10	1.59e+00	7.13e−02
{5,20,100}µM	1.38e−05	3.45e−06	1.63e+03	<1e−10	1.64e+00	8.15e−02
{5,40,100}µM	1.32e−05	3.39e−06	2.12e+02	<1e−10	3.29e−01	9.18e−02
{10,20,40}µM	2.44e−05	2.88e−06	1.03e+04	<1e−10	9.14e+00	7.19e−02
{10,20,100}µM	2.43e−05	3.03e−06	8.04e+02	<1e−10	7.17e−01	7.65e−02
{10,40,100}µM	2.57e−05	3.07e−06	2.00e+02	<1e−10	2.80e−01	8.79e−02
{20,40,100}µM	1.00e−07	4.37e−06	6.87e+02	<1e−10	7.53e−01	7.92e−02
{5,10,20,40}µM	1.43e−05	3.61e−06	2.47e+03	<1e−10	2.33e+00	7.18e−02
{5,10,20,100}µM	1.99e−05	3.22e−06	3.89e+02	<1e−10	3.81e−01	7.52e−02
{5,10,40,100}µM	2.03e−05	3.27e−06	2.06e+02	<1e−10	3.13e−01	9.07e−02
{5,20,40,100}µM	1.38e−05	3.46e−06	6.90e+02	<1e−10	7.94e−01	8.38e−02
{10,20,40,100}µM	2.35e−05	3.05e−06	4.77e+02	<1e−10	4.91e−01	7.89e−02
{5,10,20,40,100}µM	1.91e−05	3.21e−06	4.88e+02	<1e−10	5.36e−01	8.09e−02

Notes: Values shown in regular fonts correspond to the differences for data used in the fitting procedure, whereas values in bold face correspond to errors in the predicted distributions.

10.7 Conclusion

This chapter has presented a few of the phenomena that result from discrete stochastic reactions—stochastic amplifications, stochastic damping, stochastic resonance, and stochastic switching. For each of these phenomena, FSP analysis tools were used to illustrate these behaviors. The chapter showed how different mechanisms and parameters lead to different responses in the face of stochasticity, and it was illustrated how it is possible to use information about the variability of individual cells to help infer regulatory mechanisms and parameters from single cell data. For the readers' convenience, all examples included in this work can be reproduced using FSP Toolkit codes, which can be downloaded online (http://cnls.lanl.gov/~munsky) or requested from the author.

Acknowledgments

The author would like to thank Mustafa Khammash for help in the development of the FSP methodology in Section 10.3; Brooke Trinh for the experimental data in Section 10.6; Chunbo Lou, Stephen Payne, Alvin Tamsir, and other students at the fourth annual

Chapter 10

q-bio Summer Summer School for their feedback on some of the examples; and the Center for Nonlinear Studies (CNLS) at Los Alamos National Laboratory for providing a stimulating environment in which to pursue this research.

References

[1] McAdams, M., and A. Arkin. 1999. Its a noisy business! *Tren. Gen.* 15:65–69.

[2] Elowitz, M., A. Levine, E. Siggia, and P. Swain. 2002. Stochastic gene expression in a single cell. *Science.* 297:1183–1186.

[3] Thattai, M., and A. van Oudenaarden. 2001. Intrinsic noise in gene regulatory networks. *Proc. Natl. Acad. Sci.* 98:8614–8619.

[4] Hasty, J., J. Pradines, M. Dolnik, and J. Collins. 2000. Noise-based switches and amplifiers for gene expression. *PNAS.* 97:2075–2080.

[5] Ozbudak, E., M. Thattai, I. Kurtser, A. Grossman, and A. van Oudenaarden. 2002. Regulation of noise in the expression of a single gene. *Nature Genetics.* 31:69–73.

[6] Federoff, N., and W. Fontana. 2002. Small numbers of big molecules. *Science.* 297:1129–1131.

[7] Kepler, T., and T. Elston. 2001. Stochasticity in transcriptional regulation: origins, consequences, and mathematical representations. *Biophys. J.* 81:3116–3136.

[8] Becskei, A., and L. Serrano. 2000. Engineering stability in gene networks by autoregulation. *Nature.* 405:590–593.

[9] Dublanche, Y., K. Michalodimitrakis, N. Kummerer, M. Foglierini, and L. Serrano. 2006. Noise in transcription negative feedback loops: simulation and experimental analysis. *Molecular Systems Biology.* 2.

[10] Nevozhay, D., R. Adams, K. Murphy, K. Josic, and G. Balazsi. 2009. Negative autoregulation linearizes the dose-response and suppresses the heterogeneity of of gene expression. *Proc. Nat. Acad. Sci. USA.* 106:5123–5128.

[11] Thieffry, D., A. Huerta, E. Perez-Rueda, and J. Collado-Vides. 1998. From specific gene regulation to genomic networks. *Bioseeays.* 20:433–440.

[12] Tan, C., F. Reza, and L. You. 2007. Noise-limited frequency signal transmission in gene circuits. *Biophysical Journal.* 93:3753–3761.

[13] Sohka, T., R. Heins, R. helan, J. Greisler, C. Townsend, and M. Ostermeier. 2009. An externally tunable bacterial band-pass filter. *Proc. Nat. Acad. Sci.* 106:10135–10140.

[14] Paulsson, J., O. Berg, and M. Ehrenberg. 2000. Stochastic focusing: Fluctuation-enhanced sensitivity of intracellular regulation. *PNAS.* 97:7148–7153.

[15] Li, H., Z. Hou, and H. Xin. 2005. Internal noise stochastic resonance for intracellular calcium oscillations in a cell system. *Phys. Rev. E.* 71.

[16] Arkin, A., J. Ross, and M. H. 1998. Stochastic kinetic analysis of developmental pathway bifurcation in phage λ-infected *Escherichia coli* cells. *Genetics.* 149:1633–1648.

[17] Wolf, D., and A. Arkin. 2002. Fifteen minutes of fim: Control of type 1 pili expression in *E. coli. OMICS: A Journal of Integrative Biology.* 6:91–114.

[18] Munsky, B., A. Hernday, D. Low, and M. Khammash. 2005. Stochastic modeling of the Pap-pili epigenetic switch. *Proc. FOSBE.*:145–148.

[19] Tian, T., and K. Burrage. 2006. Stochastic models for regulatory networks of the genetic toggle switch. *PNAS.* 103:8372–8377.

[20] Cagatay, T., M. Turcotte, M. Elowitz, J. Garcia-Ojalvo, and G. Suel. 2009. Architecture-dependent noise discriminates functionally analogous differentiation circuits. *Cell.* 139:512–522.

[21] Kobayashi, H., M. Kaern, M. Araki, K. Chung, T. Gardner, C. Cantor, and J. Collins. 2004. Programmable cells: Interfacing natural and engineered gene networks. *PNAS.* 101:8414–8419.

[22] Munsky, B., and M. Khammash. 2010. Guidelines for the identification of a stochastic model for the genetic toggle switch. *IET Systems Biology.* 4:356–366.

[23] Raj, A., and A. van Oudenaarden. 2009. Single-molecule approaches to stochastic gene expression. *Annual Review of Biophysics.* 38:255–270.

[24] Pardue, M., and G. J. 1969. Molecular hybridization of radioactive DNA to the DNA of cytological preparation. *Proc Nat Acad Sci USA.* 64:600–604.

[25] John, H., M. Birnstiel, and K. Jones. 1969. RNA-DNA hybrids at the cytological level. *Nature.* 223:582–587.

[26] Raj, A., P. van den Bogaard, S. Rifkin, A. van Oudenaarden, and S. Tyagi. 2008. Imaging individual mRNA molecules using multiple singly labeled probes. *Nature Methods*. 5:877–887.

[27] Ghosh, I., A. Hamilton,, and L. Regan. 2000. Antiparallel leucine zipper-directed protein reassembly: Application to the green fluorescent protein. *J. American Chemical Society*. 122:5658–5659.

[28] Cabantous, S., T. Terwilliger, and G. Waldo. 2004. Protein tagging and detection with engineered self-assembling fragments of green fluorescent protein. *Nature Biotechnology*. 23:845–854.

[29] Cabantous, S., and G. Waldo. 2006. *In vivo* and *in vitro* protein solubility assays using split gfp. *Nature Methods*. 3:845–854.

[30] Shapiro, H. 2003. *Practical Flow Cytometry*, 4th Ed. Wiley-Liss.

[31] Gillespie, D. T. 1977. Exact stochastic simulation of coupled chemical reactions. *J. Phys. Chem.* 81:2340–2360.

[32] Gillespie, D. T. 2001. Approximate accelerated stochastic simulation of chemically reacting systems. *J. Chem. Phys.* 115:1716–1733.

[33] Allen, R., P. Warren, and P. Rein ten Wolde. 2005. Sampling rare switching events in biochemical networks. *Phys. Rev. Lett.* 94.

[34] Rao, C. V., and A. P. Arkin. 2003. Stochastic chemical kinetics and the quasi-steady-state assumption: Application to the gillespie algorithm. *J. Chem. Phys.* 118:4999–5010.

[35] Cao, Y., D. Gillespie, and L. Petzold. 2005. The slow-scale stochastic simulation algorithm. *J. Chem. Phys.* 122.

[36] Warmflash, A., P. Bhimalapuram, and A. Dinner. 2007. Umbrella sampling for nonequilibrium processes. *J. Chem. Phys.* 127.

[37] van Kampen, N. 2007. *Stochastic Processes in Physics and Chemistry*, 3rd Ed. Elsevier.

[38] Elf, J., and M. Ehrenberg. 2003. Fast evaluations of fluctuations in biochemical networks with the linear noise approximation. *Genome Research*. 13:2475–2484.

[39] Nasell, I. 2003. An extension of the moment closure method. *Theoretical Population Biology*. 64:233–239.

[40] Gmez-Uribe, C., and G. Verghese. 2007. Mass fluctuation kinetics: Capturing stochastic effects in systems of chemical reactions through coupled mean-variance computations. *JCP*. 126.

[41] Singh, A., and J. Hespanha. 2007. A derivative matching approach to moment closure for the stochastic logistic model. *Bulletin of Mathematical Biology*. 69:1909–1925.

[42] Sinitsyn, N., N. Hengartner, and I. Nemenman. 2009. Adiabatic coarse-graining and simulations of stochastic biochemical networks. *Proc. Nat. Acad. Sci. U.S.A.* 106:10546–10551.

[43] Walczak, A., A. Mugler, and C. Wiggins. 2009. A stochastic spectral analysis of transcriptional regulatory cascades. *Proc. Nat. Acad. Sci.* 106:6529–6534.

[44] Munsky, B., and M. Khammash. 2006. The finite state projection algorithm for the solution of the chemical master equation. *J. Chem. Phys.* 124.

[45] Burrage, K., M. Hegland, S. Macnamara, and R. Sidje. 2006. A krylov-based finite state projection algorithm for solving the chemical master equation arising in the discrete modelling of biological systems. *Proc. of The A.A.Markov 150th Anniversary Meeting.*:21–37.

[46] Munsky, B., and M. Khammash. 2008. The finite state projection approach for the analysis of stochastic noise in gene networks. *IEEE Trans. Automat. Contr./IEEE Trans. Circuits and Systems: Part 1.* 52:201–214.

[47] Munsky, B. 2008. The finite state projection approach for the solution of the chemical master equation and its application to stochastic gene regulatory networks. Ph.D. thesis, Univ. of California at Santa Barbara, Santa Barbara.

[48] Warmflash, A., and A. Dinner. 2008. Signatures of combinatorial regulation in intrinsic biological noise. *Proc. Nat. Acad. Sci. USA*. 105:17262–17267.

[49] Dunlop, M., R. Cox III, J. Levine, R. Murray, and M. Elowitz. 2008. Regulatory activity revealed by dynamic correlations in gene expression noise. *Nature Genetics*. 40:1493–1498.

[50] de Ronde, W., B. Daniels, A. Mugler, N. Sinitsyn, and I. Nemenman. 2009. Mesoscopic statistical properties of multistep enzyme-mediated reactions. *IET Syst. Biol.* 3:429–437.

[51] Mettetal, J., D. Muzzey, J. Pedraza, E. Ozbudak, and A. van Oudenaarden. 2006. Predicting stochastic gene expression dynamics in single cells. *Proc. Natl. Acad. Sci. U S A*. 103:7304–7305.

[52] Munsky, B., B. Trinh, and M. Khammash. 2009. Listening to the noise: random fluctuations reveal gene network parameters. *Molecular Systems Biology*. 5.

[53] Thorsley, D., and E. Klavins. 2010. Approximating stochastic biochemical processes with Wasserstein pseudometrics. *IET Systems Biology*. 4:193–211.

[54] Gillespie, D. T. 1992. A rigorous derivation of the chemical master equation. *Physica A*. 188:404–425.

[55] Bel, G., B. Munsky, and I. Nemenman. 2010. Simplicity of completion time distributions for common complex biochemical processes. *Physical Biology.* 7.

[56] Haseltine, E., and J. Rawlings. 2002. Approximate simulation of coupled fast and slow reactions for stochastic chemical kinetics. *J. Chem. Phys.* 117:6959–6969.

[57] Salis, H., and Y. Kaznessis. 2005. Accurate hybrid stochastic simulation of a system of coupled chemical or biological reactions. *J. Chem. Phys.* 112.

[58] Tian, T., and K. Burrage. 2004. Binomial leap methods for simulating stochastic chemical kinetics. *J. Chem. Phys.* 121:10356–10364.

[59] Cao, Y., D. T. Gillespie, and L. R. Petzold. 2005. Avoiding negative populations in explicit Poisson tau-leaping. *J. Chem. Phys.* 123.

[60] Rathinam, M., L. R. Petzold, Y. Cao, and D. T. Gillespie. 2003. Stiffness in stochastic chemically reacting systems: the implicit tau-leaping method. *J. Chem. Phys.* 119:12784–12794.

[61] Gillespie, D. T. 2000. The chemical Langevin equation. *J. Chem. Phys.* 113:297–306.

[62] Munsky, B., and M. Khammash. 2008. Transient analysis of stochastic switches and trajectories with applications to gene regulatory networks. *IET Systems Biology.* 2:323–333.

[63] Munsky, B., and M. Khammash. 2007. A multiple time interval finite state projection algorithm for the solution to the chemical master equation. *J. Comp. Phys.* 226:818–835.

[64] Munsky, B., and M. Khammash. 2006. A reduced model solution for the chemical master equation arising in stochastic analyses of biological networks. *Proc. 45th IEEE Conference on Decision and Control.*:25–30.

[65] Peles, S., B. Munsky, and M. Khammash. 2006. Reduction and solution of the chemical master equation using time-scale separation and finite state projection. *J. Chem. Phys.* 125.

[66] Munsky, B., S. Peles, and M. Khammash. 2007. Stochastic analysis of gene regulatory networks using finite state projections and singular perturbation. *Proc. 26th American Control Conference (ACC).*:1323–1328.

[67] Sidje, R. B. 1998. EXPOKIT: Software package for computing matrix exponentials. *ACM Transactions on Mathematical Software.* 24:130–156.

[68] Gardner, T., C. Cantor, and J. Collins. 2000. Construction of a genetic toggle switch in *Escherichia coli*. *Nature.* 403:339–242.

[69] Warren, P., and P. Rein ten Wolde. 2005. Chemical models of genetic toggle switches. *J. Phys. Chem. B.* 109:6812–6823.

[70] Lipshtat, A., A. Loinger, N. Balaban, and O. Biham. 2006. Genetic toggle switch without cooperative binding. *Phys. Rev. Lett.* 96.

11. PKA: Prototype for Dynamic Signaling in Time and Space

Susan S. Taylor
Alexandr P. Kornev

Chapter 11

Quantitative Biology: From Molecular to Cellular Systems edited by Michael E. Wall © 2012 CRC Press / Taylor & Francis Group, LLC. ISBN: 978-1-4398-2722-2

11.1　Evolution of Protein Kinases as Regulatory Enzymes

11.1.1　Role of Phosphates in the Regulation of Biology

The importance of phosphates in biology has been recognized since the discovery of ATP and nucleic acids. Westheimer articulated clearly in his 1988 review the importance of phosphates for information transfer and the critical role of the phosphodiester backbone for the structure of both DNA and RNA (Westheimer and Huang 1988). The recent explosion of RNA biology has defined an extraordinary role for RNA in the regulation of biology as well. In terms of bioenergetics, it is ATP that provides the energy to drive all aspects of biology, and the fact that we each turn over approximately 40 kg of ATP a day just in a resting state attests to the importance of this high energy phosphate. However, Westheimer failed to appreciate the critical role of phosphates, and, in particular, protein phosphorylation, in the regulation of biology. This role was first described in the late 1950s by the pioneering work of Krebs and Fischer (1956) on glycogen phosphorylase when they demonstrated that the activity of glycogen phosphorylase was activated by the covalent attachment of a single phosphate. Their subsequent discovery of the activating kinase, phosphorylase kinase, laid the foundations for what was eventually discovered to be one of the largest gene families encoded for by the human genome. The activating kinase for phosphorylase kinase, originally named phosphorylase kinase kinase but later re-named cAMP-dependent protein kinase or PKA, was discovered a decade later (Walsh et al. 1968).

A number of protein-kinases were discovered over the next decade by conventional biochemical techniques. One of the most important set of discoveries occurred in 1978 when the transforming protein of Rous Sarcoma Virus, Src, was shown to be a protein kinase (Collett and Erikson 1978; Levinson et al. 1978). Soon thereafter Hunter and Sefton (1980) discovered that Src phosphorylated tyrosine, not serine or threonine. Once Src was cloned in 1979 (Reddy et al. 1980), it was clear that it belonged to the same family as PKA (Barker and Dayhoff 1982), which was the first protein kinase to be sequenced using traditional methods (Shoji et al. 1981). These findings established the boundaries of the kinase superfamily. However, it was not until the genomic era that we actually began to truly appreciate the full magnitude and diversity of this family and the extent to which biology had used protein phosphorylation to regulate cellular processes. Most eukaryotic organisms use about 2% of their genome to code for protein kinases (Manning et al. 2002); in the case of plants about 4% of their genomes are devoted to protein kinases (Shiu et al. 2004). These enzymes regulate pathways as diverse as cell division, growth, differentiation, memory, metabolism, and immune response, and in many cases it is a cascade of kinases that orchestrate each event. Because the protein kinases also are associated with so many diseases, the structures of many protein kinases have been solved, and this provides us with a unique opportunity to explore the evolution of this key family of regulatory enzymes in terms of structural domains and motifs and also to define the architecture that defines an active protein kinase.

One feature that distinguishes protein kinases from most other enzymes is that they are highly regulated. They are dynamic molecular switches, often part of a large macromolecular complex, that must be rapidly and precisely turned off and on. This distinguishes them in fundamental ways from the metabolic enzymes that have been studied in such depth and have evolved to efficiently generate an ignition switch that initiates a cascade of events. Failure to properly turn off a protein kinase can have catastrophic consequences as evidenced by the many oncogenic kinases that lead to malignancies.

11.1.2 Subdomains of the Eukaryotic Protein Kinase

When the sequence of Src was first compared to PKA (Barker and Dayhoff 1982), it became clear that these two enzymes, coming from such diverse sources, were nevertheless related. The advent of DNA sequencing made cloning routine, and sequences of other kinases began to emerge rapidly in the 1980s. In comparing these sequences Hanks and Hunter (1995) realized that there were many conserved motifs that extended throughout what they defined as the kinase core. These 12 subdomains have remained as a definition of the functional motifs that comprise the eukaryotic protein kinases (EPKs) (Niedner et al. 2006). Once the crystal structure of PKA was solved in 1991 (Knighton et al. 1991a), the conserved sequence motifs acquired functional significance. Given the importance of the EPKs for biology and their close link to many diseases, many kinase structures have been solved since the first PKA structure. These provide a *structural kinome* that allows us to delve more deeply into how an active EPK evolved and how it is assembled. In all cases the conserved kinase core consists of an amino terminal lobe (N-lobe) and a carboxy terminal lobe (C-lobe), with many of the motifs important for docking to ATP and positioning the γ-phosphate for catalysis associated with the N-lobe and most of the residues essential for catalysis and peptide/protein docking associated with the C-lobe (Figure 11.1). The original Hanks and Hunter assignment of subdomains based on sequence conservation has served as a useful framework for defining the structural motifs. These subdomains and their associated motifs are summarized in Figure 11.2 and are correlated with the sequence using PKA as a frame of reference. In addition, a glossary is included here to explain some of the common nomenclature that has come to define the features of the conserved protein kinase family.

11.1.3 Evolution of the Protein Kinase Superfamily

The EPKs evolved from the simpler eukaryotic-like kinases (ELKs) that are abundant in prokaryotes. Many of the ELKs are metabolic enzymes such as choline kinase that act on small molecules (Kannan et al. 2007). Both the ELKs and the EPKs have a bilobal structure. The smaller N-lobe that positions the ATP for catalysis is very similar, and the C-lobe, through the F helix, is also conserved. The two features that are unique to the EPKs are associated with the C-lobe: (1) the activation loop

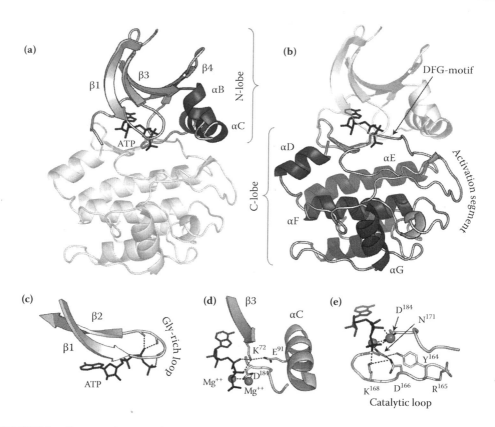

FIGURE 11.1 Conserved protein kinase core. The top panels highlight functional motifs in the N-lobe (left) and the C-lobe (right). (a) N-lobe contains five β-strands and a large αC helix. (b) C-lobe is α-helical with a large activation segment. ATP molecule is bound between the lobes (c) The phosphates are positioned by a conserved glycine-rich loop between β-strands 1 and 2 (d) Conserved residues K72 from β-strand 3 and D164 from the DFG-motif in the activation segment (e) Catalytic loop contains a set of catalytically important residues (e).

that links β-strand 9 to the F helix; and (2) the GHI helical subdomain. The activation loop is highly dynamic and is typically assembled in its active conformation by the addition of a phosphate that is sometimes mediated by either *cis* or *trans* autophosphorylation or by the action of a heterologous activating kinase (*trans*). This activation by phosphorylation is characteristic of the EPKs, and in some cases, such as RSK, four to five kinases contribute to the activation of a single kinase (Anjum and Blenis 2008). This complex regulation by phosphorylation emphasizes that the eukaryotic protein kinases have evolved to be highly dynamic *switches* initiating a downstream signaling event. They are not metabolic enzymes whose function is to efficiently turnover product. The GHI Helical subdomain has evolved most likely to be a docking site for proteins and also for introducing additional regulatory/allosteric control (Figure 11.3).

Both the GHI helical subdomain and the activation loop are anchored firmly to the hydrophobic F helix. These two novel elements of the EPKs are linked to each other by a highly conserved ion pair: Arg 280, which lies between the H and I helices, and Glu208, which is at the end of the activation loop. This site is thought to be a sensitive allosteric

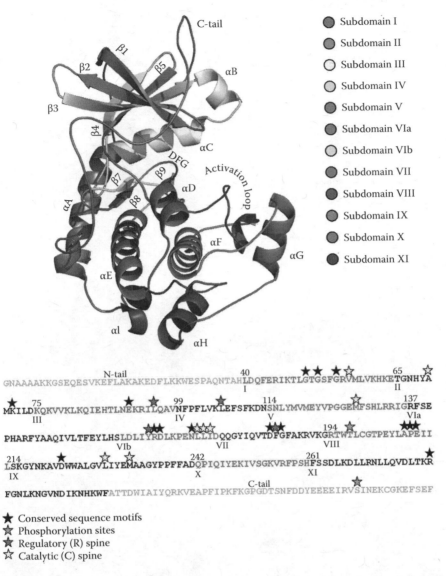

FIGURE 11.2 (See color insert.)   Subdomain assignments based on sequence conservation according to Hanks and Hunter (1995) shown on the PKA structure. Positions of the major conserved residues are shown as stars.

link to the active site and is also a site that is often mutated in disease-associated snps (Torkamani et al. 2008).

11.1.4   Link between cAMP and Kinase Activation

The realization that the protein kinases were highly regulated emerged quickly after the discovery of a phosphorylase kinase that is highly specific for glycogen phosphorylase. It is important to remember that protein kinases phosphorylate protein substrates, not peptide substrates. Glycogen phosphorylase was activated in response to glucagon, an

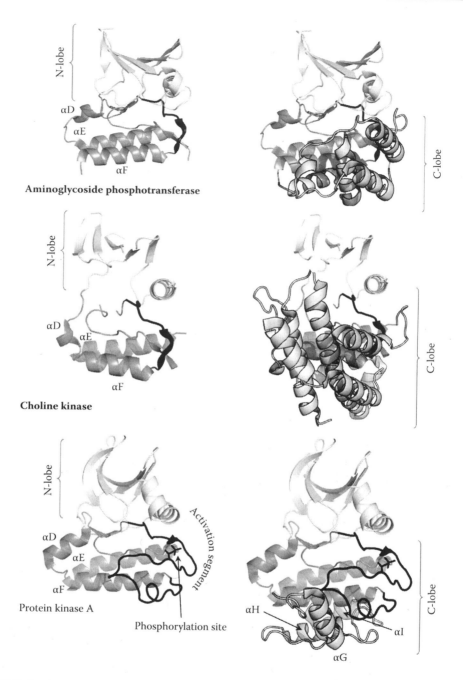

FIGURE 11.3 Activation segment and the helical GHI-subdomain distinguish EPKs from ELKs. The activation segment (shown in black) that joins the DFG-motif to the αF helix is shown on the left; the helical subdomains that follow the αF helix are shown on the right and also indicated in black. PKA, a prototype for EPKs, is compared with two eukaryote-like kinases: Aph and Ck. The extended activation segment and its regulation by phosphorylation are unique features of the EPKs. The GHI-helical subdomain is also conserved in EPKs and functions as a docking site and, most likely, as an allosteric link to the active site in EPKs. While the ELKs also have helical subdomains following the F helix they are not conserved with respect to each other or to EPKs (shown in black).

extracellular hormone. Also, about the same time that Krebs and Fischer (1956) were carrying out their pioneering work, Sutherland discovered that glucagon and epinephrin act by binding to the outside of cells and then generate an intracellular second messenger which they identified as cAMP (Robison et al. 1968). Krebs then showed that cAMP activated another kinase, cyclic AMP-dependent protein kinase (PKA), which in turn activated phosphorylase kinase (Walsh et al. 1968). Several years later Gill and Garren (1971) reported that the receptor for cAMP was a PKA regulating subunit that binds to the PKA catalytic subunit and inhibits its activity. The concept of dynamic kinase cascades and kinases functioning as molecular switches thus emerged very early on and has been mirrored in many other signaling cascades ever since. The dynamic coupling of growth factor receptors such as the insulin receptor to downstream signaling events that lead eventually to the initiation of specific gene expression in the nucleus emphasizes not only the highly dynamic nature of these pathways and these signaling molecules but also the critical importance of localization. The spatiotemporal aspect of cell signaling is as important as switching on and off the activity of the kinases and phosphatases. These pathways can also be oscillatory with kinases and phosphatases targeted to the same scaffold in proximity to a substrate protein and acting in concert. For example, in the case of PKA, two second messengers, cAMP, which activates PKA, and calcium, which activates the calcium-dependent phosphatase, calcineurin, can oscillate in a coordinated way (Ni et al. 2011). In the case of PKC, calcium and diacylglycerol (DAG) define another oscillatory cycle (Violin et al. 2003; Kunkel and Newton 2010). These oscillatory pathways emphasize that the protein kinases are dynamic switches. Also embedded within these oscillatory pathways are many feedback loops. In addition, for PKA the cAMP levels can be regulated by cyclases and phosphodiesterases that synthesize and degrade the second messenger.

11.2   Protein Kinase Structure

11.2.1   Structural Motifs in the Kinase Core

The conserved kinase core that defines the eukaryotic protein kinase superfamily consists of a bilobal protein that contains most of the essential machinery for catalysis and for scaffolding, the two functions that are essential for downstream signaling (Knighton et al. 1991a, 1991b, 1993). Each lobe is made up of helical and beta subdomains, and the active site cleft is formed when ATP binds to the active site cleft causing the two lobes to converge to form a closed conformation. The adenine ring binds to the base of the cleft, whereas the γ phosphate is positioned at the edge of the cleft. Catalysis is mediated by opening and closing of the active site cleft allowing for transfer of the phosphate and then release of the nucleotide. The conserved motifs, defined by the original classification of subdomains in Table 11.1, are distributed throughout the linear sequence, and most of these motifs cluster around the active site cleft and contribute to catalysis. The N-terminal lobe (N-lobe) contains a five-stranded antiparallel beta sheet, which is an essential part of the ATP binding mechanism. β-strands 1 and 2 (subdomain I) are joined by a glycine-rich loop, and this loop packs on top of the ATP. The adenine ring is buried beneath the β-strands while a backbone amide at the tip of the loop anchors the γ-phosphate and positions it for phosphoryl transfer (Madhusudan et al. 2002;

Chapter 11

Table 11.1   Protein Kinase Subdomains and Motifs

No.	Secondary Structure	Residues	Embedded Motifs	Common Name	C-Spine	R-Spine
I	β1–β2	40–64	GxGxxG[55]	Glycine rich loop	V[57]	
II	β3	65–75	AxK[72]	.	A[70]	
III	αB–αC	76–98	E[91]			L[95]
IV	αC–β4 loop, β4	99–113				L[106]
V	β5, Hinge, αD	114–136			M[128]	
VIA	αE	137–160				
VIB	β6–β7	161–175	HRDΦKxEN[171]	Catalytic loop	L[172], L[173], I[174]	Y[164]
VII	β8–β9	176–193	DFG[186]	Magnesium binding loop		F[185]
VIII	β9–αF loop	194–213	APE[208]	Activation loop, P+1 loop		
IX	αF	214–241	DxWxxG[225]		L[227], M[231]	
X	αG	242–260				
XI	αH–αI	261–300	R[280]			

Figure 11.1). β-strand 3 contains an essential Lys that couples to a conserved Glu in the C helix when the kinase is in an active conformation. β-strand 3 (subdomain II) also contains a highly conserved alanine that docks onto the adenine ring of ATP, which is deeply buried between the two lobes. The helical subdomain (III) in the N-lobe in PKA contains a short B helix and a long C helix. The B helix is not conserved in all kinases, but the C helix is a critical part of every kinase. The N-terminus of the C helix interacts with the activation loop phosphate, whereas the C-terminus forms part of the hinge at the base of the active site cleft. β-strands 4 and 5 do not interact directly with the nucleotide.

The C-terminal lobe (C-lobe) is mostly helical and quite stable. The E, F, and G helices, based on hydrogen deuterium exchange, are very stable and do not exchange much with solvent even after several days (Yang et al. 2005). Resting on the helical core at the bottom of the active site cleft is a four-stranded β-sheet, which contains most of the remaining catalytic residues. Between β-strands 6 and 7 lies the catalytic loop (H/YRDXKXXN). The conserved aspartate (Asp[166] in PKA) positions the peptide hydroxyl moiety for phosphoryl transfer (Adams 2001) while the conserved catalytic lysine (Lys[168] in PKA) docks to Asp[166]. In tyrosine kinases the Lys is replaced with an Arg. The asparagine (Asn171 in PKA) is one of the most highly conserved residues in the kinase superfamily and coordinates one of the magnesium ions and also bridges to the magnesium positioning loop that lies between β-strands 8 and 9. The DFG motif lies between β-strands 8 and 9 and in the nucleotide bound conformation this aspartate (Asp184 in PKA) binds to the catalytic magnesium ion. In the proposed transition state for PKA all these motifs from different parts of the molecule converge at the active site to facilitate transfer of the phosphate (Madhusudan et al. 2002).

11.2.2 Hydrophobic Spines Connect the N- and C-Lobes

Now that many kinase structures are available, they can be compared to search for conserved motifs that are in addition to the well-defined sequence motifs previously described. By rigorously analyzing the architecture of the kinase core, it was revealed that the core is built around a very stable hydrophobic core made up of three essential elements: a single hydrophobic helix that spans the large lobe (F helix); and two hydrophobic spines each made of noncontiguous residues from both lobes (Taylor and Kornev 2011). These spines are composed of noncontiguous residues and were discovered only by analyzing the structure; unlike the aforementioned linear motifs, these would never be recognized as conserved motifs based on sequence comparisons alone. The spines were first recognized when active and inactive kinases were compared using a method referred to as local spatial pattern (LSP) alignment (Kornev et al. 2006). This method allows for a rapid comparison of any two related structures and identification of spatially conserved residues. What became clear in the comparison of active and inactive kinases is that every active kinase had a contiguous hydrophobic spine that is made up of two residues from the N-lobe and two from the C-lobe. This spine is broken in most inactive kinases.

Typically the spine is assembled as a consequence of phosphorylation of the activation loop that joins β-strand 9 to the F helix (see Section 11.3.1). This causes the DFG motif that lies between β-strands 8 and 9 to flip so that the Phe is positioned to complete the spine while the Asp is positioned to interact with one of the ATP bound magnesium ions. The DFG-phenylalanine is bound to a highly conserved histidine in the HRD-motif. In PKA and some of the other AGC kinases this His is replaced with a Tyr (Tyr164), but in each case the hydrophobic residue is packed against the Phe from the DFG motif (Phe185 in PKA). In the N-lobe of PKA the two hydrophobic spine residues are Leu97 in the C helix and Leu104 in β-strand 5. These two residues not only are linked to the C-lobe through the R-spine but also serve to anchor the beta and helical subdomains together in the N-lobe (Taylor and Kornev 2011; Figure 11.4). These two spine residues are always hydrophobic, but the specific amino acid is not conserved.

Using the LSP alignment method to compare all kinases, another hydrophobic spine was found running parallel to the regulatory (R) spine. Unlike the R-spine, this second spine is completed not by the dynamic insertion of a hydrophobic amino acid side chain but rather by the adenine ring of ATP. Special attention should be paid to the alanine in β-strand 3 (Ala[70] in PKA). This Ala, along with Asn171 in the catalytic loop is one of the most invariant residues in the kinome superfamily. ATP binding thus positions the two lobes so that the catalytic residues are aligned optimally for catalysis. The dynamic effect of adding MgATP is reflected in the NMR analysis of the PKA C-subunit and shows clearly how ATP binding creates a kinase that is catalytically committed to catalysis whereas in the apo enzyme the two lobes are mostly uncoupled (Masterson et al. 2010, 2011). Both spines are anchored to the hydrophobic F helix. The R-spine is anchored to the N-terminus of the C helix by another conserved Asp (Asp220) that stabilizes the backbone of the HRD loop. The C-spine is anchored through hydrophobic residues to the C-terminus of the F helix. The catalytic loop is also firmly anchored through hydrophobic residues to the F helix (Kornev et al. 2008).

Cha ter 11

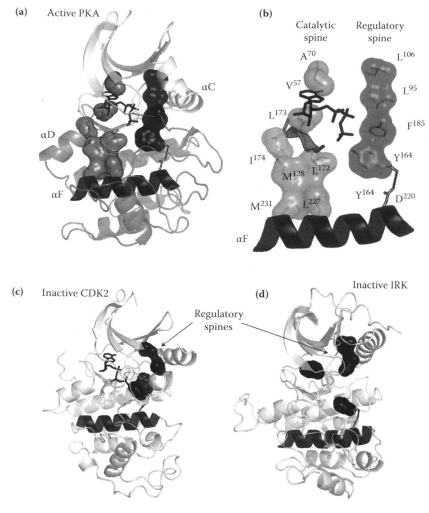

FIGURE 11.4   Internal architecture of the conserved protein kinase core is built around the αF helix. (a) Two hydrophobic spines span the molecule joining the N- and C-lobes and providing their firm but flexible connection. (b) The R-spine contains four residues from different kinase subdomains and is anchored to the αF helix by conserved D220. The C-spine is completed by ATP. Upon inactivation the R-spine is disassembled. The disassembly can be achieved in different ways: (c) Movement of the αC helix like in CDK2 (d) or transformation of the activation segment, IRK.

11.2.3   Flanking Regions Are Integrated with the Core

Although the core that defines the protein kinase superfamily contains most of the essential catalytic machinery, the core is, surprisingly, not usually sufficient to mediate optimal catalysis on its own. Instead, it is assembled into a fully active state by interactions with flanking regions or domains that are anchored to the core. In the case of PKA there is a short N-tail (39 residues) and C-tail (50 residues) that wrap around both lobes of the core (Figure 11.5). Without these tails the kinase is neither stable nor active. MAPK is similar in that it has N- and C-tails that wrap around the core in different ways but nevertheless achieve that same function of stabilizing the active kinase.

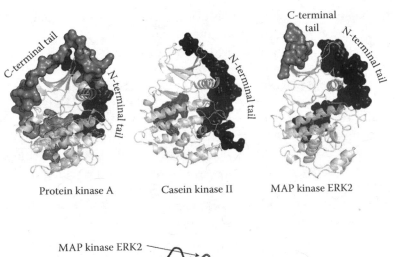

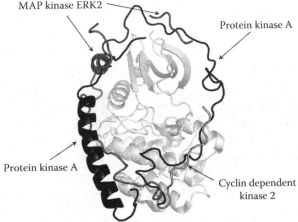

FIGURE 11.5 Integration of the flanking regions with the kinase core. Flanking regions wrap around the conserved kinase core in a kinase specific way providing its stability and allosteric mechanisms for regulation. Shown on the top are three examples: protein kinase A, casein kinase II and ERK2. Each kinase core is displayed as a ribbon. N-terminal tails are shown as black surface, and C-terminal tails are shown in gray. On the bottom, flanking regions of three different kinases are overlaid to show that they are spatially organized in similar ways to integrate the core.

The non-receptor tyrosine kinases such as Src typically have two N-terminal domains, an SH2 and an SH3 domain, and these interact with the core in the inactive conformation to keep the kinase activity turned off. Unleashing of the SH2- and SH3-domains, typically mediated by dephosphorylated a C-terminal phospho-tyrosine, activates the kinase. The receptor tyrosine kinases are the most challenging because they contain a large extracellular domain that binds typically to a growth factor. Binding of the growth factor to the extracellular domains promotes dimerization and also releases the inhibition of the cytoplasmic kinase domain. In addition to a single transmembrane helix, there is a linker region that joins the single transmembrane helix to the kinase core and also a C-terminal tail. The linker and the C-tail are filled with various phosphorylation sites that create a highly sophisticated regulatory mechanism; however, it is extremely difficult to trap these states in a stable crystal lattice for structural studies.

Cha ter 11

The highly sophisticated ways the tails can contribute to both regulation and catalysis can perhaps best be appreciated by looking more closely at the PKA catalytic subunit and the AGC subfamily. PKA is flanked at its N-terminus by an N-terminal myristyl moiety that is followed by an amphipathic helix. The hydrophobic surface of the helix is anchored to the large lobe of the kinase core and to a critical hydrophobic hinge socket that lies at the base of the active site cleft (Herberg et al. 1997). The hydrophilic surface of the helix also provides a docking site for another PKA interacting protein called A kinase interacting protein 1 (AKIP1). AKIP1 is involved in trafficking the catalytic subunit to the nucleus (Sastri et al. 2005). This N-terminal helix is not conserved in the AGC subfamily although other motifs play a similar role in docking to this surface of the core. The N-terminal myristylation site is thought to be a switch mechanism. In the free C-subunit this acyl moiety is localized to an acyl pocket similar to Abl kinase (Zhang et al. 2010), and the free C-subunit does not localize to membranes. Binding to an RII regulatory subunit, however, mobilizes the N-terminal acyl group, and this holoenzyme complex now binds well to lipid vesicles (Gangal et al. 1999).

Unlike the N-tail, the C-tail is highly conserved in all members of the AGC subfamily, such as PKC, PKG, Akt, S6K, and Rsk (Kannan et al. 2007) and is also highly regulated by phosphorylation, both *cis* and *trans*. The C-tail of PKA, and indeed of all AGC protein kinases, can be divided into three functionally distinct segments. The first segment (residues 301–316) is defined as the C-lobe tether (CLT), and in most AGC kinase structures so far this segment is firmly anchored to the kinase core by several conserved sequence motifs such as W307 and PXXP (313–316). The PXXP motif in PKCα was shown to bind to Hsp90 (Gould et al. 2009), and in Akt this motif is predicted to bind to Src (Jiang and Qiu 2003). The next segment (residues 317–336) is defined as the active site tether (AST), and this segment can be ordered or disordered. This segment contains an FDX(X)Y/F motif that is an integral part of the ATP binding site. The first Phe is part of the adenine binding pocket, and mutation of this Phe results in significant loss of activity (Phe) (Yang et al. 2009). The second aromatic residue (Phe or Tyr) lies on top of the ribose ring when the enzyme is in a closed conformation. In the absence of bound nucleotide, the AST segment is highly disordered in crystal structures while binding of the nucleotide drags the AST into the active conformation. The C-tail also has a critical phosphorylation site, Ser[338] in PKA. This has been designated as the turn motif phosphorylation site. The C-terminal portion of the C-tail (residues 336–350) is anchored to the N-lobe in the active enzyme and is referred to as the N-lobe tether (NLT). At the end of the NLT is a hydrophobic (HF) motif (FXXF), and this is anchored to the C helix in the N-lobe. PKA ends with the hydrophobic motif, but other AGC kinases continue with a phosphorylation site immediately following the HF motif and additional residues C-terminal residues.

11.3 Assembly of Regulated and Dynamic Molecular Switches

In general, the kinase core is regulated by motifs and linkers that lie outside the boundaries of the conserved core. PKA gives the advantage of having the kinase core packaged as a relatively compact enzyme with relatively short tails that flank both the N- and C-terminal ends of the kinase core. Without these tails the kinase is not active. These tails, which wrap around both lobes of the kinase core, are thus *cis* regulatory elements and are an essential part of the active enzyme. Unlike many, perhaps most, other kinases where activation is achieved by the dynamic turnover of an activation loop phosphate, the phosphates that are attached to the activation loop and the C-tail are very resistant to phosphatases. The catalytic subunit is then assembled as part of a holoenzyme complex so that activation is mediated exclusively the second messenger, cAMP.

11.3.1   Regulation by Phosphorylation of the Activation Loop

In almost every AGC kinase there are two critical sites that are regulated by phosphorylation. Between the AST and the NLT is a phosphorylation site referred to as the turn motif. In some cases this is a *cis* autophosphorylation site, perhaps even occurring cotranslationally, whereas in other cases, such as Akt, this site appears to be phosphorylated by a heterologous kinase (Oh et al. 2010, Keshwani et al. 2012). With the exception of PKA and PKG, all other AGC kinases have an additional segment that is fused to the FXXF hydrophobic motif. This segment begins with a phosphorylation site that is highly regulated. In Akt, for example, this site is regulated by mTor. Understanding the order and mechanism for regulation of the C-tail phosphorylation sites has been challenging. The other regulatory phosphorylation site in the AGC kinases is the activation loop Thr. This site is thought to be phosphorylated in mammalian cells by a heterologous kinase, PDK1 (Cheng an Weiss), but when the C-subunit is expressed in *E. coli* Thr197 can be readily phosphorylated by *trans* autophosphorylation. Although PDK1 has many features that resemble the AGC kinases, it is in fact evolutionarily unique. It can be thought of as a symbiotic kinase where its activity depends on the AGC kinase, the substrate that is to be phosphorylated (Romano et al. 2009).

11.3.2   Open and Closed Conformations

Various crystal structures define open and closed conformations with a variety of *intermediate* conformations that lie between the two. In reality there is likely a wide range of conformational dynamics that transition between open and closed conformations. Recent NMR studies of PKA revealed some surprising features of this enzyme (Masterson et al. 2010). Mapping both the fast (psec/nsec) and slow (µsec/msec) motions, Masterson et al. defined unique features of the apo enzyme versus the nucleotide bound enzyme and the peptide substrate bound enzyme. Most striking was the finding that the slow dynamics were almost nonexistent in the apo enzyme. There were a few exceptions such as Leu173, a key C-spine residue, but overall the two lobes did not show significant movements in this slow timescale range. The slow motions appeared only when ATP was added. This led to the designation of the apo enzyme as a catalytically uncommitted state and the ATP bound state being a catalytically committed state. These results seem to validate the critical importance of the catalytic spine and the importance of linking the two domains as a consequence of binding ATP. The analysis of a prominent PKA substrate, phospholamban, also emphasized that the peptide as well as the kinase was highly dynamic in the msec/µsec range, motions which are typically associated with catalysis. These results paint a different picture of catalysis where the kinase is not behaving

Chapter 11

as an efficient catalytic machine but rather as a remarkably dynamic switch where the mechanism of phosphoryl transfer might be thought of as *collision-based* catalysis rather than diffusion-based catalysis. By this mechanism, the residues that immediately flank the site that is to be phosphorylated will direct the proper alignment for positioning the P-site residue, but the docking is extremely transient. Catalysis is facilitated only when the peptide is tethered in close proximity to the active site cleft. (Diffusion within the confines of the active site cleft, but not from long distances.) The intrinsic disorder of the phosphorylation is thus a fundamental and probably essential feature of the peptide. It becomes ordered only once it is phosphorylated and bound to something else.

11.3.3   Recognition of Peptides versus Proteins

In thinking about protein kinases as phosphortranserferases we automatically think of other model enzymes where a substrate can usually be trapped in the active site by a substrate analog such as a nonhydrolysable analog of ATP. PKA was initially trapped in a conformation that resembles such a transition state by forming a complex with a peptide (IP20) derived from the heat stable protein kinase inhibitor (PKI), and this complex has served as a template for an active kinase (Knighton et al. 1993). However, PKI is not a substrate; it is instead a high affinity inhibitor of PKA that locks the enzyme into a stable complex that is absolutely dependent on ATP and two magnesium ions. It is possible to convert the inhibitor peptide into a substrate by replacing the P-site Ala with a Ser, and this peptide also has been trapped in a conformation that resembles a transition state by using ADP and aluminum fluoride as a mimetic for ATP (Madhusudan et al. 2002). Surprisingly, however, in spite of intense effort in many labs it has been remarkably difficult to trap other kinases in an *active* conformation in a crystal lattice. A recent structure of Cdk2 was trapped in a transition state complex and the second magnesium ion was essential (Bao et al. 2011).

In the case of PKI, the peptide is *trapped* at the active site cleft not only by ATP and two magnesium ions but also by the amphipathic helix that precedes the inhibitor site and docks to a hydrophobic pocket that lies between the αF and αG helices. This peptide binds with high affinity (1.0 nM), whereas in the absence of ATP and the second Mg ion the affinity is close to 1μM. It is the positioning of the second Mg^{++} that is crucial for the high affinity binding. The heptapeptide that docks to the active site cleft resembles the commonly used PKA substrate, Kemptide (LRRASLG); however, Kemptide on its own binds poorly to the C-subunit. Although its Km is approximately 20 μM, its Kd is actually closer to 300 μM (Adams and Taylor 1992). The low Km thus does not represent the true binding affinity of the peptide. Typically small peptides are extremely poor substrates. In some cases, this may be due to the fact that only the kinase core is being analyzed, and this core may lack essential catalytic features. In all cases, however, this feature of poor peptide binding and slow catalysis emphasizes that protein kinases do not work on small peptides in cells; they phosphorylate proteins. Typically these proteins are tethered to the kinase at some distal site or by binding to an accessory domain. They are not tethered tightly to the active site cleft of the core. An excellent example that demonstrates this feature of protein kinases is the structure of PKR bound to eIF2α. (Figure 11.6) This is one of the few examples of a protein complex bound to a protein kinase. This structure demonstrates how the αG helix can serve as a docking site for bringing the segment that is to be phosphorylated in proximity to the active site cleft (Dar et al. 2005). As a

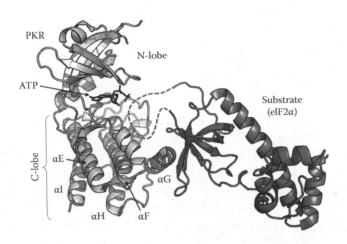

FIGURE 11.6 Tethering of a protein substrate to a protein kinase. The structure of PKR bound to its substrate eIF2α shows how the αG helix plays a major role in position of the protein for phosphorylation. The site to be phosphorylated is disordered as indicated by the dashed line.

GLOSSARY OF KEY TERMS FOR CATALYTIC AND REGULATORY SUBUNITS

CATALYTIC SUBUNIT

N-lobe: also known as *small lobe*. The lobe is composed of five β-strands that form a rigid essentially hydrophobic sheet that serves as a binding site for ATP. A conserved α helix C (subdomain III) is inserted between β-strands 3 and 4. It connects N- and C-lobes and is a regulatory element.

C-lobe: also known as *large lobe*. This mainly helical subdomain contains α helices D through I that are conserved in all eukaryote protein kinases. It serves as a docking site for peptide substrates, tethering sites for protein substrates, and contains residues that are important for catalytic activity (subdomains VIb, VII, and VIII).

AGC kinases: a group of about 60 serine/threonine kinases that share several sequence, structure and function similarities (Kannan et al. 2007). Sequence similarities include ~50 residues long C-terminal tail and 6 residues long AGC-insert (GxxxxK) in subdomain XI. The name of the group is derived from three main representatives: protein kinase A (PKA), protein kinase G (PKG), and protein kinase C (PKC).

N-tail: residues that precede subdomain I and do not form a recognized protein domain. It is not conserved in the AGC subfamily and can perform different functions or can be completely absent. In PKA it contains an α helix A that stabilizes the N-lobe and typically carries a myristylation site at the N-terminus. Exon I codes for the first 14 residues and there a number of splice variants for this exon.

C-tail: residues that follow the subdomain XI and do not form a recognized protein domain. In the case of AGC kinases there are a number of conserved sequence recognition motifs embedded within the C-Tail (Romano et al. 2009). These are divided into three conserved signatures (Kannan et al. 2007):

C-lobe tether (CLT): first part of the C-tail that docks the tail to the C-lobe surface. It includes a characteristic PxxP motif.

Active site tether (AST): middle part of the C-tail that contains FDDY (or in general $FD(x)_{1-2}Y/F$) motif. The FDDY-Phe (237 in PKA) docks into the ATP binding site of the N-lobe while the Phe/Tyr (Tyr330 in PKA docks onto the ribose ring. These hydrophobic residues are essential for activity in PKA (Yang et al. 2009, Batkin et al. 2000). This region also contains an important regulated phosphorylation site typically referred to as the *turn motif*.

N-lobe tether (NLT): C-terminal part of the tail containing a *hydrophobic motif* (FxxF) that binds into a hydrophobic pocket formed by α-C helix and $\beta 4$-strand. It stabilizes protein kinase structure and is critical for the enzyme activity.

HM extension: With the exception of PKA, most other AGC kinases extend beyond the NLT, and this region typically contains a regulated phosphorylation site that immediately follows the HM sequence.

Regulatory spine (R-spine): a stack of four hydrophobic residues that connect the two protein kinase lobes. The R-spine is a nonconsecutive structural motif as it is formed by residues from different parts of kinase sequence. Disassembly of the R-spine leads to inactivation of the kinase and thus can regulate protein kinase activity.

Catalytic spine (C-spine): a stack of eight hydrophobic residues that, together with the adenine ring of ATP, connects the N-and the C-lobe of protein kinase molecule. It provides precise and reliable positioning of ATP with respect to substrate peptides.

Glycine-rich loop: highly flexible loop in the subdomain I between β-strands 1 and 2 that contains a conserved GxxGxG motif. It positions the γ-phosphate of ATP and plays critical roles in phosphotransfer and ATP/ADP binding/release.

AxK motif: a conserved motif in the β-strand 3. Alanine (A^{70}) is a part of the C-spine and docks onto the adenine ring of ATP. Lysine (K^{72}) forms a salt bridge with a conserved glutamic acid (E^{91}) in the αC helix and coordinates the position of α and β phosphates of ATP.

C helix: α helix in the N-lobe (subdomain III). The helix is linked to β-strands 3 and 4 by flexible linkers and thus can be highly mobile. Stabilization of the C helix is critical for effective phosphotransfer. It contains two conserved residues E^{91} that forms a salt bridge with K^{72} and L^{95} that is a part of the R-spine and is always conserved as a hydrophobic residue.

Hinge region: a flexible loop in subdomain V that connects two lobes of protein kinase between the β-strand 5 in the N-lobe and the D helix in the C-lobe.

Catalytic loop: a loop between the β-strands 6 and 7 in the C-lobe (subdomain VIb) that contains several residues critical for catalysis. The general motif is HRDΦKxEN. Histidine is a part of the R-spine. In many AGC kinases it is substituted with tyrosine (Y^{164}) and is a part of the R-spine. Aspartate (D^{166}) is an important element of phosphotransfer and coordinates with one of the Mg ions that bridges the α and γ phosphates and critical for catalysis, as well as the universally conserved asparagine N^{171}, which bridges the other magnesium ion that coordinates the α and γ phosphates.

Activation segment: A large EPK-specific structural element in the C-lobe. It is comprised of the Magnesium binding loop, Activation loop, P+1 loop, APE-motif and a linker that connects APE-motif and the αF helix. The geometry of the Activation segment can vary substantially depending on phosphorylation of its Activation loop. Activity of many EPKs is regulated via such phosphorylation.

Magnesium binding loop: Part of the Activation segment that includes DFG-motif and β-strand 9.

DFG motif: Highly conserved Asp-Phe-Gly in the beginning of the activation segment. D^{184} binds to two magnesium cations in the active site and is absolutely required for catalysis. F^{185} is an important residue in the R-spine.

Activation loop: Very diverse in sequence and the most flexible part of the Activation segment between β-strand 9 and P+1 loop. It contains primary phosphorylation site and in some kinases additional phosphorylation sites. When the activation loop is phosphorylated the The(P) or Tyr(P) is anchored to the Arg in the HRD motif.

P+1 loop: A U-shaped loop in the Activation segment that accommodates side chain of the peptide substrate residue that follows the phosphorylation site (known as the P+1 site).

APE motif: highly conserved motif in the activation segment. A^{206} and P^{207} dock to a conserved tryptophan W^{222} in the F helix. E^{208} forms a salt bridge with universally conserved R^{280} in subdomain XI. These interactions connect the Activation segment to the kinase core and provide structural integrity of the C-lobe in EPKs.

F helix: A large predominantly hydrophobic helix in the C-lobe. It serves as an anchoring base for the R- and C-spines providing a foundation for the internal architecture of the protein kinase molecule.

G helix: One of the EPK-specific α helices (together with H and I helix) that binds to the activation segment right after the APE motif. It is positioned in a close proximity to the phosphorylation site and thus serves as a docking structure for substrates, regulatory proteins, phosphatases, and other interacting proteins.

H-I loop: A loop between H and I helix that contains the conserved R280. In AGC kinases it also contains an AGC-specific insert (residues XXX-YYY in PKA). It is known to be a docking site for regulating and substrate proteins.

Phosphorylation sites

Activation loop: Primary phosphorylation site is positioned at T^{197} position. It forms conserved hydrogen bonds to the HRD-arginine and the C helix thus organizing active conformation of EPKs. Activation loop can contain several other phosphorylation sites that are not conserved and dependent on activation mechanism for different EPKs.

Turn motif: Phosphorylation site in C-tail of AGC kinases. The name is related to S^{338} in PKA, the first protein kinase with known structure. It is positioned in a characteristic turn of C-tail. Structures of other AGC kinases showed that their phosphorylation sites considered to be analogous to S^{338} are positioned over the glycine-rich loop and correspond to E^{333} in PKA.

REGULATORY SUBUNIT

CNB domains: cyclic AMP (cAMP) binding domains. Ancient signaling protein domain that recognizes and binds second messengers cAMP and cGMP. It has been conserved from bacteria to humans. β-strand 8 barrel serves as a rigid motif where the cyclic phosphate docks. The α-helical subdomain is more flexible and transmits allosteric signals from the bound nucleotide to the rest of protein complex.

Dimerization/docking (D/D) domain: N-terminal part of PKA regulatory subunits (R-subunits) that serves as dimerization site. The D/D domain contains four antiparallel α helices (two from each R-subunit) and forms a groove that binds to a specific A kinase anchoring protein (AKAP) motif.

PBC: This is the signature motif of the CNB domain. It consists of a short helical insert between β-strands 6 and 7 in the β-barrel of the CNB domain and binds the phosphate of cAMP. An essential residue in this motif is an Arginine that binds to the phosphate of cAMP. Significant structural changes in the PBC are the first step in allosteric signal propagation.

N3A motif: a conserved α-helical structure in N-terminal part of CNB domains that consists of N helix, A helix and a loop between them known as 3_{10}-loop. Together with the PBC and the B/C helix this motif transfers the allosteric signal in CNB domain containing proteins.

B/C helix: a large α helix in the C-terminal part of CNB domains. In the cAMP bound conformation it is usually broken into two helices known as B helix and C helix. In the AMP free conformation it can form a single helix termed B/C helix.

Inhibitor site: flexible linker in R-subunits of PKA that contains signature sequence for PKA substrates: RRxS/TΦ. In type I R-subunits Ser/Thr is

substituted with Ala (RIα) or Gly (RIβ) so that it becomes a pseudo substrate. This sequence binds to the active site of PKA catalytic subunit
and prevents substrate binding thus inhibiting PKA activity.

N-linker: flexible linker in R-subunits of PKA that connects the dimerization
domain and the inhibitor site.

C-linker: flexible linker in R-subunits of PKA that connects the inhibitor site
and CNB domains.

consequence of this docking, the region of eIF2α that is poised to be phosphorylated has
become disordered, but it is not stabilized at the active site, suggesting that docking to
the active site and docking to the αG helix may be mutually exclusive. Once the substrate
protein is phosphorylated, it will dissociate from the kinase.

11.4 Regulation of PKA

11.4.1 Regulatory subunits of PKA

In addition to being enzymes, protein kinases are also scaffolds that serve as a docking
surface for many proteins including kinases, phosphatases, substrates, and other signaling proteins and domains as well as flexible linkers. One of the most important features
of a protein kinase is that it must be regulated. In the case of PKA, regulation is achieved
by binding of the second messenger, cAMP. To achieve this, the active phosphorylated
catalytic subunit is packaged as a holoenzyme complex with regulatory (R) subunits
that are the major receptors for cAMP in most eukaryotic cells. In mammals there are
four functionally nonredundant regulatory subunits for PKA: Iα, Iβ, IIα, and IIβ. Each
regulatory subunit is composed of a dimerization domain at the N-terminus followed by
a flexible linker that contains an inhibitor site. The two classes I and II are distinguished
by their ability to be autophophorylated. RI subunits have a pseudophosphorylation site
with an Ala at the P-site, whereas RII subunits have a Ser and are autophosphorylated.
At the C-terminus lie two tandem cyclic nucleotide binding (CNB) domains. In the
absence of cAMP each regulatory subunit dimer interacts with two catalytic subunits
forming an inactive tetrameric holoenzyme complex where the inhibitor site (IS) is
docked to the active site cleft of the catalytic subunit. The cooperative binding of cAMP
to the regulatory subunit unleashes the active catalytic subunits. To understand how
PKA is regulated, it is necessary to first understand how cAMP binds in a cooperative
manner to the regulatory subunit and then how the regulatory subunits dock onto the
catalytic subunit in the absence of cAMP and block its activity.

11.4.2 cAMP Binding to the Regulatory Subunits

Four functionally nonredundant PKA regulatory subunits can be divided into two general classes, I and II, each having α and β subtypes (McKnight 1998). All have the same
domain organization previously described. In the presence of cAMP the free R-subunit
is a stable dimer where N-terminal dimerization domain is linked by a highly flexible

linker to two stable cAMP binding domains. To understand the dynamic properties of the R-subunits one first needs to understand the CNB-domain. cAMP has been conserved as a second messenger throughout biology in both prokaryotes and eukaryotes, and the CNB domain has coevolved as a docking domain for cAMP that allows the second messenger to mediate a biological response to an extracellular signal. In mammalian cells intracellular levels of cAMP are regulated in two ways. Typically binding of a peptide hormone or transmitter or a small ligand to a G-protein coupled receptor (GPCR) will lead to the activation of Gαs. Binding of the ligand to the trimeric complex, Gαβγ, promotes the exchange of GTP for GDP on Gαs; this, in turn, promotes the dissociation of Gβγ, which allows the active GTP-bound Gαs to bind to and activate an adenylylate cyclase.

11.4.3   Cyclic Nucleotide Binding Domains

Cyclic nucleotide binding domains are widespread in biology (Kannan et al. 2007). They are conserved as signaling motifs in cyclic nucleotide gated channels (HCN), in transcription factors such as the catabolite gene activator protein (CAP), and in cyclic nucleotide exchange factors such as EPAC. Each CNB domain is composed of a helical subdomain and a beta subdomain (Figure 11.7). The beta subdomain is a contiguous eight-stranded β-barrel that harbors the docking site for the cyclic nucleotide. This docking site is referred to as the phosphate binding cassette (PBC), and this is the signature motif for each CNB domain. There is a short helix, which is part of the PBC, between β-strands 5 and 6. The PBC is flanked by two key residues that anchor the cyclic nucleotide and shield it from solvent and from PDEs. A conserved arginine (Arg[209] and Arg[333] in the CNB-A- and CNB-B domains of the RIα subunit) is located in the PBC helix and docks to the phosphate moiety. In addition, a conserved glutamate (Glu200 and Glu222 in the CNB-A- and CNB-B-domains of RIα) binds to the 2'-OH of the ribose moiety.

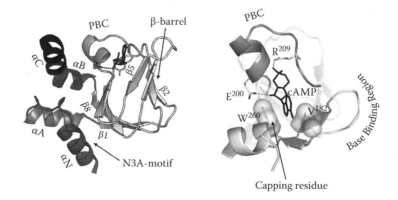

FIGURE 11.7   General organization of the cyclic-nucleotide binding (CNB) domain. The CNB-A domain of RIα is used as a prototype of this highly conserved structural domain. The left panel shows a β-barrel consisting of eight contiguous β-strands (light gray) and an α-helical subdomain including noncontiguous α helices (dark gray). The right panel shows how the cyclic nucleotide is bound to the phosphate binding cassette, which is the signature for this domain. The right panel also shows how the base of the nucleotide is bound by hydrophobic interactions.

The helical part of the CNB domain is made up of two noncontiguous parts. The N-terminal helical segments consist of an N helix linked to the A helix by a loop that contains a small 3_{10} helix. This segment precedes β-strand 1 and is referred to as the N3A motif. β-strand 8 is followed by a second helical motif referred to as the B/C helix. In the presence of cAMP the B/C helix is kinked and moves in toward the PBC, while the N3A motif moves out, where it typically interacts with other domains or proteins (Figure 11.8). The adenine ring of cAMP is capped by a hydrophobic residue that is spatially conserved in all CNB domains but that can come from different parts of the molecule. In the RPK R-subunits the capping residue for cAMP bound to CNB-A typically comes from the CNB-B domain, and the capping residue for the CNB-B domain lies in the C helix of the CNB-B domain (Su et al. 1995; Diller et al. 2000; Berman et al. 2005).

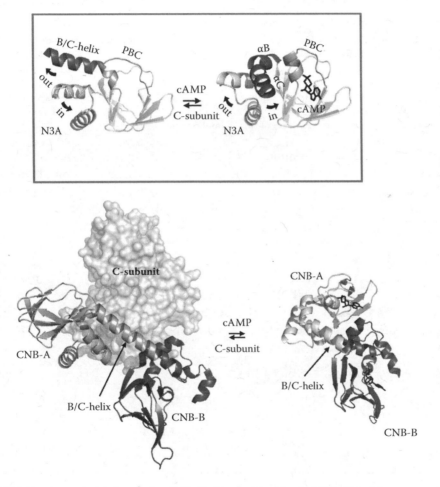

FIGURE 11.8   Structural changes in the CNB domain induced by cAMP biding. The top panel shows conserved structural changes in all CNB domains. In the absence of cAMP (left) the B/C helix moves out, while the N3A motif moves in. Upon cAMP binding the PBC moves toward the cAMP phosphate moving the B/C helix in and causing the N3A motif to move out. The bottom panel shows global cAMP-induced structural changes in PKA RIα regulatory subunit, which contains a tandem of two CNB domains. In the presence of cAMP RIα has a compact configuration with two CNB domains packed against each other (right bottom panel). In the cAMP free form RIa unfolds and forms an extensive interface with the C-subunit of PKA (left bottom panel).

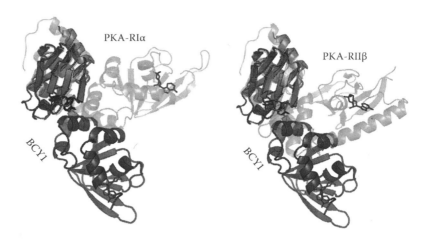

FIGURE 11.9 Organization of the two tandem CNB domains is unique to each regulatory subunit. Although the fold of each CNB domain is highly conserved the way the two domains create an interface in the cAMP-bound conformation is different. Shown in dark gray are two molecules of the PKA regulatory subunit in *Saccharomyces cerevisiae* (BCY1). The left panel compares the relative orientation of the CNB domains in BCY1 to those in the mammalian RIα regulatory subunit (light gray) after overlaying CNB-A domains. The right panel shows a similar comparison of BCY1 to the mammalian RIIβ regulatory subunit (light gray), providing another variation of this theme.

In PKA there are two contiguous CNB domains that, in the presence of cAMP, interact to form a network of allosteric contacts. Although each cAMP-bound CNB domain can be easily superimposed, the orientation of the two domains relative to each other is actually quite distinct for each regulatory subunit. This was apparent when the structure of the RIIβ subunit was solved and compared with RIα (Diller et al. 2000). The recent structure solution of BCYL, the yeast homolog of the PKA regulatory subunit (Rinaldi et al. 2010), shows yet another variation on this theme (Figure 11.9) and demonstrates why it is essential to solve each structure. Having a structure of the CNB domain of RIα is sufficient to model the CNB domains of all the different R-subunits where only the precise position of the B/C helix will vary; however, one structure is not sufficient to model the interfaces between the two CNB domains in RIα, RIIα, and BCYL.

11.4.4 Regulation of PKA by cAMP

Each regulatory subunit binds to two molecules of cAMP. In the RIα subunit the binding is highly cooperative. cAMP binds first to the C-terminal CNB domain, CNB-B, and then to the CNB-A domain. It is then docking to the A domain that unleashes the inhibitor site at the active site cleft of the C-subunit. To understand the molecular basis for inhibition of the C-subunit by the R-subunit and to understand the activation of the holoenzyme by cAMP, it was essential to solve the structure of a holoenzyme complex. The structure of an RIα subunit containing a single cAMP binding domain complexed to the C-subunit was solved first (Figure 11.10) (Kim et al. 2005). This was followed quickly by structures of RIα (Kim et al. 2007) and RIIα (Wu et al. 2007) containing both CNB domains bound to the C-subunit. These structures revealed the major conformational flexibility of the CNB domains of the regulatory subunits as they release cAMP, separate from each other, and then wrap

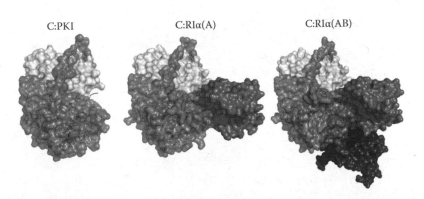

C:PKI   C:RIα(A)   C:RIα(AB)

FIGURE 11.10 (See color insert.) The catalytic subunit serves as a scaffold that interacts with multiple regulatory proteins. On the left is a representation of the catalytic subunit bound to a peptide from the heat-stable protein kinase inhibitor (PKI). The middle panel shows a catalytic subunit bound to a deletion mutant of RIα that contains a single nucleotide binding domain (CNB-A). On the right is the catalytic subunit bound to a deletion mutant of RIα that contains both single nucleotide binding domains (CNB-A and CNB-B). The catalytic subunit is shown as a space filling model with the N-lobe in white and the C-lobe in tan. PKI 1 to 24 is shown as a red ribbon. The R-subunits are also shown as space-filling models. The Inhibitor site that docks to the catalytic site is shown in red. The CNB-A domain in dark teal, CNB-B domain is in turquoise.

themselves around the large lobe of the catalytic subunit and lock the IS in the linker into the active site cleft.

When cAMP is released and the R-subunit binds to the C-subunit, each CNB domain undergoes a large change that mediates this global and dramatic change in conformation. The β-barrel remains remarkably stable during this transition although there are subtle changes in the PBC. However, the N3A motif, once the PBC is free of cAMP, moves in and forms a critical contact with the PBC. The B/C helix is also released, and, when bound to the C-subunit, it extends into a single long helix that is referred to as the out configuration. Thus, when cAMP is bound the N3A motif is out and the B/C helix is in, whereas in the holoenzyme the N3A motif is in and the B/C helix is out. A similar set of changes is seen in the CNB-B domain. The B/C helix of the CNB-A domain is fused directly to the N3A motif of the CNB-B domain. This N3A motif also moves in to contact the PBC of CNB-B while the B/C helix moves out away from the PBC. However, in this case the B/C helix does not extend out into a single long helix but rather forms a novel helical motif that is moved out and away from the PBC (Figure 11.8).

What was most surprising about the conformational changes in the CNB domains is that the cAMP binding site is distorted in such a major way. Although it is possible to appreciate what the cAMP-free PBC might look like, no one anticipated the global reorganization of the helical subdomains that occurs. In the presence of cAMP additional interactions besides the PBC are critical for anchoring the nucleotide. These form a hydrophobic cap for the adenine ring. One side of the hydrophobic cap is formed by the relatively stable β-strand 4 containing the base-binding region (BBR). The other side is donated by different elements in each R-subunit. In RIα the cap for domain A comes from the aromatic ring of Trp260, which is part of the N3A motif in CNB-B. In RIIβ the cap residue for CNB-A is provided by Arg381, which lies in the B helix of CNB-B. In BCYL the hydrophobic cap is provided by a Tyr309 in the N3A motif of CNB-B.

Although the spatial location of Trp260 in RIα and Tyr309 in BCYL is identical relative to the adenine ring, the fact that the aromatic ring is replaced by the tyrosine ring means that the orientation of the two domains is very different in these two highly homologous proteins (Figure 11.9).

11.4.5   Dynamic Features of the CNB Domains

While crystal structures have provided us with models of the different cAMP-bound conformations, the elegant nuclear magnetic resonance (NMR) studies from the Melacini laboratory have provided a dynamic picture of these motifs in RIα (Das et al. 2006, 2007; McNicholl et al. 2010). Their work defines an allosteric network that reaches out across the domain. They describe, in addition to the primary docking of cAMP to the PBC, a set of two shells that radiate out from the PBC and provide an extended allosteric network. The primary shell is composed of the Gly169 and Asp170 that lie at the beginning of β-strand 3. The secondary shell then reaches out to the B/C helix. More recently, NMR structures of cAMP bound to a deletion mutant of RIα that contains both CNB domains were solved (Byeon et al. 2010), and these structures allow us to appreciate how the extended allosteric network links the two domains.

11.5   Assembly of Tetrameric Holoenzymes

11.5.1   Isoform Differences Are Defined by SAXS/SANS: Assembly of a Tetrameric Holoenzyme

While the free R- and C-subunits allowed us to understand how cAMP binds to the R-subunits and how the C-subunit functions as a catalyst, it was not until a holoenzyme structure was solved that we could truly begin to appreciate how PKA was actually inhibited by the regulatory subunit and how it was then activated by the binding of cAMP. However, in cells each molecule of PKA is assembled as a tetrameric holoenzyme. To understand the global architecture of the holoenzymes, small-angle X-ray and neutron scattering were used (SAXS and SANS; see Chapter 6).

What is most striking about the SAXS data is that each of the four holoenzymes (RIα, RIβ, RIIα, and RIIβ) is highly distinct. Once again, the data might have predicted that having one tetrameric structure might allow all four to be modeled. The SAXS data, however, show that this will not be possible. Based on the SAXS data, the free RIα subunit is Y-shaped (Vigil et al. 2004), and recent analysis of the RIβ subunit indicates that it also has a similar conformation. However, when the RIα holoenzyme forms it expands into a relatively compact structure, although there are major changes in the structure of the CNB domains in the RIα subunit. The maximum diameter D_{max} goes from 120 Å to approximately 140 Å. In contrast, the RIβ structure is much more extended with a D_{max} of 190 Å.

The dimeric RIIα and RIIβ subunits are much more rod-like. They assume a more dumbbell shape with the CNB domains at each end. When the C-subunits bind to RIIα, the structure remains extended and rod-like. The D_{max} is approximately 180 Å and is nearly identical to the free RIIα dimer. However, when the RIIβ dimer binds to two molecules of C, the resulting holoenzyme is very compact and almost globular. Thus, each holoenzyme is quite different.

11.5.2   Quaternary Structures of the Tetrameric PKA Holoenzymes ($I\alpha$, $I\beta$, $II\alpha$, and $II\beta$) Are Distinct

To understand the spatial organization of the different PKA holoenzymes, SAXS and SANS were used (see Chapter 6). These low-resolution methods provide a picture of each holoenzyme and indicate that they are surprisingly different even though the overall domain organization is conserved. The two RII holoenzymes were especially striking. Both of the cAMP bound RII dimers were extended and highly asymmetric. When the RIIα was assembled into a tetrameric holoenzyme, the complex remained extended and showed little change in the maximum diameter (200 Å), indicating the two R:C complexes were attached to the D/D domain by a flexible linker. When the RIIβ tetramer was assembled, however, it condensed into a much more globular protein. By carrying out linker swaps It was determined that the linker region was responsible for the remarkably different quaternary structures (Vigil et al. 2006).

11.5.3   Role of the Linker Region in Defining the Quaternary Structure

In addition to SAXS and SANS analyses, the dynamic features of the RIα subunit have been mapped by scanning cysteine mutagenesis, where specific residues were replaced with Cys and then labeled with fluorescein. With this strategy, the intrinsic flexibility of each residue can be mapped. By selecting residues that were strategically positioned in each part of the RIα subunit, it was possible to confirm the flexibility of the linker region as well as the N-terminal region preceding the D/D domain in the cAMP bound conformation. In the cAMP-bound state, the protein is also known to be susceptible to proteolysis at the P-site (Arg92). When holoenzyme forms, however, this site is protected, suggesting that both sides of the linker that lie N-and C-terminal to the inhibitor site, N-linker and C-linker are ordered. The fluorescence studies confirm this. They show that two key residues in the linker region, S75C and S81C are very flexible in the free protein but become much more rigid in the holoenzyme complex (Gangal et al. 1998). From the R:C complexes where the R-subunits begin with the inhibitor site, we can appreciate how this previously disordered segment (the C-linker) becomes ordered as it wraps around CNB A domain of the regulatory subunit and reinforces the positioning of the extended B/C helix.

11.5.4   Assembly of an RIα Tetramer

This analysis of the N-linker region and the differences in flexibility and proteolysis of the N-linker led us to think that the N-linker, in addition to the D/D domain, may play a role in stabilizing the tetramer. Thus, several versions of RIα were engineered containing an extended linker as well as variants that contain the D/D domain. All were set up for crystallization, but only one gave diffraction quality crystals, RIα(75-244) (Boettcher et al. 2011). When solved, these crystals showed only a single R:C complex in the asymmetric unit; however, the symmetry mate gave a stunning view of how allostery might be achieved by the cooperative interaction of the R and C domains in the tetrameric complex. These interactions could never be appreciated from the crystal of a heterodimer containing only one R and one C-subunit (Figure 11.11).

Chapter 11

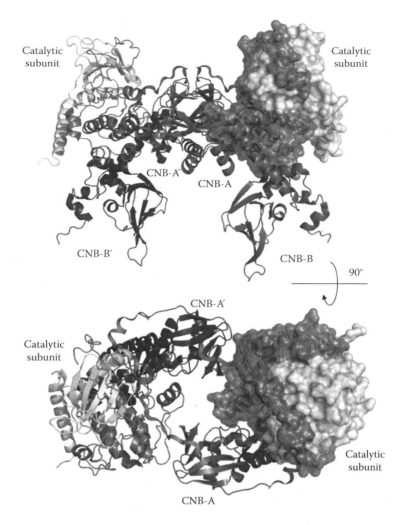

FIGURE 11.11 (See color insert.)   Model of the tetrameric RIα holoenzyme. Crystallization of a deletion mutant of the RIα subunit that contains an extended linker segment bound to a catalytic subunit revealed a tetrameric configuration of the two RC dimers. The model of the full length protein based on this tetrameric configuration is consistent with small-angle X-ray scattering and small-angle neutron scattering data. It is also consistent with single particle image reconstruction models (unpublished results). The upper panel shows the relative position of the dimerization docking domain and the well-separated locations of the catalytic subunits. Rotation of this model by 90 degrees reveals the incredible symmetry of the two active sites where the extended linker of one R-subunit is docked onto the CNB-A domain of the symmetry related dimer. Coloring of the catalytic subunit is as described in Figure 11.10, the R-subunits are shown as ribbons. In red is the linker region that becomes ordered only in the complex. The regulatory subunit of one domain is shown in turquoise, while the regulatory subunit from the symmetry related dimer is shown in dark teal.

11.6   Localization of PKA in Cells

11.6.1   A Kinase Anchoring Proteins

A major mechanism for achieving specificity in PKA signaling is through localization (see Chapter 7). These spatiotemporal aspects of PKA signaling are also unique for each isoform (Iα, Iβ, IIα, and IIβ). How do we localize PKA to specific sites in the cell? A major mechanism for targeting of PKA is through the AKAP, which contains an amphipathic helix that binds through its hydrophobic surface to the D/D domain of each regulatory subunit. We refer to this as an A kinase binding (AKB) motif. Type II R-subunits typically bind with high affinity (1–2 nM). In addition, dual specific AKAPs or D-AKAPs bind to both RI and RII subunits (Huang et al. 1997). Sphingosine kinase interacting protein was shown to be highly specific for RI subunits and localizes to the intermembrane space in mitochondria (Means et al. 2011). AKAP220 contains several AKBs, and one is very specific for RI. In the case of AKAP220, it appears as though the AKAP220 complex is present both in stimulated and unstimulated cells even though the resting cells show RI to be diffuse in the cytoplasm. Only after the C-subunit is released is targeting of the RI:AKAP220 complex to specific sites evident. This is in contrast to RII subunits, which are typically anchored via AKAPs to membrane targets. AKAP220 establishes another novel paradigm for signaling by RIα holoenzymes. While it is long recognized that binding of cAMP unleashes the activity of the catalytic subunit, in the case of AKAP220 cAMP also frees up the targeting potential of the AKAP-bound R-subunit. The D/D domains of RIα are also disulfide bonded, and there is evidence that targeting of RIα can be mediated by oxidative stress that leads to the formation of these disulfide bonds.

11.6.2   Molecular Basis for Anchoring

The D/D domain of each regulatory subunit forms an antiparallel four-helix bundle. The first two antiparallel helices form the docking site for the AKAP helix. The initial structure of a RIIα D/D domain was solved by NMR (Newlon et al. 1999), and this was followed by the NMR structure of an RIα D/D domain (Banky et al. 2000). The NMR structure of the RIIα D/D domain bound to the AKAP peptide from AKAP79 was then solved, and this complex defined for the first time the molecular features of this docking mechanism (Newlon et al. 2001). Subsequent crystal structures were solved of RIIα bound to the AKB from D-AKAP2 and to an engineered peptide that has high affinity and high specificity for binding to RIIα (Gold et al. 2006; Kinderman et al. 2006). The recent structure of the RIα D/D domain bound to the same AKAP peptide from D-AKAP2 paved the way for understanding how an AKAP can achieve specificity for RI versus RII and also provides a foundation for elucidating the molecular codes that define not only an AKAP but also whether it will bind preferentially to RIα or RIIα (Sarma et al. 2010). The rules for docking of amphipathic helices have not been clearly articulated, but such helices are widespread in biology. Just for PKA, several amphipathic helices are found to be important. The AKAP helices determine localization of PKA, whereas PKI, which is mostly disordered in solution, contains two stable amphipathic helices. Near the C-terminus is the nuclear export signal that

Chapter 11

orchestrates exit from the nucleus by binding to exportin. At the N-terminus is the helix that determines high affinity binding of PKA so that it is taken along as cargo and delivered to the cytoplasm. The spatial features of each of these amphipathic helices are quite distinct and clearly specify important signaling codes.

11.6.3   Assembly of Macromolecular Complexes

Most RII-subunits or Type II holoenzymes appear to be targeted to specific sites on a membrane such as a channel or a transporter. Often the PKA is located in close proximity to a phosphatase. The PKA complex targeted to a channel or a transporter can thus function as a single oscillatory molecular circuit that regulates the transient opening and closing of channels (Ni et al. 2011). AKAP79/150 is one such example, where PKA and calcineurin, a calcium-activated phosphatase, are colocalized (Coghlan et al. 1995). The opening of the channel, which is facilitated by PKA phosphorylation of the C-terminus of the channel–transporter, is thus regulated by two different second messengers: cAMP, through its activation of PKA; and calcium, through its activation of calcineurin. PKA is also a metabolic enzyme that can stay on for a long time and not only regulate the phosphorylation state of cytoplasmic proteins but also mediate gene transcription by translocating into the nucleus (Montminy et al. 2004). The persistent signal, in principle, is achieved by inhibition of the phosphatase inhibitors such as PP1-In and DDARP-32 (Nimmo and Cohen 1978, Hemmings et al. 1984). Both of these proteins are potent inhibitors of PP1 when they are phosphorylated by PKA so that the phosphatase activity is muted as long as PKA is active. This mechanism, in principle, allows the cell to maintain a persistent PKA signal.

11.7   Closing Perspectives and Future Challenges

Advances in genome sequencing have allowed us to define the boundaries of the protein kinase superfamily and also to elucidate how the protein kinases have evolved in unique ways from their prokaryotic precursors, the ELKs. By delving deeply into the sequences of each family, we can also decipher features that distinguish the different families (Kannan and Neuwald 2005; Kannan, Haste, et al. 2007; Kannan, Wu, et al. 2007). Having crystal structure of PKA allowed us for the first time to define the conserved sequence motifs as functional motifs. We now have a structural kinome where at least the kinase cores of most of the subfamilies are represented. While we have delved deeply into the structure and function of PKA, having a structural kinome and developing methods to compare active and inactive kinases, we can begin to decipher some of the fundamental rules that govern kinase structure and function. We can also begin to understand how mutations lead to defects that cause diseases such as cancer.

However, to understand how a kinase is regulated in its physiological state cannot be deciphered from the core alone. It will require crystallizing full-length kinases to show how the kinase core is either turned on or off by the dynamically regulated domains and linkers that flank the tail. Unfortunately, we have only a few examples so far. We have an understanding of Src regulation, for example, because we have structures of the SH2 and SH3 domains fused to the kinase core (Xu et al. 1999). This shows how Src can exist in two fundamentally different conformations. In the off conformation the C-terminal tail is phosphorylated by another kinase, Csk, and this phosphate is bound to its own SH2 domain. Removal of the C-terminal phosphate generates and active kinase that can be autophosphorylated in *trans* on its activation loop. Regulation of Cdk is another example where the kinase alone is inactive. Activation is achieved by binding of cyclin, which then allows for phosphorylation of three sites. The activation loop Thr is phosphorylated and is thus in an active conformation, but the additional phosphorylation of adjacent Thr and Tyr residues in the glycine-rich loop keeps the enzyme in an inactive state. Removal of these phosphates by a phosphatase, Wee1, is then the key activation step. The recent structure of the dodecameric CAM kinase II with its attached docking domain and its inhibitory calmodulin binding domain provide us for the first time with an appreciation of how this kinase is regulated in complex ways by calcium and calmodulin (Chao et al. 2011).

Obtaining crystals of these larger complexes has been extremely challenging in part because the proteins are intrinsically so dynamic. Coupling crystallography with SAXS and SANS provides a mechanism to understand at low resolution how domain organization changes as a function of activation. Cryo-electron microscopy is also providing insights into how activation of full-length kinases such as the receptor tyrosine kinases is achieved. However, these are also still low resolution. Clearly, this will be a highly interdisciplinary challenge.

To understand the dynamic features of these enzymes will require other techniques such as NMR, fluorescence methods, and spin resonance methods such as DEER. Bringing all of these hybrid methods together to understand the regulation of one kinase is also a computational challenge. Computing is becoming much faster, and the design of computers such as Aton (Shaw et al. 2010), which can carry out molecular dynamic simulations in the msec time range, are opening up new opportunities, as are accelerated molecular dynamics simulation methods. Reaching such long timescales will allow us to confirm our simulations with experimental validations. However, it will be necessary to develop tools that will allow for the use of these different types of data in an iterative way. Ultimately, we need to focus on bringing together all of these approaches to achieve a molecular understanding of signaling events. We are left in awe of the remarkable sophisticated ways that have evolved to precisely regulated biological events.

References

Adams, J. A. (2001). "Kinetic and catalytic mechanisms of protein kinases." *Chem Rev* 101(8): 2271–2290.

Adams, J. A. and S. S. Taylor (1992). "Energetic limits of phosphotransfer in the catalytic subunit of cAMP-dependent protein kinase as measured by viscosity experiments." *Biochemistry* 31(36): 8516–8522.

Anjum, R. and J. Blenis (2008). "The RSK family of kinases: emerging roles in cellular signalling." *Nat Rev Mol Cell Biol* 9(10): 747–758.

Banky, P., M. G. Newlon, et al. (2000). "Isoform-specific differences between the type Iα and IIα cyclic AMP-dependent protein kinase anchoring domains revealed by solution NMR." *J Biol Chem* 275(45): 35146–35152.

Bao, Z. Q., D. M. Jacobsen, et al. (2011). "Briefly bound to activate: transient binding of a second catalytic magnesium activates the structure and dynamics of CDK2 kinase for catalysis." *Structure* 19(5): 675–690.

Barker, W. C. and M. O. Dayhoff (1982). "Viral *src* gene products are related to the catalytic chain of mammalian cAMP-dependent protein kinase." *Proc Natl Acad Sci U S A* 79(9): 2836–2839.

Chapter 11

Batkin, M., Schvartz, I., and S. Shaltiel. (2000). "Snapping of the carboxyl terminal tail of the catalytic subunit of PKA onto its core: characterization of the sites by mutagenesis." *Biochemistry* 39(18), 5366–5373 (2000).

Berman, H. M., L. F. Ten Eyck, et al. (2005). "The cAMP binding domain: An ancient signaling module." *Proceedings of the National Academy of Sciences of the United States of America* 102(1): 45–50.

Boettcher, A. J., J. Wu, et al. (2011). "Realizing the Allosteric Potential of the Tetrameric Protein Kinase A RIα Holoenzyme." *Structure* 19(2): 265–276.

Byeon, I. J., K. K. Dao, et al. (2010). "Allosteric communication between cAMP binding sites in the RI subunit of protein kinase A revealed by NMR." *J Biol Chem* 285(18): 14062–14070.

Chao, L. H., M. M. Stratton, et al. (2011). "A mechanism for tunable autoinhibition in the structure of a human Ca2+/calmodulin-dependent kinase II holoenzyme." *Cell* 146(5): 732–745.

Cheng, X., Y. Ma, et al. (1998). "Phosphorylation and activation of cAMP-dependent protein kinase by phosphoinositide-dependent protein kinase." *Proc Natl Acad Sci U S A* 95(17): 9849–9854.

Coghlan, V. M., B. A. Perrino, et al. (1995). "Association of protein kinase A and protein phosphatase 2B with a common anchoring protein." *Science* 267(5194): 108–111.

Collett, M. S. and R. L. Erikson (1978). "Protein kinase activity associated with the avian sarcoma virus *src* gene product." *Proc Natl Acad Sci U S A* 75(4): 2021–2024.

Dar, A. C., T. E. Dever, et al. (2005). "Higher-order substrate recognition of eIF2α by the RNA-dependent protein kinase PKR." *Cell* 122(6): 887–900.

Das, R., M. Abu-Abed, et al. (2006). "Mapping allostery through equilibrium perturbation NMR spectroscopy." *J Am Chem Soc* 128(26): 8406–8407.

Das, R., V. Esposito, et al. (2007). "cAMP activation of PKA defines an ancient signaling mechanism." *Proc Natl Acad Sci U S A* 104(1): 93–98.

Diller, T. C., N. H. Xuong, et al. (2000). "Type IIβ regulatory subunit of cAMP-dependent protein kinase: purification strategies to optimize crystallization." *Protein Expr Purif* 20(3): 357–364.

Gangal, M., T. Clifford, et al. (1999). "Mobilization of the A-kinase N-myristate through an isoform-specific intermolecular switch." *Proc Natl Acad Sci U S A* 96(22): 12394–12399.

Gangal, M., S. Cox, et al. (1998). "Backbone flexibility of five sites on the catalytic subunit of cAMP-dependent protein kinase in the open and closed conformations." *Biochemistry* 37(39): 13728–13735.

Gill, G. N. and L. D. Garren (1971). "Role of the receptor in the mechanism of action of adenosine 3':5'-cyclic monophosphate." *Proc Natl Acad Sci U S A* 68(4): 786–790.

Gold, M. G., B. Lygren, et al. (2006). "Molecular basis of AKAP specificity for PKA regulatory subunits." *Mol Cell* 24(3): 383–395.

Gould, C. M., N. Kannan, et al. (2009). "The chaperones Hsp90 and Cdc37 mediate the maturation and stabilization of protein kinase C through a conserved PXXP motif in the C-terminal tail." *J Biol Chem* 284(8): 4921–4935.

Hanks, S. K. and T. Hunter (1995). "Protein kinases 6. The eukaryotic protein kinase superfamily: kinase (catalytic) domain structure and classification." *FASEB J* 9(8): 576–596.

Hemmings, H. C., Jr., P. Greengard, et al. (1984). "DARPP-32, a dopamine-regulated neuronal phosphoprotein, is a potent inhibitor of protein phosphatase-1." *Nature* 310(5977): 503–505 (1984).

Herberg, F. W., B. Zimmermann, et al. (1997). "Importance of the A-helix of the catalytic subunit of cAMP-dependent protein kinase for stability and for orienting subdomains at the cleft interface." *Protein Science* 6(3): 569–579.

Huang, L. J., K. Durick, et al. (1997). "D-AKAP2, a novel protein kinase A anchoring protein with a putative RGS domain." *Proc Natl Acad Sci U S A* 94(21): 11184–11189.

Hunter, T. and B. M. Sefton (1980). "Transforming gene product of Rous sarcoma virus phosphorylates tyrosine." *Proc Natl Acad Sci U S A* 77(3): 1311–1315.

Jiang, T. and Y. Qiu (2003). "Interaction between Src and a C-terminal proline-rich motif of Akt is required for Akt activation." *J Biol Chem* 278(18): 15789–15793.

Kannan, N., N. Haste, et al. (2007). "The hallmark of AGC kinase functional divergence is its C-terminal tail, a cis-acting regulatory module." *Proc Natl Acad Sci U S A* 104(4): 1272–1277.

Kannan, N. and A. F. Neuwald (2005). "Did protein kinase regulatory mechanisms evolve through elaboration of a simple structural component?" *J Mol Biol* 351(5): 956–972.

Kannan, N., S. S. Taylor, et al. (2007). "Structural and functional diversity of the microbial kinome." *PLoS Biol* 5(3): e17.

Kannan, N., J. Wu, et al. (2007). "Evolution of allostery in the cyclic nucleotide binding module." *Genome Biol* 8(12): R264.

Keshwani, M. M., C. Klammt, et al. (2012). "Cotranslational cis-phosphorylation of the COOH-terminal tail is a key priming step in the maturation of cAMP-dependent protein kinase." *Proc Natl Acad Sci U S A* 109(20): E1221–E1229.

Kim, C., C. Y. Cheng, et al. (2007). "PKA-I holoenzyme structure reveals a mechanism for cAMP-dependent activation." *Cell* 130(6): 1032–1043.

Kim, C., N. H. Xuong, et al. (2005). "Crystal structure of a complex between the catalytic and regulatory (RIα) subunits of PKA." *Science* 307(5710): 690–696.

Kinderman, F. S., C. Kim, et al. (2006). "A dynamic mechanism for AKAP binding to RII isoforms of cAMP-dependent protein kinase." *Mol Cell* 24(3): 397–408.

Knighton, D. R., S. M. Bell, et al. (1993). "2.0 Å refined crystal structure of the catalytic subunit of cAMP-dependent protein kinase complexed with a peptide inhibitor and detergent." *Acta Crystallogr D Biol Crystallogr* 49(Pt 3): 357–361.

Knighton, D. R., J. H. Zheng, et al. (1991a). "Crystal structure of the catalytic subunit of cyclic adenosine monophosphate-dependent protein kinase." *Science* 253(5018): 407–414.

Knighton, D. R., J. H. Zheng, et al. (1991b). "Structure of a peptide inhibitor bound to the catalytic subunit of cyclic adenosine monophosphate-dependent protein kinase." *Science* 253(5018): 414–420.

Kornev, A. P., N. M. Haste, et al. (2006). "Surface comparison of active and inactive protein kinases identifies a conserved activation mechanism." *Proc Natl Acad Sci U S A* 103(47): 17783–17788.

Kornev, A. P., S. S. Taylor, et al. (2008). "A helix scaffold for the assembly of active protein kinases." *Proc Natl Acad Sci U S A* 105(38): 14377–14382.

Krebs, E. G. and E. H. Fischer (1956). "The phosphorylase *b* to *a* converting enzyme of rabbit skeletal muscle." *Biochim Biophys Acta* 20(1): 150–157.

Kunkel, M. T. and A. C. Newton (2010). "Calcium transduces plasma membrane receptor signals to produce diacylglycerol at Golgi membranes." *J Biol Chem* 285(30): 22748–22752.

Levinson, A. D., H. Oppermann, et al. (1978). "Evidence that the transforming gene of avian sarcoma virus encodes a protein kinase associated with a phosphoprotein." *Cell* 15(2): 561–572.

Madhusudan, P. Akamine, et al. (2002). "Crystal structure of a transition state mimic of the catalytic subunit of cAMP-dependent protein kinase." *Nat Struct Biol* 9(4): 273–277.

Manning, G., D. B. Whyte, et al. (2002). "The protein kinase complement of the human genome." *Science* 298(5600): 1912–1934.

Masterson, L. R., C. Cheng, et al. (2010). "Dynamics connect substrate recognition to catalysis in protein kinase A." *Nat Chem Biol* 6(11): 821–828.

Masterson, L. R., L. Shi, et al. (2011). "Dynamically committed, uncommitted, and quenched states encoded in protein kinase A revealed by NMR spectroscopy." *Proc Natl Acad Sci U S A*.

McKnight, G. S. (1998). "Cyclic AMP, PKA, and the physiological regulation of adiposity." *Recent Prog Horm Res* 53: 139–159.

McNicholl, E. T., R. Das, et al. (2010). "Communication between tandem cAMP binding domains in the regulatory subunit of protein kinase A-Iα as revealed by domain-silencing mutations." *J Biol Chem* 285(20): 15523–15537.

Means, C. K., B. Lyngren, et al. (2011). "An entirely specific type I A-kinase anchoring protein that can sequester two molecules of PKA holoenzyme." *Proceedings of the National Academy of Sciences of the United States of America* (accepted).

Montminy, M., S. H. Koo, et al. (2004). "The CREB family: key regulators of hepatic metabolism." *Ann Endocrinol (Paris)* 65(1): 73–75.

Newlon, M. G., M. Roy, et al. (2001). "A novel mechanism of PKA anchoring revealed by solution structures of anchoring complexes." *EMBO J* 20(7): 1651–1662.

Newlon, M. G., M. Roy, et al. (1999). "The molecular basis for protein kinase A anchoring revealed by solution NMR." *Nat Struct Biol* 6(3): 222–227.

Ni, Q., A. Ganesan, et al. (2011). "Signaling diversity of PKA achieved via a Ca2+-cAMP-PKA oscillatory circuit." *Nat Chem Biol* 7(1): 34–40.

Niedner, R. H., O. V. Buzko, et al. (2006). "Protein kinase resource: an integrated environment for phosphorylation research." *Proteins* 63(1): 78–86.

Nimmo, G. A. and P. Cohen. (1978). "The regulation of glycogen metabolism. Purification and characterisation of protein phosphatase inhibitor-1 from rabbit skeletal muscle." *Eur J Biochem* 87(2): 341–351.

Nirula, A., M. Ho, et al. (2006). "Phosphoinositide-dependent kinase 1 targets protein kinase A in a pathway that regulates interleukin 4." *J Exp Med* 203(7): 1733–1744.

Oh, W.J., C. C. Wu, et al. (2010). "mTORC2 can associate with ribosomes to promote cotranslational phosphorylation and stability of nascent Akt polypeptide." *EMBO J* 29(23): 3939–3951.

Reddy, E. P., M. J. Smith, et al. (1980). "Nucleotide sequence analysis of the transforming region and large terminal redundancies of Moloney murine sarcoma virus." *Proc Natl Acad Sci U S A* 77(9): 5234–5238.

Rinaldi, J., J. Wu, et al. (2010). "Structure of yeast regulatory subunit: a glimpse into the evolution of PKA signaling." *Structure* 18(11): 1471–1482.

Robison, G. A., R. W. Butcher, et al. (1968). "Cyclic AMP." *Annu Rev Biochem* 37: 149–174.

Romano, R. A., N. Kannan, et al. (2009). "A chimeric mechanism for polyvalent trans-phosphorylation of PKA by PDK1." *Protein Sci* 18(7): 1486–1497.

Sarma, G. N., F. S. Kinderman, et al. (2010). "Structure of D-AKAP2:PKA RI complex: insights into AKAP specificity and selectivity." *Structure* 18(2): 155–166.

Sastri, M., D. M. Barraclough, et al. (2005). "A-kinase-interacting protein localizes protein kinase A in the nucleus." *Proc Natl Acad Sci U S A* 102(2): 349–354.

Shaw, D. E., P. Maragakis, et al. (2010). "Atomic-level characterization of the structural dynamics of proteins." *Science* 330(6002): 341–346.

Shiu, S. H., W. M. Karlowski, et al. (2004). "Comparative analysis of the receptor-like kinase family in Arabidopsis and rice." *Plant Cell* 16(5): 1220–1234.

Shoji, S., D. C. Parmelee, et al. (1981). "Complete amino acid sequence of the catalytic subunit of bovine cardiac muscle cyclic AMP-dependent protein kinase." *Proc Natl Acad Sci U S A* 78(2): 848–851.

Su, Y., W. R. Dostmann, et al. (1995). "Regulatory subunit of protein kinase A: structure of deletion mutant with cAMP binding domains." *Science* 269(5225): 807–813.

Taylor, S. S. and A. P. Kornev (2011). "Protein kinases: evolution of dynamic regulatory proteins." *Trends Biochem Sci* 36(2): 65–77.

Torkamani, A., N. Kannan, et al. (2008). "Congenital disease SNPs target lineage specific structural elements in protein kinases." *Proc Natl Acad Sci U S A* 105(26): 9011–9016.

Vigil, D., D. K. Blumenthal, et al. (2004). "Conformational differences among solution structures of the type Iα, IIα and IIβ protein kinase A regulatory subunit homodimers: role of the linker regions." *J Mol Biol* 337(5): 1183–1194.

Vigil, D., D. K. Blumenthal, et al. (2006). "Solution scattering reveals large differences in the global structures of type II protein kinase A isoforms." *J Mol Biol* 357(3): 880–889.

Violin, J. D., J. Zhang, et al. (2003). "A genetically encoded fluorescent reporter reveals oscillatory phosphorylation by protein kinase C." *J Cell Biol* 161(5): 899–909.

Walsh, D. A., J. P. Perkins, et al. (1968). "An adenosine 3',5'-monophosphate-dependant protein kinase from rabbit skeletal muscle." *J Biol Chem* 243(13): 3763–3765.

Westheimer, F. H. and S. Huang (1988). "Rates and mechanisms of hydrolysis of esters of phosphorus-acid." *J Am Chem Soc* 110(1): 181–185.

Wu, J., S. H. Brown, et al. (2007). "PKA type IIα holoenzyme reveals a combinatorial strategy for isoform diversity." *Science* 318(5848): 274–279.

Xu, W., A. Doshi, et al. (1999). "Crystal structures of c-Src reveal features of its autoinhibitory mechanism." *Mol Cell* 3(5): 629–638.

Yang, J., S. M. Garrod, et al. (2005). "Allosteric network of cAMP-dependent protein kinase revealed by mutation of Tyr204 in the P+1 loop." *J Mol Biol* 346(1): 191–201.

Yang, J., E. J. Kennedy, et al. (2009). "Contribution of non-catalytic core residues to activity and regulation in protein kinase A." *J Biol Chem* 284(10): 6241–6248.

Zhang, J., F. J. Adrian, et al. (2010). "Targeting Bcr-Abl by combining allosteric with ATP-binding-site inhibitors." *Nature* 463(7280): 501–506.

12. Stochastic Simulation of the Phage Lambda Gene Regulatory Circuitry

John W. Little
Adam P. Arkin

Chapter 12

Quantitative Biology: From Molecular to Cellular Systems edited by Michael E. Wall © 2012 CRC Press / Taylor & Francis Group, LLC. ISBN: 978-1-4398-2722-2

12.1   Introduction

This chapter describes a stochastic simulation of the phage λ gene regulatory circuitry, one of the best-characterized circuits in biology. This simulation is stochastic because the crucial regulatory events occur when the cell contains small numbers of regulatory molecules, and because many of these regulatory events occur in only a small fraction of the cell's and can be thought of as rare events. The primary goals of the modeling are to determine the critical features that control the behavior of the circuitry; to test specific mechanistic models for its operation; to facilitate productive interplay between modeling and experiment, allowing refinement of the model and the parameter set; and ultimately to develop a quantitative description of the λ circuitry and its interaction with the host, to supplement the qualitative wiring diagrams currently used to describe the circuitry.

The cycle of modeling, experiment, and refinement described here is a detailed exemplar of how models are used to organize knowledge, understand the limitations of data, test hypotheses, and design experiments to discriminate efficiently among competing hypotheses. Few systems provide the mechanistic depth available in the λ literature; thus, this example is a vanguard for ways that modeling in other systems may develop. Further, this system represents a model of decision making of a heterologous (viral) circuit operating in a defined host. It is, in many ways, the key model system to demonstrate how to characterize and model host–pathogen interaction and host–context interactions for *synthetic biological* circuits (many of which are made out of components of this bacteriophage). Indeed, the ability to use models of λ to understand how the particulars of its circuit design lead to decision making and stable dormant states has inspired us to apply many of the same methodologies described herein to characterize and model a conceptually similar *decision* perhaps underlying establishment of HIV latency (Weinberger et al., 2005).

This chapter first explains why λ is an excellent model system for systems analysis and why it sets a gold standard for the quantitative and mechanistic modeling of biological systems. The λ circuitry is then reviewed, focusing both on the systems behavior and on the mechanistic events that underlie this behavior. Then the goals of the modeling are further detailed, and the need for a *dedicated* model is discussed. The mechanistic model underlying the simulation is described, and the behavior of the simulation is given. Examples are given in which the predictions of the model fit experimental observations. The chapter concludes by describing our view of the value of simulations, both in constructing an accurate model of a circuit and in providing an opportunity to think hard about what is and what is not known about regulatory circuits. The chapter also discusses the likelihood of discovering principles of system behavior beyond the specific system being modeled and describes a specific example of this kind of discovery.

For numerous reasons, λ is an excellent subject for simulation. First, it shows a range of regulatory decisions, stable regulatory states, and epigenetic switches. Second, most of the critical components and interactions have been identified, and these have been extensively analyzed at the molecular and mechanistic level so that reasonable parameter values are available or can be measured. These mechanisms are diverse and span a wide range of fundamental and common mechanisms in cellular regulatory networks. Third, a wide variety of mutants are available that affect regulatory processes, and again these have been analyzed. Importantly, this analysis has taken place both in simplified subsystems, allowing dissection of the mechanisms, and more recently in the context of the intact virus, to test their effects on the overall behavior of the circuit. This allows a principled approach both to construction of models and testing their predictions. Finally, λ offers the opportunity to relate decisions made at the molecular level to the behavior of populations and eventually to evolutionary strategies for optimizing viral behavior, particularly when taken together with the behavior of other related viruses. Experimental characterization of λ is deep at multiple levels, making it an excellent candidate for developing multiscale models that range from dynamics of regulatory networks to population genetics. Few biological systems offer this prospect.

12.2   Phage λ Gene Regulatory Circuitry

Phage λ is a bacterial virus, or bacteriophage, that infects host cells (Hendrix et al., 1983; Ptashne, 2004). After infection, most phages follow a so-called lytic pathway, in which a temporal pattern of gene expression occurs, leading to synthesis of new copies of the viral genome, followed by packaging these copies into mature virus particles, which are then released from the host cell, usually by lysis of the cell. Although λ can follow this same lytic pathway, it has an alternative mode of existence, the lysogenic state, in which the viral genome is physically integrated into the genome of the host, and expression of the lytic genes is blocked by the action of a master regulatory protein termed λ repressor or CI ("see-one," not "see-eye") (Figure 12.1).

The lysogenic state is extremely stable. It can switch to the lytic pathway, but in the absence of perturbations it does so at an almost undetectably low rate, probably $< 10^{-8}$ per cell generation (Little and Michalowski, 2010). Switching can be extremely efficient, however, when the host SOS response is triggered by treatments that damage DNA or inhibit DNA replication (Little and Mount, 1982; Little, 1993). A key part of this host response is activation of the host RecA protein to a form that can mediate the proteolytic cleavage of the host LexA repressor. Activated RecA can also mediate cleavage of λ CI, inactivating it and allowing expression of lytic genes; the lytic pathway ensues.

Hence, λ has three different regulatory events (Figure 12.1) that are well understood and make fruitful subjects for modeling. First, the lysis–lysogeny decision controls the developmental fate of the infected cell. Second is the stability of the lysogenic state. The lytic state can also be reasonably stable when the lytic functions are blocked by mutation so that the cell is not killed by this alternate pattern of gene expression; this "anti-immune" state has no normal role in λ biology. Third, prophage induction is an

The Phage λ Life Cycle

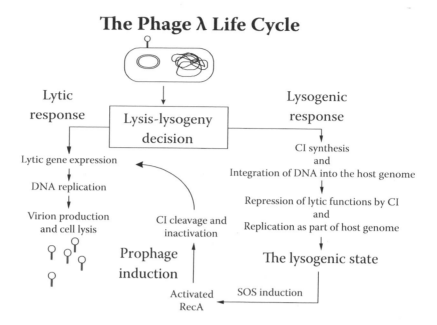

FIGURE 12.1 Lambda life cycle. At the top is a cell newly infected with λ. The infected cell can follow either of two developmental pathways, one of which culminates in a stable epigenetic state, the lysogenic state. This state can switch to the lytic pathway in the process of prophage induction. See text for details. (From Little, 2006. With permission.)

epigenetic switch of regulatory state and can occur in nearly 100% of treated cells. Hence, the regulatory circuitry is balanced in such a way that it is almost completely stable in the absence of perturbations and almost completely destabilized by treatments that induce the host SOS regulatory system. The λ system presents the clearest, best-characterized examples of dynamic decision making and stabilization of fate and is a crucible for testing theories of biological switching.

12.3 Mechanisms Underlying the Regulatory Circuitry

The mechanisms controlling the regulatory circuitry can best be understood with reference to several maps of the regulatory region, showing the locations of *cis*-acting sites and the genes for *trans*-acting factors. Maps at several scales are shown in Figure 12.2.

12.3.1 Lysis–Lysogeny Decision

At a mechanistic level, this decision depends on the level of a viral regulatory protein termed CII (Herskowitz and Hagen, 1980; Oppenheim et al., 2005). CII is a transcriptional activator. When the CII level is high, it activates transcription from three λ promoters. These act in various ways to favor the lysogenic pathway. The most direct mechanism is by stimulating the expression of high levels of CI from the P_{RE} promoter (Figure 12.2b). Expression of massive amounts of CI shuts off the lytic promoters P_L and P_R, favoring the lysogenic pathway. CII also controls expression of the Int protein

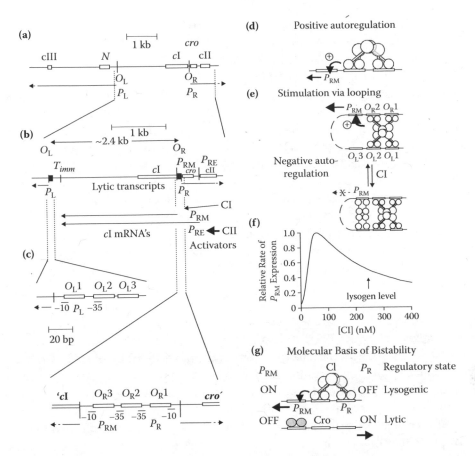

FIGURE 12.2 Lambda maps, CI activities. A, B, C. Maps of λ. Maps are to scale, with the scale indicated. (a) Map of the early region of λ, showing about 10% of the λ genome. Locations of the *cII* and *cIII* genes are indicated; these play a role in the lysis-lysogeny decision and are expressed from the P_R and P_L promoters, respectively, as shown. Only the P_L and P_R transcripts are shown; these continue beyond the region shown, as indicated by the dashed line. The *N* gene product acts to prevent transcription termination of these transcripts. (Oppenheim et al., 2005) (b) More detailed map of this region. This region is often termed the *immunity region* since it contains the *cis*-acting sites (O_L and O_R) and genes for *trans*-acting factors (CI and Cro) that dictate immunity specificity. Below the map are shown two transcripts that encode CI; these are transcribed from different promoters and are subject to different regulatory controls, as indicated. Both terminate at T_{imm}. C. Location of *cis*-acting sites in the O_L and O_R regions. The O_L region contains a single promoter, P_L, and three binding sites for both CI and Cro. Spacing between O_L2 and O_L3 is smaller than between the other operator pairs. The O_R region has the P_{RM} and P_R promoters, overlapping three binding sites for CI and Cro. (d–e) Activities of CI. (d) Positive autoregulation. Reversible dimerization of CI is not shown. CI dimers bind cooperatively to O_R1 and O_R2. When bound to O_R2, CI stimulates P_{RM} about 8- to 10-fold. (e) Loop formation and regulatory effects. When tetramers (pairs of dimers) are bound at O_L and O_R, usually at the sites indicated, they can form a long-range loop, making a CI octamer. In the form shown, expression of P_{RM} is further stimulated about 2-fold. When more molecules of CI bind, O_R3 is occupied and P_{RM} is repressed. Detailed configurations of these complexes are not known. (f) Rate of P_{RM} expression as a function of CI concentration. The output of the simulation is shown; similar data have been obtained experimentally (Dodd et al., 2001, 2004). (g) Molecular basis of bistability. Alternative occupancy patterns of the O_R operators at moderate concentrations of CI and Cro are shown; each pattern perpetuates itself. At higher Cro concentrations, Cro binds to O_R2 or O_R1, turning off P_R; at higher CI concentrations, looped forms (e) lead to occupancy of O_R3, turning off P_{RM}. (From Little, 2006, and Little and Michalowski, 2010. With permission.)

(not shown), which catalyzes site-specific integration of the λ genome into that of the host, and of the P_{aQ} promoter (not shown), which blocks expression of later stages of the lytic pathway (Kobiler et al., 2005; Little, 2005).

The level of CII is therefore crucial in determining the fate of the infected cell. CII is metabolically unstable, being degraded by the host FtsH protein (also known as HflB) (Ito and Akiyama, 2005). The virus makes another protein, CIII, that antagonizes the activity of FtsH. The lysogenic response is favored when cells are infected with multiple phages, perhaps due to higher levels of CII and CIII in multiply-infected cells. Aside from the effects of CIII, the mechanisms controlling the level of FtsH activity are not well understood. FtsH is membrane-bound, and its cytoplasmic domain has the catalytic activity. Other host factors, HflKC and HflD, are also involved in the lysis–lysogeny decision, and mutants lacking these factors have high frequencies of lysogenization (Ito and Akiyama, 2005). Possibly they modulate FtsH activity. Both HflKC and HflD are also membrane bound, making detailed biochemical analysis a challenge. In addition, the lysis–lysogeny decision is believed to depend on *cell physiology*, in ways that are not understood but may involve modulation of FtsH activity. For these reasons, and because Little's primary research interests lie in the remaining two aspects of λ regulation, the lysis–lysogeny decision is not modeled in the work described here.

However, Arkin et al. (1998) did so previously, using a different approach, in one of the first stochastic simulations of a complex biological system. This work successfully predicted the known response of lysogenization frequency to the number of phages infecting a cell, but it did not include numerous features of the present simulation, including cell division, DNA replication, and (perforce) more recently discovered aspects of λ regulation, primarily those involving looping and its effects on *cI* expression. In principle this simulation could be updated, incorporating these features and several other advances, including studies (St-Pierre and Endy, 2008) describing the effect of cell volume on lysogenization frequency; single-cell studies suggesting that the decision is made on a phage-by-phage basis, not a cell-by-cell basis (Zeng et al., 2010); and advances in understanding of CII degradation (Ito and Akiyama, 2005). Ideally, it would also include limited replication of infecting phage DNA, since this occurs prior to the lysis–lysogeny decision (Herskowitz and Hagen, 1980).

12.3.2 Stability of Gene Regulatory States

Most, but not all, of the regulatory events involved in stabilizing regulatory states occur in a complex control region termed the O_R region (Figure 12.2c) (Ptashne, 2004). This region contains two promoters and three binding sites, termed O_R operators, to which phage regulatory proteins bind. Initiation at the lysogenic promoter, P_{RM}, leads to expression of the *cI* gene and is subject to multiple regulatory controls. Initiation at the early lytic promoter P_R leads to expression of the early rightward lytic genes, including another regulatory protein, Cro, whose action antagonizes that of CI. Indeed, both CI and Cro bind to the three O_R operators, but with differing affinities and consequences.

The simpler of the two proteins, Cro, binds most tightly to O_R3 (Figure 12.2g) (Darling et al., 2000b). When bound to O_R3, Cro represses P_{RM} without affecting expression of P_R. At higher levels of Cro, it binds as well to O_R2 and O_R1, giving partial repression of P_R in a simple example of negative autoregulation. Hence, a cell containing Cro and no

CI continues to express Cro and no CI, giving the aforementioned anti-immune state. During infection, this simple rule (the state "Cro but no CI" is self-perpetuating) can be overridden by expression of CI from the P_{RE} promoter. In addition, as stated already the anti-immune state has no normal role in λ biology, and lethal lytic functions must be blocked by mutation to enable this state to persist for many generations (Eisen et al., 1970; Neubauer and Calef, 1970).

CI is far more complicated, and the complexities are introduced in stages. CI binds tightly to O_R1 and O_R2, shutting off P_R expression. It also stimulates its own expression from P_{RM} (Figure 12.2d). As is the case with Cro, a cell containing CI and no Cro perpetuates this pattern of gene expression (Figure 12.2g). Hence, this circuit has the wiring diagram of a bistable switch with double-negative feedback; CI represses Cro and vice versa.

However, for a bistable switch to function, the system needs at least one other component, that is, some form of nonlinearity, for the states to be stable (Ferrell, 2002). Cro DNA binding is nonlinear, because Cro dimerizes weakly ($K_{dimer} \approx 1 \mu M$) (Darling et al., 2000a; LeFevre and Cordes, 2003), so that its binding to DNA is a nonlinear function of [Cro]. CI binding is nonlinear for two reasons—it dimerizes rather weakly ($K_{dimer} \approx 5$–10 nM) (Koblan and Ackers, 1991b), and its binding to adjacent binding sites, such as O_R1 and O_R2, is highly cooperative (cooperativity parameter ≈ 100) (Johnson et al., 1979; Koblan and Ackers, 1992). By itself, O_R1 is a strong CI binding site, while O_R2 is a weak binding site, but cooperativity promotes occupancy of O_R2. CI can bind cooperatively to O_R2 and O_R3, but the usual binding pattern is at sites 1 and 2 because site 1 is a tight binding site. In addition, the fact that CI stimulates its own expression from P_{RM} means that the system has positive feedback, another feature important for bistable circuits since it helps drive the system away from an intermediate undecided state.

CI also has a higher-order form of cooperative DNA binding (Figure 12.2e). In addition to the O_R regulatory region, CI and Cro bind to a second region, the O_L region (Figure 12.2b), located ~2.4 kb from O_R. As at O_R, there are three binding sites for each protein at O_L but only one promoter, the early lytic promoter P_L (Figure 12.2c). CI also binds cooperatively at O_L; again, O_L1 is the tightest binding site, while O_L2 and O_L3 are weaker sites; cooperative binding usually leads to occupancy of O_L1 and O_L2, leaving O_L3 free.

When CI is bound as a tetramer (dimers at two adjacent sites) both to O_R and O_L, cooperative contacts between the two tetramers lead to formation of a long-range loop containing an octamer of CI molecules (Figure 12.2e) (Dodd et al., 2001). In addition, when this loop is formed, further cooperative interactions can lead to occupancy of O_L3 and O_R3, forming a dodecameric structure. Initial loop formation is evidently only slightly favored (Dodd et al., 2004), but subsequent binding of two more CI dimers is strongly favored.

Formation of looped structures has two opposing effects on expression of P_{RM}. First, if O_R3 is not occupied, looping affords a further stimulation of $\approx$ twofold on P_{RM}, giving another form of positive feedback (Anderson and Yang, 2008a). Second, if O_R3 is occupied, P_{RM} is repressed, giving negative feedback of CI expression. At the level of CI present in a lysogen, this promoter is expressed at perhaps 40–50% the maximal rate (Dodd et al., 2004; Anderson and Yang, 2008a).

As a consequence of all these regulatory controls, expression of P_{RM} follows a complex response to [CI] (Figure 12.2f). In the absence of CI, the promoter is nearly silent. As [CI] increases, non-linearity due to weak dimerization and cooperative binding give a sigmoid binding curve at O_R2; bound CI stimulates P_{RM}. Looping gives further

Cha ter 12

stimulation, until the curve reaches a peak. At higher levels, the rate of expression declines due to negative autoregulation.

This curve illustrates how the regulatory circuitry is able to maintain a high degree of stability in the absence of perturbations (Dodd et al., 2001) and to become destabilized in the presence of an active mechanism (proteolytic cleavage) for removing CI. At the [CI] found in a lysogen, negative autoregulation helps to maintain a constant level of CI by counteracting random fluctuations in its level. As CI becomes depleted due to cleavage, repression of P_{RM} is relieved, counteracting the effects of cleavage, but below a critical level (the peak in the curve) the positive autoregulation begins to decline, leading to diminished expression. Finally, as the level further declines, P_R is derepressed, leading to expression of Cro, which in turns represses P_{RM} by binding to O_R3. At this point, the switching process is complete, and the cell can follow the lytic pathway.

12.3.3 Prophage Induction

This process is triggered by activation of the host SOS response (Little and Mount, 1982), so called because it responds to a distress signal. The host does not support this response for the benefit of λ; rather, the SOS response controls a suite of genes (termed SOS genes) that are involved in repairing DNA damage and coping in other ways with the effects of blocking DNA synthesis. It is believed that λ has evolved to take advantage of this signal that a cell is in trouble and may not survive, affording the virus a way to escape a doomed host.

At the level of systems behavior, prophage induction exhibits threshold behavior (Little et al., 1999) (Figure 12.3). At low doses of DNA damage, usually provided in our experimental work by ultraviolet irradiation, little or no switching occurs. At a particular dose, switching becomes efficient, and at higher doses nearly 100% of cells become

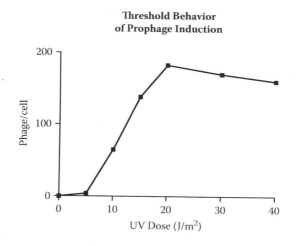

FIGURE 12.3 Threshold behavior of prophage induction. Experimental data are from Michalowski et al. (2004) for wild-type λ. In this experiment, exponentially growing cells are given graded doses of ultraviolet light, as indicated, followed by growth for 2 hr and assay of phage produced. This procedure does not measure switching directly, but it is much more convenient than techniques that measure switched cells. Direct measurements of switched cells (JWL, unpublished data) suggest that, except at the lowest ultraviolet doses, the yield of phage per switched cell is constant, indicating that this indirect measure accurately reports fraction of switched cells, except at low doses.

induced. The dose giving 50% the maximal phage yield is defined as the *set-point* for induction, and one of the goals of the modeling is to reproduce threshold behavior and the effects of mutations on the set-point.

Prophage induction requires the action of the host RecA protein. RecA plays multiple roles in cellular metabolism (Cox et al., 2000; Schlacher et al., 2006). It catalyzes an early step (strand invasion) in genetic recombination. It also plays important roles in DNA repair and mutagenesis. In addition, it has a regulatory role in the SOS system. The form that catalyzes cleavage, termed RecA*, can form in vitro by binding of RecA to single-stranded DNA (ssDNA) in the presence of ATP or a poorly hydrolyzable analogue like ATP-γ-S (Craig and Roberts, 1980). In this complex, RecA forms a helical filament on the ssDNA (Chen et al., 2008). Formation of this filament is reversible, especially when (as in the cell) it contains ATP. Though it is believed that this filament is the activated form of RecA, it is not known with certainty whether ssDNA is the only activating molecule or what DNA structures (e.g., stalled replication forks, daughter strand gaps) contain ssDNA after DNA damage in vivo. These limits to present knowledge complicate the modeling of this process.

Mechanistically, specific cleavage is an unusual reaction, in that CI and LexA actually catalyze their own cleavage in an intramolecular reaction, and RecA somehow stimulates this reaction (Little, 1984, 1993). Put another way, RecA* acts indirectly as a *co-protease* to catalyze self-cleavage. An attractive model, based mostly on structural analysis (Luo et al., 2001), is that RecA* stabilizes a reactive form of these proteins. However, this feature does not complicate modeling of RecA-mediated cleavage; RecA* still acts as a catalyst of the reaction, and the reaction can be treated using standard enzyme kinetics.

Although details of λ regulation may contribute to the threshold behavior of prophage induction, we believe that it results primarily from the kinetics with which RecA is activated. It takes roughly 30 min to cleave all the CI; at low doses of DNA damage, RecA* does not persist long enough to complete this process. One goal of the modeling is to test this kinetic hypothesis.

12.3.4   Methods for Analyzing λ

There exist a number of other phages, related to λ, termed lambdoid phages (Campbell, 1994; Weisberg et al., 1999; Juhala et al., 2000). Where tested, they have all the regulatory decisions that λ does. Some of the mechanisms resemble those of λ, and the organization of *cis* regulatory regions is the usually the same or very similar to λ. It is likely, however, that some of the underlying mechanisms differ in detail. It would eventually be of interest to compare the behavior of the λ circuitry with that of other lambdoid phages. To this end, we indicate here the ways many of the λ features were discovered and analyzed, as a way of assessing how much work it would be to attain a comparably detailed mechanistic model for other phages. As a historical note, many of these experimental approaches were developed originally for use in λ studies.

Existence of the various *cis*-acting sites is relatively easy to establish, by a combination of sequence analysis, DNase I footprinting, and mapping the location of transcripts and regulatory regions. The latter can be done by primer extension of transcripts or by the use of reporter genes to delimit the location. This information is available for many lambdoid phages (e.g., see Bushman, 1993; Degnan et al., 2007).

The use of uncoupled systems to analyze the response of promoters to regulatory proteins, originally developed for use in λ (Maurer et al., 1980; Meyer and Ptashne, 1980; Meyer et al., 1980), is widely used throughout molecular biology. In this approach, the expression of a regulatory protein is uncoupled from the level of that protein, abolishing any self-regulation. For instance, CI can be expressed from a *lacP::cI* fusion, and graded amounts of CI can be provided with graded amounts of IPTG. Then the consequences of varying CI levels on P_{RM} expression are assessed by using $P_{RM}::lacZ$ reporters; typically several different versions of the reporter are used to assess detailed features of CI action (e.g., the effect of looping). This approach gave some of the first indications for positive and negative autoregulation of λ P_{RM} and the existence of looping-mediated effects. In sum, this approach allows an input–output relationship to be measured between CI level and P_{RM} activity; this is not possible in the coupled system, in which self-regulation of CI prevents experimental control of the level of CI.

The affinities of the various protein–DNA interactions can be determined using DNase I footprinting, which can also yield cooperativity parameters (Koblan and Ackers, 1992). This is a labor-intensive undertaking; for instance, it has not been done rigorously for the O_L region in λ. And, for the data to be readily interpretable, dimer dissociation constants for CI and Cro need to be determined so that the concentration of dimers in the reactions are known. In addition, this analysis should be done under conditions resembling as closely as possible the in vivo conditions, a constraint not yet realized for λ.

Parameters for RNAP interactions with the promoters, at least for P_R and P_{RM}, have been analyzed by the abortive initiation method developed by McClure (1980). This again is labor-intensive, although it may suffice for modeling to simplify the treatment of transcription initiation in the simulation so that it is not necessary to know these values.

Details of housekeeping functions such as transcription and translation can likely be assumed to be the same as for λ, and can be simplified for simulation purposes.

These approaches would provide the bulk of the parameter set needed for modeling. However, for refining the parameter set, the use of mutant versions of the circuit is extremely helpful. Although some mutations can be made in *cis* acting sites, studying the properties of mutations in proteins that affect subtle aspects of function will demand the isolation of such mutations. It is unlikely that such mutations can be predicted from the sequence of the proteins, particularly since these are weak interactions stabilized by the chelate effect. Specifically, *cI* mutants affecting cooperativity (Beckett et al., 1993; Benson et al., 1994; Whipple et al., 1994) or positive autoregulation (Hochschild et al., 1983) were isolated in genetic screens and selections, mostly involving reporter genes. These efforts could be adapted to other CI genes and have been to some extent, such as with cooperativity mutants of phages HK022 (Mao and Little, 1998) and P22 (Valenzuela and Ptashne, 1989). The examples and tools provided by previous efforts will make this easier, but it is still not trivial.

As a final cautionary note, despite many years of intensive study of λ, features of the circuitry were missed until recently, specifically the negative and positive effects of looping on P_{RM} expression. This means that a casual search for features found in λ is not sufficient. Also, one must be alive to the possibility that another lambdoid phage will contain features not present in λ, or qualitatively different from λ. Four examples come from phage HK022 (Weisberg et al., 1999). First, its CI also binds to another operator, termed O_{FR}, located downstream of *cro*, that has no counterpart in λ (Carlson and

Little, 1993); O_{FR} is likely involved in some type of looping. Second, expression of P_L is not essential for lytic growth of HK022. Third, its anti-termination mechanism (not a part of the current modeling effort for λ) does not involve the use of a protein like λ N protein but is based on modification of RNAP by the transcript itself. Finally, there is no genetic material between the end of CI and the O_L region, in contrast to λ and most other lambdoid phages (Degnan et al., 2007). Other examples of variations among lambdoid phages are given in Campbell (1994).

12.4   Previous Modeling Efforts on Lambda

There is a long history of modeling the λ circuit. The first λ model (Thomas et al., 1976) was based on Boolean logic. The first detailed mechanistic model (Shea and Ackers, 1985) used a deterministic approach, based on measured free energies of interaction with CI and Cro with the various operators. Another study (Reinitz and Vaisnys, 1990) identified missing levels of cooperativity in the λ circuit; later work identified DNA looping and weak dimerization of Cro as examples of this. A later approach (Aurell et al., 2002) attempted to estimate the stability of the lysogenic state, using a similar deterministic approach to evaluate operator occupancies, and a first-exit approach to estimate stabilities. Still other efforts (Zhu et al., 2004; Lou et al., 2007) have used a range of mathematical approaches to assess stabilities. All of these efforts did not include looping between O_L and O_R and hence were missing an important regulatory feature, one that confers additional positive and negative autoregulation on the circuit.

More recently, a stochastic simulation (Morelli et al., 2009) was developed, independently of the work described here, that does include looping (although not the stimulation of P_{RM} by looping). This effort highlighted the importance of looping for stability of regulatory states. It included the use of a method called forward flux sampling (FFS; Allen et al., 2009), which is designed to speed the simulation of rare events such as switching of wild-type; this approach was adopted here.

These modeling efforts have had a diverse range of goals, as previously discussed for the general case (Arkin, 2001). Many have sought to further our understanding of λ and to identify deficiencies in our understanding at the mechanistic level. Others have sought to illustrate design principles of cellular regulatory circuits (e.g., Savageau and Fasani, 2009), such as bistability or extreme stability. Still others are intended to analyze the relevance of certain physical principles in the operation of cellular circuits, such as noise. Finally, since part of the λ circuit have commonly been used in a wide range of synthetic biology studies, simulating the behavior of these working parts has often been done as part of simulating the behavior of constructed circuits. The goals of the work described in this chapter are somewhat different, as described next.

12.5   Stochastic Model of the λ Circuitry

12.5.1   Need for a Stochastic Model

The focus of this modeling effort is to simulate the stability of the lysogenic state and the process of prophage induction. The lysogenic state is highly stable and serves as the paradigm for a stable epigenetic state in a system lacking features, such as chromatin architecture, that lock down regulatory decisions irreversibly. Prophage induction is likewise a long-standing paradigm for an epigenetic switch. Each case involves a switching event that can be rare and, more importantly, occurs at small numbers of CI molecules. This demands the use of a stochastic simulation (Gillespie, 1977, 1992; McAdams and Arkin, 1997; Rao et al., 2002) (see also Chapters 3 and 10).

12.5.2 Goals of the Modeling

The first goal is to discover the critical parameters that control the high degree of stability of the lysogenic state and that govern the switching process. A second is to attain a fit between a wide set of experimental observations, on one hand, and the predictions of the simulation on the other hand. This will make the model broadly applicable and provide evidence that the mechanistic model underlying the simulation is a reasonable representation of the biological system. A third goal is to make predictions from the simulation and test them experimentally. A fourth is to test mechanistic models for the systems behavior of the circuitry, such as threshold behavior of prophage induction. A fifth and related goal is to test the importance of various mechanistic features in determining the behavior, particularly those that have been proposed as a result of detailed mechanistic work. This can be done only in the context of a detailed mechanistic model such as the present one; examples are given below.

12.5.3 General Approach

To relate the output of the simulation to the results of experiments, it is necessary to look at observable endpoints. There are two primary endpoints for which abundant experimental data are available, namely, the rate of switching from the lysogenic state to the lytic state in mutants less stable than WT, and the fraction of cells that switch during the process of prophage induction. In principle, the stability of the lytic or anti-immune state could be analyzed. However, this is not emphasized in our modeling efforts because we have not studied it experimentally and the limited available data are hard to relate to switching rates.

Although the output of the simulation could be taken to be the system state as a function of time, in practice the primary output is considered to be the concentration of the two regulatory proteins CI and Cro as a function of time. The program makes a running plot of this output. Hence, switching is generally defined as a transition from a high-CI/low-Cro state to a low-CI/high-Cro state.

Since the simulation is stochastic, in general each simulation takes a different time course. To obtain a measure of switching rates, many trajectories of the system, typically 1000, are therefore simulated, and the kinetics and overall event probabilities of the switching events are characterized statistically. Hence, to measure stability of the system, we start it in the lysogenic state (high [CI] and no Cro), follow the time course, and tabulate the times at which switching occurs to a low-CI/high-Cro state and the fraction that do not switch in a specified time interval. Similarly, for measuring prophage induction, we start the system in the lysogenic state; then allow CI cleavage to occur for a specified length of time, and tabulate the fraction of runs that switch (by the same criteria) and the time at which this occurs. Again, these data are characterized statistically.

This system is simulated in the framework of the chemical master equation, which models the chemistry as a spatially homogeneous discrete state (molecule number-based) continuous time Markov process and is, in most ways, a more rigorous description of chemistry than the deterministic/continuous chemical kinetics that are the more common representation (see Chapters 3 and 10). The role of space and mechanical processes are abstracted in this picture, and, of course, the chemical mechanisms we choose are often themselves coarse approximations of the real underlying physics. But experience has shown that this level of analysis is acceptable for a wide array of systems such as this.

The simulation of system behavior uses a version of the exact stochastic simulation algorithm, now commonly called the Gillespie algorithm (Gillespie, 1977) (see also Chapters 3 and 10). The simulation consists of two parts: a mechanistic model of the circuitry, which underlies the simulation, and the simulation itself. The simulation runs as a loop (Figure 12.4a). At the start of each iteration of the loop, the system is in a defined state, and that state allows certain events to occur. Each of these has a given probability of occurring. The Gillespie algorithm uses specific random numbers to choose which event will occur and when. The chosen event occurs at the chosen time, the state of the system is changed, and the next iteration begins.

The mechanistic model is one of the most highly detailed ones currently being developed. The degree of detail lies not in the number of regulatory components, in contrast, for instance, to models of *Drosophila* embryogenesis (Von Dassow et al., 2000; Manu et al., 2009), but in the depth of detail in which regulatory interactions are described. This complexity is necessary, in part, because the regulation of CI expression is so complicated. It is feasible because the depth of knowledge about λ permits a detailed model

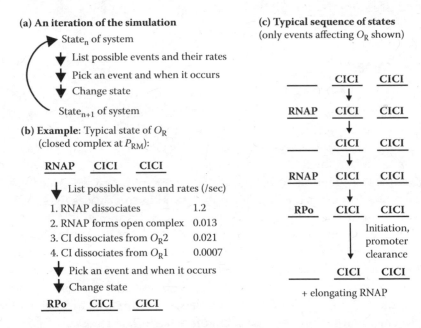

FIGURE 12.4 State machine and operation of the simulation. (a) The simulation runs in a loop, as indicated; at the end of each iteration, the state of the system changes. (b) An example, in which only the O_R region of one prophage is indicated. (c) Typical sequence of changes at a single site; generally, many steps (not shown) would be taking place at other O_R or O_L regions, or in solution, so that the changes shown would usually not occur in successive turns of the loop.

to be made that has a chance of being reasonably accurate. In addition, inclusion of features such as DNA replication both makes the model more realistic and uncovers unsuspected roles for such features. This model breaks down each reaction into a number of elementary mass-action steps, each of which is assumed to have exponentially distributed reaction firing times, as required to use the Gillespie algorithm.

Due to the complexity of the underlying mechanistic model and the conditional nature of several mechanistic features, we chose to write our own simulation code rather than to use standard software. In particular, complexities introduced by CI-mediated looping make it necessary to assess probabilities of certain events depending on detailed features of the system state. The mechanistic model embodies all the aspects of the regulatory circuitry outlined above, and simplified versions of several general housekeeping functions involved in gene expression such as transcription and translation. As is standard, a *parameter set* is provided to the model so that the probabilities can be assessed.

12.5.4   States of the System

The state of the system at each point in a simulation is defined by several features. First are the numbers of molecules of each of several forms of the two DNA-binding proteins CI and Cro, including free monomers, free dimers, and dimers bound nonspecifically to DNA. Second are the states of the O_L and O_R regions for each of several prophages, typically four; in addition to the local states of each region, a CI-mediated loop can form between these two sites on a prophage, creating additional possible states of these regions. Third are other features such as the cell volume, the positions of transcribing RNAPs and of ribosomes on mRNAs, and a flag indicating whether DNA replication has occurred.

12.5.5   Parameter Set

An important part of any modeling effort is the choice of parameters. A wide body of knowledge about parameters in the λ system is available, resulting from a combination of in vivo and in vitro work. Parameters based on this work have been used in most previous modeling efforts; with advances in knowledge and understanding, the parameter set has become more refined.

In the course of evaluating available parameter sets, however, we came to realize that estimates of the parameters are both incomplete and in some cases likely to be inaccurate. In many cases the measured values are not directly comparable, being measured under different conditions and often under conditions that do not likely reflect the in vivo situation. We expect this is a general issue. The next section describes an example that is applicable to the λ case and is likely a more general concern for modeling.

Parameter sensitivity tests are used to identify the parameters critical to the proper operation of the circuitry. In these tests, the values of parameters are systematically varied, and the effects on various aspects of circuit behavior are assessed. To date, we have used three different tests. First, the system starts with an amount of CI and Cro that yields about 50% each of the lysogenic and lytic state, and the fraction of each outcome is measured. This is not meant to mimic the lysis–lysogeny decision but to assess the balance between CI and Cro. Second, the stability of a relatively unstable mutant is assessed, typically one with reduced k_f values for P_{RM}. The test is whether the half-life of the circuit is affected. Third, a dose-response curve is carried out for prophage induction, asking whether the set-point for switching is affected. These tests are ongoing, but they do reveal both that the exact values for many parameters are not critical, and that the values for several parameters, such as affinity of Cro for O_R3 and of CI for O_R1 and O_R2, are critical; the critical parameters are usually those that have been expected from qualitative descriptions to be critical. In addition, the tests allow assessment of features such as rates of binding and unbinding of RNAP or regulatory proteins, kinetic parameters for transcription, and events related to CI-mediated looping. Other uses are indicated below.

12.5.6   Affinities of Specific DNA–Binding Proteins

Affinities of Cro and CI for their operators have been measured meticulously for the O_R region, mostly by the Ackers laboratory (Koblan and Ackers, 1991b; Koblan and Ackers, 1991a; Koblan and Ackers, 1992; Darling et al., 2000b). Parameters based on these detailed analyses are used in current modeling efforts. However, this analysis is incomplete, and the values obtained may not accurately reflect the in vivo values.

When these studies were done, the importance of O_L for regulation at O_R was not yet known, and only a cursory study (Senear et al., 1986) of CI binding to O_L was done. This did not measure the affinities of CI for the individual binding sites, so the value of ω, the cooperativity parameter, could not be measured unambiguously. This is a particular concern, since the spacing between O_L2 and O_L3 is only 3 bp, unlike the 6 or 7 bp spacing between O_R2 and O_R3 or O_R2 and O_R1, respectively, which have been shown to have about the same level of cooperativity. With another CI repressor, that of phage HK022, reducing the spacing has a substantial effect on the value of ω (Mao et al., 1994) and changes qualitatively the nature of cooperative interactions (Liu and Little, 1998). Hence, it is plausible that the value of ω for binding to O_L2 and O_L3 differs from the value currently used in most modeling efforts. In turn, this may influence the values of the individual affinities and of the cooperativity parameter for binding to O_L1 and O_L2, and which looped forms are favored.

More importantly, the in vitro data may not reflect the in vivo values. Many prokaryotic specific DNA-binding proteins bind more tightly in buffers containing the anion glutamate (the major anion in *E. coli*) than in those with acetate or chloride (Leirmo et al., 1987). Typical increases in affinity are 10- to 40-fold. Yet most affinity measurements have been done in buffers containing chloride ion. How should we deal with this disconnect?

One protein that is known to bind more tightly in glutamate is RNAP (Leirmo et al., 1987). A standard value used for the concentration of free RNAP is 30 nM (McClure, 1985, but also see Bremer et al., 2003, who estimate a value > 1 µM). With this value of RNAP and the standard affinities measured in chloride-containing buffers, we are unable to attain sufficient expression of P_{RM} to attain the CI level found in a lysogen. Raising [RNAP] to 300 nM allows enough CI to be made but also uses much CPU time in binding and unbinding events. Instead, we have chosen to retain the 30 nM value and reduce the off-rates for RNAP dissociation, which is justified due to the glutamate effect.

This issue is perhaps harder to address for CI and Cro. Again, the measured affinity values for CI and Cro are in chloride buffers. We have found (JWL, unpublished data) that CI binds perhaps eightfold more tightly to O_R2, a weak binding site, in 200 mM

K-glutamate than in 200 mM KCl; this value is tentative because it is unclear whether the dimer dissociation constant also changes in glutamate, but the increase in affinity is almost certainly in the range of 5- to 10-fold.

Although CI and Cro may bind more tightly in glutamate, their affinity for nonspecific DNA may increase as well in glutamate, with the result that the concentrations of free proteins will be very low, and occupancies of specific sites will not change markedly from the values measured in vitro.

One way to evaluate whether affinities measured in vitro correspond to the in vivo values is to measure occupancies in vivo as a function of CI or Cro concentration. In principle, reporter genes could be used to measure repression and activation curves as a function of in vivo CI or Cro concentration. Workers have measured reporter activities for P_{RM}::*lacZ* and P_R::*lacZ* reporters in response to various levels of CI (Dodd et al., 2001, 2004) and of P_R::*lacZ* in response to Cro (Reinitz and Vaisnys, 1990, analyzing unpublished data of Pakula et al., 1986), provided in both cases by cells carrying plasmids with *lacP*::*cI* or *lacP*::*cro* fusions, growing cells in the presence of varying levels of IPTG to afford graded amounts of CI or Cro. However, there are several difficulties in interpreting these data.

First, there must be an independent measure of nonspecific DNA binding, to allow calculation of the free protein concentration. This measure is not available for CI and Cro; minicell measurements with two other prokaryotic DNA-binding proteins, Lac repressor and LexA, suggest that only a small fraction of these proteins is free in solution (Kao-Huang et al., 1977; Sassanfar and Roberts, 1990), and the remainder is inferred to be bound nonspecifically. With λ, two separate computational analyses have used reporter data to evaluate nonspecific binding. Importantly, both studies assumed that the in vitro affinity values are correct. The first study (Bakk and Metzler, 2004a) used the data cited above and concluded that the ΔG for nonspecific binding for both CI and Cro was about −4.2 kcal/mole. The second (Dodd et al., 2004) calculated nonspecific binding for CI, using both P_{RM}::*lacZ* and P_R::*lacZ* reporters (repression only), and obtained a value of −6.3 kcal/mole. In each case, the fits of the data to the model are reasonably good; however, for CI the first study relied mainly on P_{RM} data obtained at low CI concentrations, which is problematic due to stochastic variation (see below). Both groups concluded that the fits between modeling and data give a consistent parameter set. However, it remains possible that the same or very similar fit could be obtained if specific binding is tighter, as the glutamate effect suggests, and a compensating increase in nonspecific binding offsets that. We conclude that, without an independent measure of nonspecific binding, this approach does not by itself provide strong evidence that the in vitro data set obtains in the cell as well.

A second concern is that the measurement of CI or Cro levels is done on a bulk culture, and it must be assumed that the level of CI in all cells is the same to interpret these data. However, gene expression is stochastic, and the level of CI will vary from cell to cell. This variability will be compensated to some extent in the normal λ circuitry by negative autoregulation (Becskei and Serrano, 2000; Dodd et al., 2001), but this feature is not present in the plasmid system. This concern is especially important at low levels of CI, where noise should play a larger role, a particular concern in evaluating the early part of the CI dose-response curve for P_{RM}.

A third concern is an inconsistency (Maurer et al., 1980; Dodd et al., 2001) in the data obtained by this approach, namely, that the apparent CI activity obtained in the plasmid-based system differs from that observed when CI is provided by a prophage; the latter appears to be considerably more active than the CI provided from a plasmid. A similar disparity was observed (I. Dodd, personal communication) with the P_R::*lacZ* fusion, where CI made from a prophage gave a greater degree of repression than plasmid-provided CI at the same level. Dodd et al. speculated that this disparity might result from heterogeneity in plasmid copy number. As an alternative approach, we (C.B. Michalowski and JWL) have made a set of λ prophages that supply variable levels of CI. This will remove this copy-number issue, and negative autoregulation should reduce, but not eliminate, cell-to-cell variation. This approach cannot, however, be used for low CI values, since lysogens will switch to the lytic state. In sum, we conclude that evaluating in vivo occupancies by use of reporter genes is not as straightforward as it might appear.

12.5.7   Treatment of Protein–DNA Interactions

In the previous stochastic simulation (Arkin et al., 1998), and in most deterministic λ modeling efforts, the partition function developed by Shea and Ackers (1985) was used to assign probabilities that a given regulatory region is in a particular state. The underlying assumption is that protein binding and unbinding occurs on a much faster timescale than other events such as open complex formation by RNAP. This approach has two flaws. It allows what appear to be biochemically implausible transitions (e.g., those involving several sequential dissociation and binding events) to occur in a single step, on the timescale of much less than a second. In addition, the complexes formed upon CI-mediated looping (not yet discovered in 1998) are presumably held together by a large number of weak interactions, and it is plausible that they dissociate relatively slowly. If so, the assumption that all states of the regulatory region are in rapid equilibrium may be invalid.

Hence, we have chosen to implement a state machine. In this approach, the state of each regulatory region is made explicit at each stage of the simulation (Figure 12.4), and each state of the regulatory region can change to a limited number of possibilities (Figure 12.4). For instance, in the configuration shown in Figure 12.4b, four possible transitions of the O_R region are allowed, and each has a probability given by the parameter set. Other transitions (such as CI binding to O_R1) involve the concentrations of CI, Cro, or RNAP. For a given prophage, our standard model specifies 43 and 84 different states of O_L and O_R, respectively; in all, there are 2583 allowable O_L–O_R combinations. With multiple prophages, the number of possible states becomes astronomical. To avoid specifying a vast number of possible states, the states of O_L and O_R are specified separately, but looped forms must form and dissociate coordinately at both sites when a transition of this type is chosen. On- and off-rates are assigned for each binding reaction. The on-rates are assumed to be diffusion-limited, and they are set at 10^8 mole^{-1} sec^{-1}; off-rates are then dictated by measured dissociation constants.

12.5.8   Nonspecific DNA Binding

Most if not all specific DNA-binding proteins bind nonspecifically to any duplex DNA. Though nonspecific binding is far weaker, the great abundance of nonspecific sites (~3×10^7 in a cell with four chromosomes) means that a substantial fraction of CI and Cro are likely bound nonspecifically (Kao-Huang et al., 1977; Bakk and Metzler, 2004b).

Chapter 12

Since nonspecific binding is weak, equilibration between free dimers and nonspecifically bound dimers are assumed to be instantaneous, and this is reequilibriated at the end of each iteration of the loop. We typically assume that, of these dimers, 2/3 are bound nonspecifically, but these parameters can be changed.

12.5.9 CI-Mediated Looping

Looping complicates the simulation for several reasons. First, looping occurs only or preferentially in *cis* (I.B. Dodd, personal communication). Accordingly, each prophage must be treated separately; O_L from one prophage can interact only with O_R of that same prophage. Hence, the state of the system specifies the states of O_L and O_R for each prophage.

Second, many different looped forms are possible, and current knowledge about which are allowed is limited. The simplest looped form is an octamer with CI dimers bound cooperatively to adjacent operators. Since binding to O_L1 and O_L2, and to O_R1 and O_R2, are energetically favored, the predominant octamer likely contains CI bound to these four sites. Alternate occupancy patterns (O_L2 and O_L3, O_R2 and O_R3) are possible, giving four looped octamer forms. In addition, the two tetramers can approach one another in two ways (Anderson and Yang, 2008a, 2008b) (Figure 12.5), since DNA is quite flexible over the distance (~2.4 kbp) separating O_L and O_R. As there is no information about this or its functional consequences, for the sake of simplicity these two configurations are not distinguished.

Third, looping has different mechanistic consequences depending on which sites are occupied at O_L and O_R. If O_L3 and O_R3 are occupied, P_{RM} is repressed. If both sites are free, P_{RM} is activated ~twofold (Anderson and Yang, 2008a, 2008b); it is unclear at present whether this occurs in both configurations (Figure 12.5) of the octamer. Since these configurations are not distinguished, the strength of P_{RM} is simply increased by a value "act2." Anderson and Yang also included another factor, "act3," which could differ from act2 in their earlier paper (Anderson and Yang, 2008a) but which they set equal to act2 in later work (Anderson and Yang, 2008b). In this latter model, the effect of CI bound to O_L3 depends on the octamer configuration. Again, for simplicity, we use act3 whenever O_L3 is occupied; this does not accord exactly with their model, but more experimental work, likely difficult, will be required to validate these features of their model. For the moment, we allow the parameter values for act2 and act3 to differ from each other, but normally set them to be equal. Though the values of these factors affect expression of *cI*, and thereby

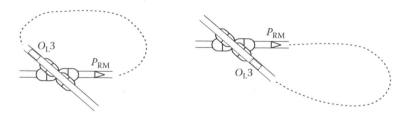

FIGURE 12.5 Two looped forms of the CI octamer. These forms differ in the orientation of O_L relative to O_R, resulting either in a bight-like or "parallel" (left panel) or in a hairpin-like or "antiparallel" (right panel) configuration of the loop, as indicated by the dotted lines (not to scale). (Adapted from Anderson and Yang, 2008b.)

the level of CI in the cell, it is unclear at present how to validate particular choices since we cannot readily control the distribution of octamer configurations in vivo.

Fourth, the program allows an O_R or O_L region to form an octamer loop if it has CI bound to two adjacent sites, but if and only if its counterpart site is also able to loop for the same reason. Otherwise the probability of changing the state of O_R or O_L to a looped form is zero.

12.5.10 Transcription and Translation

Initiation of transcription by RNAP occurs in at least two stages (Figure 12.6a) (McClure, 1985) (see also Chapter 9). The first, specific binding to the promoter, forms an initial *closed complex* (RPc). RPc can freely dissociate. We simulate this step as a simple binding event, ignoring the much more complex nature of this process (Weindl et al., 2009). In a step that is generally rate-limiting for most promoters, RPc then undergoes an *isomerization* to form an open complex, RPo, in which the two DNA strands are separated near the start-point of transcription. Formation of RPo is often irreversible, and though a parameter is provided for RPo → RPc we normally set its rate at zero. RPo then initiates transcription, forming an initiation complex RPi. This moves at a specified rate down the DNA, and after one sec or so it clears the promoter, leaving the free promoter and an elongation complex (Figure 12.6a).

Elongation complexes then proceed down the DNA to the end of the gene being transcribed. Here only transcription of *cI* and *cro* is treated, not other viral genes, since they are the only viral proteins known to control the regulatory events modeled here. This would not be adequate in modeling the lysis–lysogeny decision, which also involves the action of the viral N, CII and CIII proteins (Arkin et al., 1998) as well as host proteases. Transcription of a gene is not instantaneous, and a small time delay between initiation and mRNA completion occurs. This time delay has been incorporated, but, for the sake

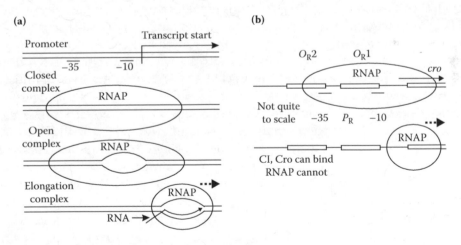

FIGURE 12.6 Promoter clearance and transient exposure of CI and Cro binding sites. (a) Forms of RNA polymerase (RNAP) bound to a promoter, shown schematically. The oval covers the regions protected by RNAP in footprinting experiments; the footprint becomes much smaller once elongation begins. (b) Exposure of O_R2 and O_R1 during promoter clearance. Until RNAP moves farther, it occludes the binding site for another RNAP molecule, presenting CI and Cro with a transient advantage in binding to the operator sequences. At an earlier stage, O_R2 but not O_R1 might be available.

of computational speed, elongation is treated as occurring in a small number of steps instead of one nt at a time; typically RNAP moves in steps of 30 nt, and the average time it takes to complete the transcript is the same as it would be for 1-nt steps. In the parameter sensitivity tests previously described, no significant difference was seen whether RNAP moved in 1-nt or 30-nt steps. For simplicity, care is also not taken to ensure that one RNAP cannot overtake another; this occurs occasionally but should occur rarely enough to affect neither the quantitative nor the qualitative dynamics of the model significantly.

The model does not include recent evidence that transcription occurs not as a Poisson process but in bursts (Golding et al., 2005) (see also Chapter 3). The mechanistic basis of this feature is not understood (Mitarai et al., 2008), so it is difficult to model. Including such a feature is expected to have substantial effects on the stability of the lysogenic state, since it would provide larger pulses of Cro protein on occasion.

When RNAP changes to elongation mode, its contacts with the promoter are released. During this process of promoter clearance, it takes a few seconds for RNAP to move far enough to allow another RNAP molecule to bind to the promoter. At this stage (Figure 12.6b) there is a transient opportunity for CI or Cro to bind before another molecule of RNAP can bind, since the operators that overlap the promoter are available for binding (Lanzer and Bujard, 1988). This feature is incorporated as an option for running the simulation. In limited tests, though, little difference has been found from the outcome when this feature is not included, so it is generally not included for the sake of simplicity. We note, however, that testing the importance of this mechanistic feature is possible only in the context of a highly detailed model such as this one.

In prokaryotes, the initial translation of an mRNA can occur in a coupled fashion with transcription; that is, the ribosome can follow closely behind the RNAP. For simplicity, this complication has been ignored, so that an mRNA becomes available for translation only once it is completed; all translation events are equivalent. Again, for the sake of computational speed translation is allowed to occur in a small number of steps, in each of which the ribosome moves 10 codons; this provides a time delay, but the average time to make the protein is the same as with 1-codon steps. In the parameter sensitivity tests, no significant difference is seen when ribosomes moved in 1-codon steps rather than in 10-codon steps. Since mRNAs are unstable, a pathway is included for mRNA degradation. As in the previous simulation (Arkin et al., 1998), prior to translation initiation, a given mRNA is offered a choice, dictated by a random number, between being translated and being degraded. For simplicity, only one ribosome on an mRNA at a time is allowed. This likely makes it take longer to translate a given mRNA than is actually the case.

The computational speed increase gained by allowing RNAP and ribosomes to move in steps larger than a single added polymeric unit is roughly fourfold when the system is in the lysogenic state. In our view, this advantage offsets the inaccuracy of this simplification.

As a final mechanistic detail, a fraction of newly synthesized Cro monomers is in an inactive conformation, and proline isomerization is necessary to convert these to an active form. It has been suggested that this creates an additional time delay before Cro can exert its effects in vivo (Satumba and Mossing, 2002). This feature has been incorporated into the present mechanistic model. It is assumed that only mature monomers can dimerize, and parameters are provided for the fraction of newly made molecules in an inactive form and a rate constant for conversion to the active form (maturation rate).

With the range of values measured in vitro for the maturation rate, in the parameter sensitivity tests the balance of the circuitry was shifted slightly toward CI, but the effects are modest. Since it is plausible that the maturation rate is increased in vivo by proline isomerases, we have not explored this feature further. At the same time, this feature may be more important in the lysis–lysogeny decision, which can be tested computationally only when the model is expanded to include this process. Once again, although our simulation does not support this hypothesis, it could be tested computationally only in the context of a highly detailed model.

12.5.11   DNA Replication

Cells growing exponentially in rich medium usually have four copies of the chromosomal site at which λ integrates (Nielsen et al., 2007). Hence, four copies of the prophage are present at the outset of the simulation. The chromosome is replicated, doubling the number of prophages. This occurs at about 80% of the time between successive cell divisions; hence, when the volume of the cell is 1.8x its initial volume, the DNA is replicated, and the number of prophages is doubled.

The fate of DNA-bound proteins when the fork passes is poorly understood. We assume that the bound proteins simply dissociate. Hence, all the regulatory regions are initially free of bound proteins after replication. For simplicity, we do not knock off transcribing RNAP's upon passage of the fork. Passage of the fork offers a transient opportunity for RNAP to bind to P_R, which is almost always repressed in the lysogenic state. It is plausible that this could destabilize the lysogenic state. Indeed, the simulation reveals that relatively unstable mutants are markedly destabilized by this effect (see below).

12.5.12   Cell Growth and Division

To analyze stability of regulatory states, we need to be able to follow a cell for many generations. Accordingly, features have been incorporated to allow this.

We assume that the volume of the cell increases linearly with time. Previous work (Arkin et al., 1998) suggested that exponential growth of volume had a negligible effect on the outcome of the simulation. The initial volume is chosen to be 1.6×10^{-15} liter, to give 1 molecule/cell at a concentration of 1 nM. The rate of cell growth is chosen to give a doubling time of 35 min, but this parameter can be varied.

When the volume has increased to twice its initial value, cell division takes place. Half of the prophages, with their bound proteins, are discarded, and the free and non-specifically bound Cro and CI are distributed to the two daughter cells according to a binomial distribution (Berg, 1978; Huh and Paulsson, 2011), as independently done previously (Rosenfeld et al., 2005). Proteins bound to the retained prophages remain bound during cell division, as occurs in vivo.

12.5.13   CI and Cro Monomer–Dimer Equilibria

Both CI and Cro undergo readily reversible equilibration between monomers and dimer forms (Koblan and Ackers, 1991b; Darling et al., 2000a). CI can also form higher-order oligomers in solution at high concentration, probably due to contacts the same or

Chapter 12

similar to those that allow cooperative DNA binding, but we have ignored these, since their biological relevance is unclear and they only occur at concentrations higher than those present in vivo. Equilibration of Cro is rather slow (Jia et al., 2005), while that of CI is assumed in the model to be diffusion controlled. In some simulations, the rates of equilibration are slowed 10-fold to save computational time.

12.5.14   RecA–Mediated Cleavage

Several aspects of this reaction lead to our treatment. First, only monomers of CI are efficient substrates for cleavage (Cohen et al., 1981; Crowl et al., 1981; Gimble and Sauer, 1989). Although parameters for the rate of cleavage of free dimers and DNA-bound dimmers are included, they are set to zero. Since the dimer dissociation constant for CI is in the range of 5–10 nM, the concentration of monomers seen in the simulation at the CI level in a lysogen ($\approx$250 nM) is about 20 nM. Second, although RecA acts indirectly to catalyze self-cleavage, it does act as a catalyst, and we can treat it as the enzyme, and CI monomers as the substrate, in a standard enzyme kinetics treatment.

Third, uninduced cells contain roughly 7,000 molecules of RecA (Sassanfar and Roberts, 1990), and studies by Sassanfar and Roberts suggest that most or all of this RecA is activated by ultraviolet irradiation. Accordingly, the concentration of activated RecA (RecA*) is in vast excess of CI monomers; that is, enzyme is in excess over substrate. Hence, the concentration of enzyme (RecA*) is scarcely depleted by formation of the E:S complex. The rate equation is

$$\text{Rate} = k_2 \times [S] \times \{[E_t]/([E_t] + K_s)\}$$

where k_2 is the rate constant for $E{:}S \rightarrow E + P$, $[S]$ is the monomer concentration, $[E_t]$ is the concentration of RecA*, and K_s is the dissociation constant, or $[E][S] / [E{:}S]$. Hence, at a fixed [RecA*], the equation simplifies to

$$\text{Rate} = \text{constant} \times [S]$$

giving first-order kinetics of monomer degradation. Since monomers are in equilibrium with free dimers, however, the kinetics will be more complicated.

After ultraviolet irradiation, RecA is activated within 1 min, as judged by studies of LexA cleavage (Little, 1983; Sassanfar and Roberts, 1990). The level of activated RecA is maximal early after ultraviolet irradiation and declines progressively with time thereafter. This decline is faster at low ultraviolet doses than at higher doses. The initial level of cleavage activity, again as judged by analysis of LexA cleavage, is about the same over a range of ultraviolet doses.

The cleavage reaction is simulated by the following treatment, which is undoubtedly oversimplified but represents these features. First, we mimic choosing a cell at random from the population so that the result of many simulations will approximate the effect of treating a mass culture. In an unsynchronized culture, the relative number of cells of a given age x ($0 \leq x \leq 1$) is given (Ingraham et al., 1983) by $2^{(1-x)}$, or $2 \cdot 2^{-x}$. To choose a random cell from the population, we determine the fraction of cells of or less than a given age x, then set the age x of that cell using a random number (RND). The integral of 2×2^{-x} is $F(x) = -2 \times 2^{-x}/\ln(2)$; the desired expression is $[F(x) - F(0)]/[F(1) - F(0)] = (2^{-x} - 1)/(0.5 - 1) = (1 - 2^{-x})/0.5$. Setting this equal to RND and solving for x gives $x = \ln(1 - 0.5 \times \text{RND})/\ln(0.5)$. Cleavage begins when the cell volume is $1 + x$.

Second, when cleavage begins, RecA is activated instantaneously to its maximum level. A dose-response curve is given by leaving the [RecA*] at this high level for differing lengths of time, after which it abruptly returns to zero (Figure 12.7). The simulation has the option of allowing the level of RecA* to decline more gradually with time, with different kinetic regimes, but this has not been explored in detail as yet; in limited trials we see threshold behavior. Then the fate of the simulation is followed, either switching or not, and the results of many runs are tabulated to give the fraction switching for a given time at which RecA is active.

This treatment is almost certainly oversimplified in several ways. First, it is not known whether the level of RecA* is the same in all cells. It seems unlikely that this is the case. DNA damage can occur either on the leading or lagging strand, and it is unclear whether the same types of structures are formed when forks encounter damage on each of these strands (Heller and Marians, 2006; Yao and O'Donnell, 2009); if not, the amount of RecA* might differ in the two cases. Moreover, it is unclear whether RecA is invariably activated at a stalled fork and whether the amount of RecA* is the same per stalled fork. Second, as the cells recover from DNA damage, stalled forks are restarted, and presumably the amount of RecA* declines in some type of stochastic fashion as the forks are able to traverse DNA for increasing lengths of time without encountering sites of damage. Nothing is known about these issues, and it is unclear how to simulate this.

Finally, there are complications and uncertainties in incorporating two aspects of the biological system in modeling prophage induction. The first is that DNA damage leads to a block in DNA replication, because replication forks have stalled at sites of damage (Cox et al., 2000; Yao and O'Donnell, 2009). Hence, a realistic simulation should include blocking replication after DNA damage. The other complication is that when the SOS system is induced, cell division soon ceases due to the induction of the SulA

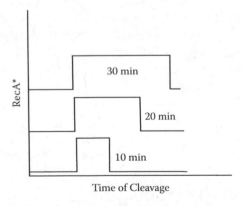

FIGURE 12.7 Sample time courses for the level of activated RecA (i.e., the input signal) during prophage induction. Three time courses for activation of RecA are shown, separated for clarity; the basal level is zero in each case. Many separate simulations for each RecA* time course are run, and the fraction switching to the lytic state is tabulated as an output.

Chapter 12

protein, which blocks cell division (Mukherjee et al., 1998); this feature that should also be included.

Among the difficulties in adding these features are the following. First, it is unclear how to model fork progression after DNA damaging events; forks resume synthesis once they are reassembled and may halt again at remaining downstream sites of damage. Hence, the kinetics of fork progression is likely to be complex and to differ with different forks and on different chromosomes. Second, new initiations can apparently occur normally at *oriC* after ultraviolet damage (Rudolph et al., 2007). When these pass the point at which the previous forks first stalled, perhaps they can move faster than forks that had stalled, because the downstream forks will have aided in removal of damage. Third, it is unclear how to model resumption of cell division. Do the long cells arising in SOS divide in several stages, making daughter cells that are longer than newly born cells during normal growth, or do they divide to make several cells of the normal size? Is this process stochastic? And how are chromosomes divided up in this process? Fourth, how are resumption of replication and cell division inter-related? When the cells grow in the absence of DNA replication, an imbalance between DNA content and cell volume arises and becomes progressively worse; how is this balance restored during the return to normal growth? Answers to these and similar questions may result from ongoing research. For instance, time-lapse microscopy of individual cells during the recovery phase should set constraints on how to model both recovery processes. Efforts to model this recovery phase are in progress.

For the moment, we have chosen to model prophage induction by blocking DNA replication and cell division once induction begins. We believe that this choice most accurately represents the events taking place during the time the regulatory decision (whether to switch or not) is usually made. Later inaccuracies resulting from not resuming these processes are expected to have a small effect on the regulatory decision. At the same time, the simulation does offer the user the choice of whether to allow replication and cell division once cleavage has begun, allowing one to assess the effect of blocking one or both processes.

These issues provide a striking example of the challenges facing modelers who consider the context of the whole cell, particularly in its response to cell stress. They also highlight one powerful aspect of modeling, namely, its ability to pinpoint gaps in our knowledge of cellular processes.

12.5.15   Output of the Model

The primary output of the model is the levels of CI and Cro as a function of time. Other aspects of the system states can, of course, be enumerated if desired. CI and Cro levels as a function of time are usually presented graphically. This output is used in various ways.

12.6   Behavior of the Model

To date, the model has been used to simulate two different aspects of λ regulation, namely, stability of regulatory states, and prophage induction. These are described in turn. In each case, we discuss fits of the simulation to experimental data.

12.6.1 Stability of the Lysogenic State

The lysogenic state of wild-type λ is highly stable; in the absence of the SOS system, its rate of switching is probably $< 10^{-8}$/generation (Little and Michalowski, 2010), a value that is below the mutation rate of a typical gene. Hence, it is uncertain that even the rare events we observed occurred in a cell that was otherwise wild-type, and the stability may be greater than this. It is a mystery what forces make this state so enormously stable. One approach to this question is to analyze mutants that reduce stability. A number of mutants have been characterized that are less stable (Little and Michalowski, 2010), switching at a readily detectable rate. Some of these mutants involve multiple changes and may be hard to model accurately; an example ($\lambda O_R 323$) is given. Others have a better understood mechanistic basis. An example is the loss of positive autoregulation, which presumably works by reducing the level of CI in the cell and allowing more frequent excursions of the system to a point at which switching becomes likely. Hence, one goal of the modeling is to reproduce the reduction of stability in existing mutants.

To assess the stability of the lysogenic state computationally, the simulation begins with enough CI to maintain that state, typically 250 nM. As the simulation runs, the levels of CI and Cro change. The system is judged to switch its state when a criterion such as Cro > 150 nM CI < 10 nM is met; this criterion is chosen by the ability of the system to remain stable in the anti-immune state when begun with these values of CI and Cro. We do many simulations of duration Tsim, and the times of switching are tabulated. Examples of switching events for a relatively unstable circuit, including one simulation that did not switch, are in Figure 12.8a; the first panel is labeled to illustrate the conventions in these plots. Half-lives are then determined either by a first-order exponential decay curve, plotting time as a function of fraction unswitched (Figure 12.8b), or by the equation $-0.693 * \text{Tsim}/\ln U$, where U is the fraction unswitched. This equation is generally used only when $0.1 \leq U \leq 0.9$; outside these limits large numbers of simulations must be done to obtain adequate statistics.

The lysogenic state is so stable, both in the cell and in the simulation, that with our standard parameter set we are unable to run the simulation long enough for switching to occur. To estimate its half-life, we implemented the forward-flux sampling (FFS) method, which samples rare events far more efficiently (Allen et al., 2006, 2009). This approach affords an estimate of the half-life for the wild-type of about 2×10^{13} generations. This value is much greater than the value determined experimentally, about 10^8 generations (Little and Michalowski, 2010), but, as stated already, the latter value is a lower bound.

The next section describes a multiple mutant, λJL516, which lacks cooperative DNA binding and carries two suppressor mutations that restore nearly normal behavior for prophage induction (Babiç and Little, 2007). We used the FFS method to assess its intrinsic stability *in silico* and found that it has a half-life of about 4×10^5 generations. Experimentally, the half-life is almost certainly $> 5 \times 10^6$ generations, and this is a lower bound as well. Evidently the simulation underestimates the stability of this mutant.

The circuit can be destabilized computationally by adjusting various parameters, including affinities of CI for operators, promoter strength, or efficiency of P_{RM} mRNA translation, and we can simulate the behavior of real mutants known to be unstable in cases where mutational effects on parameters can be assessed. In these cases, the half-lives are assessed directly as discussed already.

Chapter 12

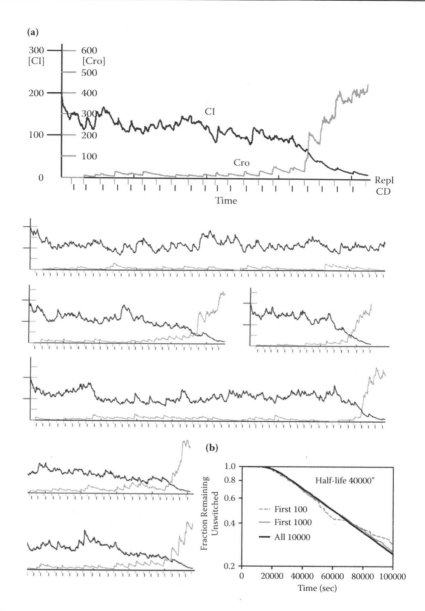

FIGURE 12.8 Switching of an unstable variant. In these simulations, P_{RM} k_f values were set at 0.25x their canonical values, resulting in a half-life of about 40,000 sec. (a) Seven simulations from a single run. The top panel is labeled to show the conventions for data representation. The x-axis is time of simulation; each simulation was 100,000 sec, and each tick above the line represents 10% of the total time. Most graphs are cropped at the time of switching. Below the axis, gray and black ticks indicate the times of DNA replication and cell division, respectively. The y-axis represents the concentrations of CI (black ticks to the left of the axis) and Cro (gray ticks to the right of the axis); in each case, ticks indicate 100 nM increments. Curves showing running CI and Cro concentrations are in black and gray, respectively. (b) Fraction of simulations remaining unswitched as a function of time. In this run, 10,000 simulations were done; of these the first 100, first 1,000, or all 10,000 were plotted. Smaller numbers of simulations give noisier plots, as expected, but the calculated half-life is similar. The half-life was calculated by fitting the data to a first-order decay curve with a plateau phase before decay begins. The plateau results because the initial CI concentration was 200 nM, and switching was called when CI < 10 nM; therefore, even if the system switches immediately it takes a period of time for CI levels to drop due to dilution.

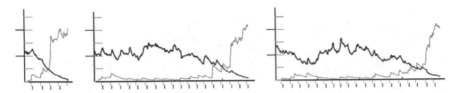

FIGURE 12.9 Bursts of Cro upon replication. In these simulations, RNAP bound 7x more weakly to P_{RM}, to mimic a mutant with this property; this variant has a half-life of 30,000 sec (Table 12.1). Plots are labeled as in Figure 12.8, and the timescale is the same. In many cases, bursts of Cro are seen at the time of replication, and a gradual accumulation of Cro often occurs prior to switching.

This approach has revealed an interesting effect that is quite likely to be of general importance. As noted, DNA replication has the potential to destabilize the lysogenic state by opening the lytic promoters transiently. We find this to be the case with a variety of unstable variants. In the simulation, bursts of Cro synthesis frequently occur soon after replication. Figures 12.8 and 12.9 show the output for two different unstable "mutants" (gray and black ticks below the x-axis indicate the times of replication and cell division, respectively). Several bursts of Cro are evident at the time of replication.

To test whether transient expression of P_R, and hence of Cro, destabilizes a circuit, we compare the effect of blocking RNAP binding to P_R for various periods of time after replication occurs. If it is blocked between 500 and 550 sec after replication, the stability is about the same as it is if RNAP binding to P_R is not blocked at all (Table 12.1). In striking contrast, if RNAP binding is blocked between 0 and 50 sec after replication, a regime that should prevent bursts of Cro synthesis caused by replication, we find marked stabilization of circuits that have been destabilized by several different mechanisms. Specifically, various parameters or sets of related parameters have been changed to give a half-life of ~30–40K sec (~15–20 generations), which revealed that for most conditions tested the effect was substantial (Table 12.1). The $\lambda O_R 323$ mutant is the exception, and is discussed below.

Tests are ongoing to determine whether the magnitude of this stabilization depends on the half-life of the system. Though the magnitude is similar in the first four cases shown (Table 12.1), it is unclear whether this is significant. The important point is that the destabilizing effect on stability is substantial. It may be difficult to devise an in vivo test of this observation.

The output traces in Figures 12.8 and 12.9 show several notable features. First, as with all stochastic simulations (including those of wild-type, not shown), the levels of CI (and Cro, when present) constantly change rather than reaching a steady state value as in a deterministic model. Second, there are frequent instances in which no CI or Cro synthesis occurs for a substantial fraction of a generation, so that the protein level drops in a curve similar to a first-order decay curve. Third, a pattern frequently seen with unstable mutants is that Cro levels gradually build up over the course of several generations; often, Cro is made mostly or completely at the time of DNA replication. Frequently, the synthesis of CI eventually ceases, and CI is diluted out, leading to a large burst of Cro and subsequent switching. This same pattern is also seen (not shown) for two other unstable mutants, listed in Table 12.1 as lines C and D. The implication is that, for these four variants at least, switching is not an abrupt process but is the culmination of events occurring over several generations. Fourth, this gradual accumulation of Cro does not invariably lead to switching; in some cases the Cro level returns to a value near zero.

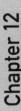

Chapter 12

Table 12.1 Destabilization of Unstable Variants by Replication

Destabilizing condition	No RNAP@P_R 0–50	No RNAP@P_R 500–550	No blocking of RNAP@P_R	Ratio (0–50/ 500–550)
	Half-life (sec)			
A. Weak P_{RM} K_B – off-rates 7x faster	0.9×10^6	3.3×10^4	2.9×10^4	28
B. Weak P_{RM} k_f – k_f's 0.25 × canonical	1.1×10^6	4.4×10^4	3.9×10^4	25
C. No positive autoregulation at O_R2 without looping	0.7×10^6	3.2×10^4	3.0×10^4	22
D. Inefficient CI translation (1.2 CI/mRNA)	1.7×10^6	3.5×10^4	3.1×10^4	50
E. λO_R323 with 9 CI/ mRNA and 50% free dimers bound nonspecifically	1.25×10^5	5×10^4	4.2×10^4	2.5

Notes: Wild-type parameters were changed as indicated to destabilize the wild-type lysogenic state. For each condition, RNAP binding to P_R was either not blocked or blocked either at 0–50 or 500–550 sec after replication, indicated by "No RNAP@P_R"; the ratio of the latter two values is given in the last column. The generation time in the simulation is about 2000 sec. A,B: Several values are provided in the model for P_{RM} off-rates and rates of open complex formation, since these values vary somewhat depending on whether RNAP is located at P_R (Fong et al., 1993). C: the model has three parameters for stimulation of P_{RM}, and values relative to unstimulated P_{RM} are given in parentheses: stimulation by CI bound at O_R2, no looping (1); stimulation by looping with CI bound at O_R2 and O_L1, with O_L3 free (3.7) or with CI bound to O_L3 (3.7); these latter terms correspond to "act2" and "act3" (see text), respectively; these values had to be increased to stabilize this mutant. D: The average number of CI monomers per transcript is decreased from 5 to 1.2. E. See text.

As a more complex example of an unstable mutant, we described (Little et al., 1999) a mutant, λO_R323, in which the O_R1 site is altered to the sequence of O_R3. This change affects only the binding of CI and Cro to the mutated site; CI binds more weakly, and Cro binds more strongly. This mutant can maintain a stable lysogenic state in vivo. Previous attempts to model its stability were either unsuccessful (Aurell et al., 2002), or were successful (Zhu et al., 2004) only when parameters were greatly altered. In both cases, the underlying models did not include CI-mediated looping and positive and negative effects on P_{RM} expression. A more recent stochastic model (Morelli et al., 2009) that included looping (but not activation of P_{RM} by looping) also predicted stability of λO_R323 under certain conditions.

The properties of this mutant in our simulation suggest that a modest amount of parameter adjustment can stabilize it. We typically allow expression of an average of 5 CI molecules/mRNA, and have 2/3 of the dimers bound nonspecifically to DNA. With these parameters, the mutant is unstable—it switches within a few generations. However, if we have 9 CI/mRNA and half the dimers bound nonspecifically, the mutant has a half-life of roughly 75 generations, stable enough that a lysogen could be isolated in the lab but far less stable than the actual mutant. The lysogenic state of this mutant can also be stabilized by making CI bind five times more tightly to its operators, provided we do not at the same time increase the strength of nonspecific binding, or by increasing the degree to which looping between O_L and O_R stimulates P_{RM}. Although more exploration of parameter space is needed, these preliminary efforts suggest that it should be possible to find parameters that stabilize this mutant.

In marked contrast to other unstable mutants, $\lambda O_R 323$ was only modestly stabilized by preventing Cro synthesis right after DNA replication (Table 12.1). It is unclear as yet why this difference occurs. Repression of P_R is less complete in this case; hence, we surmise that that removing CI from O_R during replication would have less effect on stability.

As discussed earlier, in certain mutants the λ circuitry can also exist stably in an anti-immune state. Although we can also simulate this regulatory state, we have not focused on it, since it is not a normal part of λ biology and there is not a large body of experimental data with which to compare the output of the simulation. The FFS method (Allen et al., 2009) provides an estimate for the half-life of the anti-immune state of about 2×10^6 generations in the simulation. There are not reliable experimental estimates of this value; published data suggest that this state is relatively unstable, but selective pressure against anti-immune cells likely reduced their apparent stability in these studies (Harvey Eisen, personal communication). Nonetheless, it would be an additional and somewhat independent check on the parameter set to develop experimental data for this purpose.

12.6.2   Prophage Induction and Threshold Behavior

Experimental data constrain the time at which switching occurs after ultraviolet irradiation. Switching generally becomes irreversible 20–30 min after treatment (Tomizawa and Ogawa, 1966; Bailone et al., 1979), and measurement of CI activity (Bailone et al., 1979) shows that little activity remains after 30 min. In addition, cell lysis occurs about 30 minutes later after ultraviolet induction than after infection, suggesting that lytic growth after induction begins after a delay of about 30 min. Accordingly, cleavage rates in the simulation have been adjusted such that 50% of the simulations have switched by 30 min when cleavage activity persists indefinitely.

The output of several simulations is shown in Figure 12.10. Two types of predominant patterns are shown. Some simulations switch (Figure 12.10a); CI levels drop, Cro synthesis begins, and the system abruptly switches. Others (Figure 12.10b) do not switch; CI levels drop during cleavage (shown by the hatched bar) and eventually return to the previous level when cleavage stops. These simulations took place with an intermediate time of cleavage (20 min); at shorter times, the nonswitching patterns are typical, while at longer times of cleavage many or most of the runs resemble those in Figure 12.10b, except as discussed next.

In these simulations, the CI level declines rapidly at the outset of cleavage, followed by a slower decline. Roughly similar kinetics are observed experimentally on bulk cultures using Westerns (CB Michalowski and JWL, unpublished data). Often the level of CI reaches a pseudo-steady state (e.g., Figure 12.10b), in which degradation is balanced by new synthesis; this is often disrupted upon synthesis of a burst of Cro.

Chapter 12

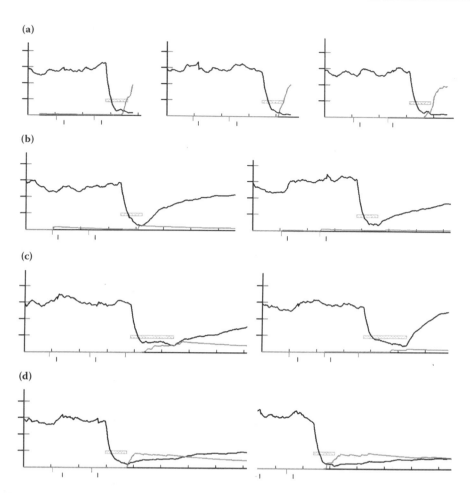

FIGURE 12.10 Simulation of prophage induction. As in Figures 12.8 and 12.9, the x-axis represents time. Simulations were for 15,000 sec; ticks above the axis are at 1500 sec intervals. The gray hatched box represents the time at which cleavage is active (20 min in these simulations except as indicated). (a) Simulations that switch. Usually a burst of Cro occurs, prompting a switch. Generally, the decision is a crisp one, particularly if it occurs while cleavage is active. In all cases, CI levels drop rapidly and then level off in the range of about 50 nM. (b) Simulations that do not switch. (c) Examples of simulations in which replication occurs before the onset of cleavage; these examples do not switch. Note, especially in the left panel, that CI levels reach a pseudo-steady state during cleavage. Cleavage was for 40 min. (d) Examples of indecisive behavior after cleavage has ceased.

Importantly, the simulation displays threshold behavior in the ultraviolet dose response (Figure 12.11a); that is, when RecA is activated for various lengths of time, little switching occurs until this time exceeds a certain value. The dose giving 50% switching is taken as the set-point. This is consistent with the model in which threshold behavior arises because RecA does not remain active long enough at low doses to cleave all the CI.

To test the effects of blocking DNA replication and cell division once RecA is activated, we allow one or both processes. If small compensating decreases are made in the cleavage rate, very similar dose-response curves result (Figure 12.11a). When cell division is allowed, the shape of the curve is the same whether replication is allowed or not.

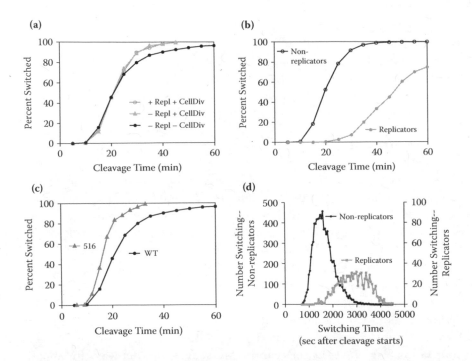

FIGURE 12.11 Threshold behavior of prophage induction for wild-type and λJL516. (a) Dose-response curve for WT. The responses to varying lengths of cleavage time are illustrated; the percentage of simulations that switch is plotted. The curves represent three cases: when DNA replication (Repl) and cell division (CellDiv) are allowed once cleavage begins; when cell division but not replication is allowed; and when neither is allowed; for each time point, 1,000 simulations were done for the first two curves, and 10,000 for the third. The first two curves overlap. The cleavage rate constant (see second equation in section 12.5.14) was 0.062/sec, 0.070/sec, and 0.075/sec, respectively, and was adjusted to achieve the same setpoint (cleavage time giving 50% switched) for all three curves. (b) Dose-response curves for simulations in which replication has or has not occurred shortly before cleavage begins. Once cleavage begins, neither replication nor cell division is allowed. Data from (a) are sorted into that those that have not replicated and those that have, shown as nonreplicators and replicators, respectively; about 13% of the simulations were in the latter class. (c) Dose-response curves for WT and λJL516. In each case, replication and cell division are not allowed after cleavage begins; WT data are from (a). (d) Distribution of switching times for WT. The times at which 10,000 simulations switched to the lytic state after the onset of cleavage for 60 min (replication and cell division were not allowed after cleavage began) were tabulated and binned into 50 sec bins. The number of switches in each bin is shown for nonreplicators and replicators. The curve for replicators is presumably truncated on the right because cleavage ceased at 3,600 sec.

When both replication and cell division are blocked, a small nonswitching fraction is seen at high doses when compared with the other two curves.

Strikingly, closer examination of this nonswitching fraction reveals a feature that, once recognized, seems self-evident. If cleavage begins after replication but before cell division has occurred, the cells are much more resistant to switching (Figure 12.11b). A reasonable model for this is that doubling the gene dosage markedly increases the capacity to replenish the CI in the face of cleavage. Examples are seen in Figure 12.10c, from simulations with cleavage for 40 min. In these cases, again CI levels reach a transient plateau, in which degradation and new synthesis are roughly balanced; the CI levels at the plateau are higher than in the absence of prior replication.

Chapter 12

In this view, then, an asynchronous population of cells is intrinsically heterogeneous in its response to DNA damaging treatments. Those that have replicated the λ prophage have a higher set-point than those that have not. One experimental prediction this makes is that, if we move a λ prophage to a location closer to the origin of replication, the apparent set-point for prophage induction should be higher, because there will be a higher proportion of cells with replicated prophages. Alternatively, synchronized cells might rather abruptly become more resistant to induction as they progress through the cell cycle, becoming resistant at the time at which the prophage replicates.

The simulation also predicts that switching times are broadly distributed (Figure 12.11d), both for those that have replicated at the time cleavage begins and those that have not. For the latter, after the peak in the distribution, the fraction remaining unswitched decays roughly exponentially. Experimental evidence with GFP expressed from plasmids also suggests a distribution of switching times (Amir et al., 2007), but this result may have been skewed by a large number of plasmid-borne CI-binding sites acting as sinks.

In simulations without replication and cell division, we frequently see cases in which both regulatory proteins are present at substantial levels long after cleavage has ceased, and their amounts change very slowly with time—the system has become "indecisive" (e.g., Figure 12.10d). Although one protein will eventually win out (CI in the examples given), the decline in the level of the other regulator (Cro in this case) due to dilution slows more and more, since the cells cannot divide. Put another way, this apparent sluggishness is partly an artifact of the way we have modeled the recovery phase. If we allow cell division to occur throughout the induction process (not shown), the system becomes far more responsive, moving to one or the other stable state more promptly. This points up the need to model the course of events during the return to a normal growth state. In addition, as the cells grow larger, the stochastic effects become less important because the number of protein molecules increases; presumably for this reason, the fluctuations in CI and Cro levels become smaller and smaller. We return to this issue of indecisiveness in the following section.

We have begun testing whether the simulation fits with the properties of mutants that are found experimentally to affect the set-point; here are two examples. First, our experimental data (Michalowski et al., 2004) indicate that, with P_{RM} alleles of various strengths, there is a reasonable correlation between P_{RM} strength and the set point (Figure 12.12b). Similarly, in the simulation, when we alter the strength of P_{RM} by changing the value of kf, the rate at which open complexes form, the set-point changes too, such that there is a roughly linear relationship between P_{RM} strength and set-point (Figure 12.12a).

As a more complex example, we have analyzed experimentally a cI mutation, cI Y210N, which blocks cooperative CI binding. A phage carrying this mutation, λ cI Y210N, cannot lysogenize. When two other mutations, a change in O_R2 that allows ~5x tighter CI binding, and an up-promoter mutant in P_{RM}, are added to this phage, the resulting triple mutant, termed λJL516, can form stable lysogens and has a set point for prophage induction near that of wild-type (Babiç and Little, 2007). Strikingly, in the simulation, the behavior of the triple mutant is likewise close to that of wild-type (Figure 11C), with two differences. First, the curve is steeper. Although such curves generally become steeper at lower set points (e.g. Figure 12A), this one is steeper yet. We surmise that one contributor to the steeper curve is that switching occurs at much

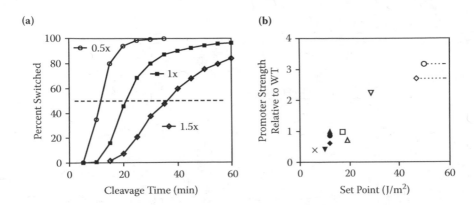

FIGURE 12.12 Correlation between strength of P_{RM} and set-point for prophage induction. (a) Dose-response curves in the simulation, using three different values of k_f for open-complex formation at P_{RM}; these are 0.5, 1, and 1.5 times the canonical value as indicated. The intersection of the dashed line and a curve represents the set-point for that curve. Data for the canonical value are from Figure 12.11a. (b) Experimental correlation between P_{RM} strength (y-axis) and the set-point. The dashed lines for the two right-most mutants indicate uncertainty in assessing the set-point. (From Michalowski et al., 2004. With permission.)

higher CI levels in this mutant. In consequence, its behavior is less stochastic and could be viewed as approaching a deterministic model. We expect that a deterministic model of prophage induction would have switch-like behavior, going from no induction to 100% induction at a discrete value for the time of cleavage. Experimentally, in contrast, the curves for WT and this mutant have about the same steepness.

The second difference has to do with the criteria for deciding if a switch has occurred. At the doses giving intermediate levels of switching, a sizable fraction of simulations give indecisive behavior, as described for wild-type but more frequent for this mutant. It is uncertain how to relate this to λ biology. If a significant amount of P_R expression occurs, other downstream events might occur that would either make switching irreversible or lead to unproductive states that would kill the host cell or lead to loss of the prophage. For instance, λ DNA replication requires genes expressed from P_R, and prophage replication in an unexcised chromosome might be lethal to the cell. Excision of the prophage due to expression of excision functions under control of P_L would remove the prophage (Weisberg and Gallant, 1967). These processes are not included in the mechanistic model, so their impact is hard to assess. This consideration does not affect the conclusion that this triple mutant has threshold behavior; its behavior is robust at the doses giving close to 0 or 100% switching, and this concern does not come into play at those doses.

Despite the complications just discussed, these two examples of correspondence between experiment and simulation are encouraging. We conclude that this simplified model for prophage induction fits well with a rather limited set of data with mutants.

12.6.3 Decisiveness of the Regulatory Circuit

As noted above, after prophage induction, the wild-type system often does not resolve promptly to one or the other stable state, in part because the mechanistic model does not

Chapter 12

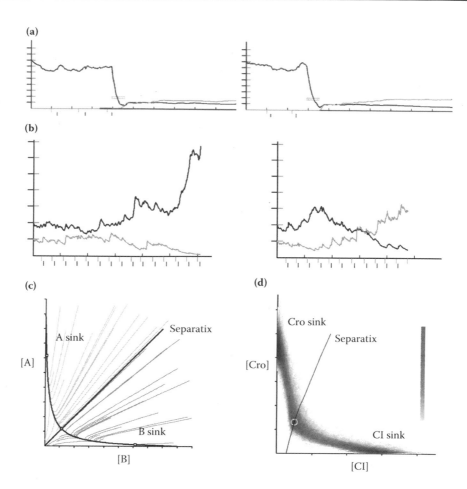

FIGURE 12.13 Indecisiveness of λJL516. (a) Representative simulations of λJL516 prophage induction in which a crisp decision was not made during or after cleavage. Cleavage took place for 16 min, at time indicated by the gray hatched box. CI is black, Cro is gray. Following cleavage, a long period of time ensued in which the regulatory system was in a state of limbo. (b) Representative simulations without cleavage, starting from a condition with CI = 200 Cro = 100, which also did not resolve quickly. Gray and black ticks below the x-axis represent the times of DNA replication and cell division, respectively. (c) Phase plot for a simple deterministic model in which each of two regulators (A and B, y- and x-axes, respectively) represses the other and activates its own expression. In each case, the regulator binds more tightly to the site that represses the other's promoter than it does to the site which activates its own promoter. Black circles at upper left and lower right are stable sinks; that at the lower left is an unstable point from which the simulation moves only if one value changes slightly. The black diagonal line is the separatrix; simulations starting to its left (right) reach the A (B) sink. Curves starting at several points in phase space are plotted; lighter (darker) gray curves reach the A (B) sink. Trajectories of these points first move to the black arc and then follow the arc to a sink. (d) Phase plot for stochastic simulation of wild-type λ, starting at CI = 50 Cro = 130 (white circle). The x- and y-axes are concentrations of CI and Cro, respectively; ticks are multiples of 100 nM. The program saved CI/Cro pairs every 20 sec for simulations of 50,000 sec (1,000 simulations); 49% of these simulations resulted in CI wins. The resulting data for the first 15,000 sec of each simulation were binned (for each axis, the bin size was 1/250 the maximum observed value for that protein) and the number of points in bins is plotted, using the gray scale on the right. Values higher in the scale represent CI/Cro values that are occupied more frequently; the top of the scale shown is 10% of the maximum observed value, and higher values appear darker gray. Sinks and the separatrix are indicated; the region near the unstable point surrounds the white circle.

currently permit resumption of replication and cell division. This indecisive behavior is more frequently seen in prophage induction of the triple mutant phage λJL516, just described (Figure 12.13a). The difference between λJL516 and wild-type is more apparent when replication and cell division are not blocked at the outset of induction (not shown), but that version of the model does not properly describe the events occurring at the time the decision is usually reached.

To test whether the decisiveness of the mutant differs from that of the wild-type, we used an alternative approach not involving prophage induction. Instead, the wild-type and the mutant were simulated by starting with intermediate values of CI and Cro and letting the system run until either of the two stable states was reached. This approach confirmed that the mutant is much less decisive than wild-type; many mutant simulations took a long time to reach a decision (Figure 12.13b), whereas the wild-type was generally more prompt (not shown).

To depict the results of many simulations, a phase plot with CI plotted against Cro was used. Before describing this analysis, we first consider the appearance of the phase plot from a much simpler, deterministic model of a bistable switch with two regulators A and B (Figure 12.13c). Two stable sinks correspond to A and B wins. These sinks are connected by an arc representing the pathway followed by the A/B plots as a function of time after starting at any point in phase space (Figure 12.13c). There is a separatrix, a line dividing the region in which A would win from that for the B winners. Where the separatrix meets the arc lies an unstable fixed point (circle at lower left), from which the system would not move unless slightly perturbed away from the point. Movement away from this point is initially slow, and for a while it accelerates with increasing distance from the point.

For a stochastic model, phase plots of multiple simulations have features that roughly correspond to the sinks, the separatrix, the arcs, and the unstable point (Figure 12.13d). The sinks correspond to the lysogenic state and the anti-immune state. They are not discrete points but stable basins in which the system meanders about for long periods of time, as seen for instance in time courses of simulations in the lysogenic state (e.g., Figure 12.8). The separatrix is a line representing 50% wins by each stable fate; moving away toward a given sink, the percentage of wins by that sink increases, eventually reaching 100%. The arc followed by individual trajectories is no longer a single line, but a cloud of reasonably likely point pairs, which we will term the *arc cloud*. No unstable point can exist, since the system is constantly in flux, but near the intersection of the separatrix and the arc cloud the system moves relatively slowly on average, eventually escaping toward one or the other stable state. The decisiveness of the system should then dictate how rapidly, on average, it moves from this region toward a stable state.

Consider now the behavior of the mutant phage and of the WT in the stochastic simulation. We start the system near a point corresponding to the unstable point, do many simulations, and plot the frequency of CI–Cro pairs as a function of time in successive 2,000 sec intervals (several time slices are depicted in Figure 12.14). The wild-type system moves toward the sinks more rapidly than the mutant system, which tends to remain near the starting point (Figure 12.14) for a longer period of time. The impetus to move away from this region might be termed a *driving force* or *pressure*, and this pressure is substantially lower for this particular mutant than for wild-type. To date, this is basically a heuristic idea, and we have not analyzed it mathematically.

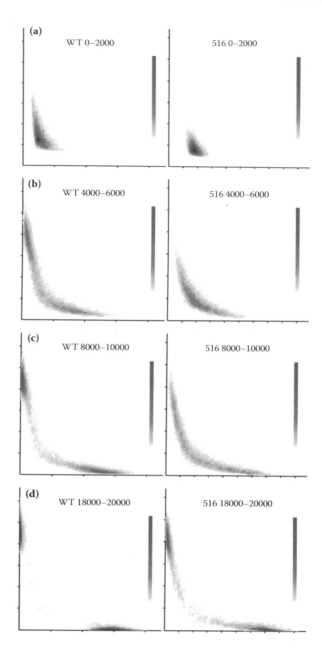

FIGURE 12.14 Phase plots for WT and λJL516 as a function of time. For wild-type and λJL516, 1,000 simulations of 50,000 sec were done, starting at CI = 50 Cro = 130 or CI = 180 Cro = 80, respectively, giving 510 and 503 Cro wins, respectively. CI/Cro pairs were binned for 2,000 sec intervals. The gray scale is as in Figure 12.13; the top end represents the bins most highly occupied during the first 20,000 sec, not during the depicted time interval. WT moves away from the starting point towards the sinks about twice as rapidly as λJL516. By 20,000 sec, both systems have largely reached one or the other stable state. Times after the start of the simulations: (a) 0–2,000 sec. (b) 4,000–6,000 sec. (c) 8,000–10,000 sec. (d) 18,000–20,000 sec.

Nonlinear features of a circuit would likely contribute to the driving force moving it away from an undecided state, pushing it toward a stable state. At the mechanistic level, perhaps two different features of this particular mutant contribute to this behavior by weakening or removing nonlinearity. First, it has no cooperative DNA binding. Second, the concentration of CI near the starting point is much higher than it would be for the corresponding state in the wild-type. Accordingly, a much greater fraction of the CI is in the form of dimers than in the wild-type case.

12.7   Opportunities Presented by Modeling

This work exemplifies several benefits of modeling. One obvious and general goal is to make a model that is consistent with experimental data. Success in such a fit lends confidence that we understand the system well, and hopefully will lead to a more complete description than one afforded by wiring diagrams, descriptions of connections in uncoupled systems, and the like. Similarly, discrepancies between experiment and modeling can lead to refinements of the parameter set, or recognition that something is missing. In addition, parameter sensitivity tests can guide experimenters to focus on parameters whose precise values are critical for proper behavior. It also helps us realize where the accepted parameter set is weak and in need of refinement, as in the case of the glutamate effect.

We close by mentioning another valuable aspect of modeling: It makes the modeler think hard about what is actually happening in the cell and try to recognize current limitations to our knowledge. Even with our extensive experience with the λ system, several such limitations became apparent only once we became deeply immersed in developing the underlying model, as illustrated by four interesting examples.

The first example is activation of RecA, which is specific to our system. Although it has long been a puzzle as to what the activating signal is, workers in the field have believed it to be some form of single-stranded DNA. Early work on DNA repair suggested that the fork could continue beyond the site of damage, leaving a daughter-strand gap in its wake (Rupp and Howard-Flanders, 1968). Later work focused on events that resulted in fork stalling (Cox et al., 2000). The current and emerging view is that a range of different events can occur when the fork encounters a noncoding lesion (Yao and O'Donnell, 2009) This leaves up in the air the issue of whether RecA is invariably activated at the site of damage, and whether the amount of activated RecA varies widely from cell to cell. It points up the need for single-cell studies to quantify RecA*, a real experimental challenge.

The second example is the uncertainty in how to model the recovery phase after SOS induction, particularly the resumption of DNA replication and cell division. We surmise that this is a general issue with cellular stress responses: The initial responses to a range of stresses have been much better studied experimentally than the recovery phases. One likely reason is that the recovery phase is heterogeneous on a cell-by-cell basis and therefore hard to analyze. An example from our own work, mentioned already, has to do with recovery from SOS induction; until we analyzed LexA cleavage during this phase (Little, 1983), no work had been done on this phase. Still, this work dealt with mass cultures and gives no insight into the behavior of individual cells, as would be needed for modeling.

Chapter 12

The third example has to do with the intrinsic heterogeneity of the cell population in its response to SOS activation. This is due to the fact that some cells have a replicated prophage, while some do not. While this may seem obvious in hindsight, it was development of a detailed model, including DNA replication, that led us to this realization.

The final example is DNA replication, a very general feature of cellular behavior. It is our view that DNA replication should be a standard part of models that extend over multiple cell generations. Indeed, based on the present work, Arkin's group has begun including DNA replication in models of other systems. Events that occur during replication almost certainly have an impact on the behavior of regulatory circuits, yet little is known about the fate of DNA-bound proteins as the fork passes. It is known that long arrays of bound Lac or Tet repressors can stall replication forks (Possoz et al., 2006); perhaps bound CI, especially, can briefly stall the fork. If the fork stalls at O_R, due to the presence of a DNA-bound CI tetramer, it might give time for CI to rebind to O_L, perhaps speeding reassembly of a dodecamer. Are there mechanisms for maintaining a high local concentration? In the special case of λ, perhaps the CI-containing loops would provide such a mechanism. In any case, our work points to the need for further investigation of this issue.

Acknowledgments

Work described here is supported by National Institutes of Health (NIH) grant GM24178. We are grateful to Ian Dodd and Harvey Eisen for describing unpublished results.

References

Allen, R. J., D. Frenkel, and P. R. Ten Wolde. 2006. Simulating rare events in equilibrium or nonequilibrium stochastic systems. *J. Chem. Phys.* 124:024102.

Allen, R. J., C. Valeriani, and P. R. Ten Wolde. 2009. Forward flux sampling for rare event simulations. *J. Phys.: Condens. Matter* 21:463102.

Amir, A., O. Kobiler, A. Rokney, A. B. Oppenheim, and J. Stavans. 2007. Noise in timing and precision of gene activities in a genetic cascade. *Mol. Syst. Biol.* 3:71–80.

Anderson, L. M. and H. Yang. 2008a. DNA looping can enhance lysogenic CI transcription in phage lambda. *Proc. Natl. Acad. Sci. U. S. A.* 105:5827–5832.

Anderson, L. M. and H. Yang. 2008b. A simplified model for lysogenic regulation through DNA looping. *Conf. Proc. IEEE Eng Med. Biol Soc.* 1:607–10.

Arkin, A., J. Ross, and H. H. McAdams. 1998. Stochastic kinetic analysis of developmental pathway bifurcation in phage lambda-infected *Escherichia coli* cells. *Genetics* 149:1633–1648.

Arkin, A. P. 2001. Synthetic cell biology. *Curr. Opin. Biotechnol.* 12:638–644.

Aurell, E., S. Brown, J. Johanson, and K. Sneppen. 2002. Stability puzzles in phage λ. *Phys. Rev. E* 65:051914.

Babiç, A. C. and J. W. Little. 2007. Cooperative DNA binding by CI repressor is dispensable in a phage λ variant. *Proc. Natl. Acad. Sci. U. S. A.* 104:17741–17746.

Bailone, A., A. Levine, and R. Devoret. 1979. Inactivation of prophage λ repressor *in vivo*. *J. Mol. Biol.* 131:553–572.

Bakk, A. and R. Metzler. 2004b. Nonspecific binding of the O_R repressors CI and Cro of bacteriophage lambda. *J. Theor. Biol.* 231:525–533.

Bakk, A. and R. Metzler. 2004a. In vivo non-specific binding of lambda CI and Cro repressors is significant. *FEBS Lett.* 563:66–68.

Beckett, D., D. S. Burz, G. K. Ackers, and R. T. Sauer. 1993. Isolation of lambda repressor mutants with defects in cooperative operator binding. *Biochemistry* 32:9073–9079.

Becskei, A. and L. Serrano. 2000. Engineering stability in gene networks by autoregulation. *Nature* 405:590–593.

Benson, N., C. Adams, and P. Youderian. 1994. Genetic selection for mutations that impair the co-operative binding of lambda repressor. *Mol. Microbiol.* 11:567–579.

Berg, O. G. 1978. A model for the statistical fluctuations of protein numbers in a microbial population. *J. Theor. Biol.* 71:587–603.

Bremer, H., P. Dennis, and M. Ehrenberg. 2003. Free RNA polymerase and modeling global transcription in *Escherichia coli*. *Biochimie* 85:597–609.

Bushman, F. D. 1993. The bacteriophage 434 right operator. Roles of O_R1, O_R2 and O_R3. *J. Mol. Biol.* 230:28–40.

Campbell, A. 1994. Comparative molecular biology of lambdoid phages. *Annu. Rev. Microbiol.* 48:193–222.

Carlson, N. G. and J. W. Little. 1993. A novel antivirulence element in the temperate bacteriophage HK022. *J. Bacteriol.* 175:7541–7549.

Chen, Z., H. Yang, and N. P. Pavletich. 2008. Mechanism of homologous recombination from the RecA-ssDNA/dsDNA structures. *Nature* 453:489–494.

Cohen, S., B. J. Knoll, J. W. Little, and D. W. Mount. 1981. Preferential cleavage of phage λ repressor monomers by recA protease. *Nature* 294:182–184.

Cox, M. M., M. F. Goodman, K. N. Kreuzer, D. J. Sherratt, S. J. Sandler, and K. J. Marians. 2000. The importance of repairing stalled replication forks. *Nature* 404:37–41.

Craig, N. L. and J. W. Roberts. 1980. *E. coli* recA protein-directed cleavage of phage λ repressor requires polynucleotide. *Nature* 283:26–30.

Crowl, R. M., R. P. Boyce, and H. Echols. 1981. Repressor cleavage as a prophage induction mechanism: hypersensitivity of a mutant λ cI protein to RecA-mediated proteolysis. *J. Mol. Biol.* 152:815–820.

Darling, P. J., J. M. Holt, and G. K. Ackers. 2000a. Coupled energetics of lambda *cro* repressor self-assembly and site-specific DNA operator binding I: analysis of *cro* dimerization from nanomolar to micromolar concentrations. *Biochemistry* 39:11500–11507.

Darling, P. J., J. M. Holt, and G. K. Ackers. 2000b. Coupled energetics of lambda *cro* repressor self-assembly and site-specific DNA operator binding II: cooperative interactions of *cro* dimers. *J. Mol. Biol.* 302:625–638.

Degnan, P. H., C. B. Michalowski, A. C. Babiç, M. H. J. Cordes, and J. W. Little. 2007. Conservation and diversity in the immunity regions of wild phages with the immunity specificity of phage λ. *Mol. Microbiol.* 64:232–244.

Dodd, I. B., A. J. Perkins, D. Tsemitsidis, and J. B. Egan. 2001. Octamerization of λ CI repressor is needed for effective repression of P_{RM} and efficient switching from lysogeny. *Genes Dev.* 15:3013–3022.

Dodd, I. B., K. E. Shearwin, A. J. Perkins, T. Burr, A. Hochschild, and J. B. Egan. 2004. Cooperativity in long-range gene regulation by the λ CI repressor. *Genes Dev.* 18:344–354.

Eisen, H., P. Brachet, L. Pereira da Silva, and F. Jacob. 1970. Regulation of repressor expression in λ. *Proc. Natl. Acad. Sci. U. S. A.* 66:855–862.

Ferrell, J. E., Jr. 2002. Self-perpetuating states in signal transduction: positive feedback, double-negative feedback and bistability. *Curr. Opin. Cell Biol.* 14:140–148.

Fong, R. S. C., S. Woody, and G. N. Gussin. 1993. Modulation of P_{RM} activity by the lambda P_R promoter in both the presence and absence of repressor. *J. Mol. Biol.* 232:792–804.

Gillespie, D. T. 1977. Exact stochastic simulation of coupled chemical reactions. *J. Phys. Chem.* 81:2340–2361.

Gillespie, D. T. 1992. A rigorous derivation of the chemical master equation. *Physica A* 188:404–425.

Gimble, F. S. and R. T. Sauer. 1989. Lambda repressor mutants that are better substrates for RecA-mediated cleavage. *J. Mol. Biol.* 206:29–39.

Golding, I., J. Paulsson, S. M. Zawilski, and E. C. Cox. 2005. Real-time kinetics of gene activity in individual bacteria. *Cell* 123:1025–1036.

Heller, R. C. and K. J. Marians. 2006. Replication fork reactivation downstream of a blocked nascent leading strand. *Nature* 439:557–562.

Hendrix, R. W., J. W. Roberts, F. W. Stahl, and R. A. Weisberg. 1983. *Lambda II*. Cold Spring Harbor Laboratory, Cold Spring Harbor, New York.

Herskowitz, I. and D. Hagen. 1980. The lysis-lysogeny decision of phage λ: explicit programming and responsiveness. *Annu. Rev. Genet.* 14:399–445.

Hochschild, A., N. Irwin, and M. Ptashne. 1983. Repressor structure and the mechanism of positive control. *Cell* 32:319–325.

Huh, D. and J. Paulsson. 2011. Non-genetic heterogeneity from stochastic partitioning at cell division. *Nat. Genet.* 43:95–100.

Ingraham, J. L., O. Maaloe, and F. C. Neidhardt. 1983. *Growth of the Bacterial Cell.* Sinauer Associates, Inc., Sunderland, MA., p. 8.

Ito, K. and Y. Akiyama. 2005. Cellular functions, mechanism of action, and regulation of FtsH protease. *Annu. Rev. Microbiol.* 59:211–231.

Jia, H. F., W. J. Satumba, G. L. Bidwell, III, and M. C. Mossing. 2005. Slow assembly and disassembly of lambda Cro repressor dimers. *J. Mol. Biol.* 350:919–929.

Johnson, A. D., B. J. Meyer, and M. Ptashne. 1979. Interactions between DNA-bound repressors govern regulation by the λ phage repressor. *Proc. Natl. Acad. Sci. USA* 76:5061–5065.

Juhala, R. J., M. E. Ford, R. L. Duda, A. Youlton, G. F. Hatfull, and R. W. Hendrix. 2000. Genomic sequences of bacteriophages HK97 and HK022: pervasive genetic mosaicism in the lambdoid bacteriophages. *J. Mol. Biol.* 299:27–51.

Kao-Huang, Y., A. Revzin, A. P. Butler, P. O'Conner, D. W. Noble, and P. H. Von Hippel. 1977. Nonspecific DNA binding of genome-regulating proteins as a biological control mechanism: measurement of DNA-bound *Escherichia coli lac* repressor *in vivo*. *Proc. Natl. Acad. Sci. U. S. A.* 74:4228–4232.

Kobiler, O., A. Rokney, N. Friedman, D. L. Court, J. Stavans, and A. B. Oppenheim. 2005. Quantitative kinetic analysis of the bacteriophage λ genetic network. *Proc. Natl. Acad. Sci. U. S. A.* 102:4470–4475.

Koblan, K. S. and G. K. Ackers. 1992. Site-specific enthalpic regulation of DNA transcription at bacterio-phage lambda O_R. *Biochemistry* 31:57–65.

Koblan, K. S. and G. K. Ackers. 1991a. Cooperative protein-DNA interactions: effects of KCl on lambda cI binding to O_R. *Biochemistry* 30:7822–7827.

Koblan, K. S. and G. K. Ackers. 1991b. Energetics of subunit dimerization in bacteriophage lambda cI repressor: linkage to protons, temperature, and KCl. *Biochemistry* 30:7817–7821.

Lanzer, M. and H. Bujard. 1988. Promoters largely determine the efficiency of repressor action. *Proc. Natl. Acad. Sci. U. S. A.* 85:8973–8977.

LeFevre, K. R. and M. H. J. Cordes. 2003. Retroevolution of lambda Cro toward a stable monomer. *Proc. Natl. Acad. Sci. USA* 100:2345–2350.

Leirmo, S., C. Harrison, S. Cayley, R. R. Burgess, and M. T. Record, Jr. 1987. Replacement of potassium chloride by potassium glutamate dramatically enhances protein-DNA interactions in vitro. *Biochemistry* 26:2095–2101.

Little, J. W. 2006. Gene regulatory circuitry of phage λ. *In The Bacteriophages.* R. Calendar, editor. Oxford University Press, New York.

Little, J. W. 1984. Autodigestion of lexA and phage lambda repressors. *Proc. Natl. Acad. Sci. U. S. A.* 81:1375–1379.

Little, J. W. 1983. The SOS regulatory system: control of its state by the level of RecA protease. *J. Mol. Biol.* 167:791–808.

Little, J. W. 2005. Threshold effects in gene regulation: when some is not enough. *Proc. Natl. Acad. Sci. U. S. A.* 102:5310–5311.

Little, J. W. 1993. LexA cleavage and other self-processing reactions. *J. Bacteriol.* 175:4943–4950.

Little, J. W. and C. B. Michalowski. 2010. Stability and instability in the lysogenic state of phage lambda. *J. Bacteriol.* 192:6064–6076.

Little, J. W. and D. W. Mount. 1982. The SOS regulatory system of *Escherichia coli*. *Cell* 29:11–22.

Little, J. W., D. P. Shepley, and D. W. Wert. 1999. Robustness of a gene regulatory circuit. *EMBO J.* 18:4299–4307.

Liu, Z. and J. W. Little. 1998. The spacing between binding sites controls the mode of cooperative DNA-protein interactions: Implications for evolution of regulatory circuitry. *J. Mol. Biol.* 278:331–338.

Lou, C., X. Yang, X. Liu, B. He, and Q. Ouyang. 2007. A quantitative study of lambda-phage SWITCH and its components. *Biophys. J.* 92:2685–2693.

Luo, Y., R. A. Pfuetzner, S. Mosimann, M. Paetzel, E. A. Frey, M. Cherney, B. Kim, J. W. Little, and N. C. J. Strynadka. 2001. Crystal structure of LexA: a conformational switch for regulation of self-cleavage. *Cell* 106:585–594.

Manu, S. Surkova, A. V. Spirov, V. V. Gursky, H. Janssens, A. Kim, O. Radulescu, C. E. Vanario-Alonso, D. H. Sharp, M. Samsonova, and J. Reinitz. 2009. Canalization of gene expression in the *Drosophila* blastoderm by gap gene cross regulation. *PLoS. Biol.* 7:e1000049.

Mao, C., N. G. Carlson, and J. W. Little. 1994. Cooperative DNA-protein interactions: effects of changing the spacing between adjacent binding sites. *J. Mol. Biol.* 235:532–544.

Mao, C. and J. W. Little. 1998. Mutations affecting cooperative DNA binding of phage HK022 CI repressor. *J. Mol. Biol.* 279:31–48.

Maurer, R., B. J. Meyer, and M. Ptashne. 1980. Gene regulation at the right operator (O_R) of bacteriophage λ I. O_R3 and autogenous negative control by repressor. *J. Mol. Biol.* 139:147–161.

McAdams, H. H. and A. Arkin. 1997. Stochastic mechanisms in gene expression. *Proc. Natl. Acad. Sci. USA* 94:814–819.

McClure, W. R. 1980. Rate-limiting steps in RNA chain initiation. *Proc. Natl. Acad. Sci. U. S. A.* 77:5634–5638.

McClure, W. R. 1985. Mechanism and control of transcription initiation in prokaryótes. *Ann. Rev. Biochem.* 54:171–204.

Meyer, B. J., R. Maurer, and M. Ptashne. 1980. Gene regulation at the right operator (O_R) of bacteriophage λ II. O_R1, O_R2, and O_R3: their roles in mediating the effects of repressor and cro. *J. Mol. Biol.* 139:163–194.

Meyer, B. J. and M. Ptashne. 1980. Gene regulation at the right operator (O_R) of bacteriophage λ III. λ repressor directly activates gene transcription. *J. Mol. Biol.* 139:195–205.

Michalowski, C. B., M. D. Short, and J. W. Little. 2004. Sequence tolerance of the phage λ P_{RM} promoter: implications for evolution of gene regulatory circuitry. *J. Bacteriol.* 186:7988–7999.

Mitarai, N., I. B. Dodd, M. T. Crooks, and K. Sneppen. 2008. The generation of promoter-mediated transcriptional noise in bacteria. *PLoS. Comput. Biol.* 4:e1000109.

Morelli, M. J., P. R. Ten Wolde, and R. J. Allen. 2009. DNA looping provides stability and robustness to the bacteriophage lambda switch. *Proc. Natl. Acad. Sci. U. S. A.* 106:8101–8106.

Mukherjee, A., C. N. Cao, and J. Lutkenhaus. 1998. Inhibition of FtsZ polymerization by SulA, an inhibitor of septation in *Escherichia coli*. *Proc. Natl. Acad. Sci. USA* 95:2885–2890.

Neubauer, Z. and E. Calef. 1970. Immunity phase-shift in defective lysogens: non-mutational hereditary change of early regulation of λ prophage. *J. Mol. Biol.* 51:1–13.

Nielsen, H. J., B. Youngren, F. G. Hansen, and S. Austin. 2007. Dynamics of *Escherichia coli* chromosome segregation during multifork replication. *J. Bacteriol.* 189:8660–8666.

Oppenheim, A. B., O. Kobiler, J. Stavans, D. L. Court, and S. Adhya. 2005. Switches in bacteriophage lambda development. *Annu. Rev. Genet.* 39:409–429.

Pakula, A. A., V. B. Young, and R. T. Sauer. 1986. Bacteriophage lambda *cro* mutations: effects on activity and intracellular degradation. *Proc. Natl. Acad. Sci. U. S. A.* 83:8829–8833.

Possoz, C., S. R. Filipe, I. Grainge, and D. J. Sherratt. 2006. Tracking of controlled *Escherichia coli* replication fork stalling and restart at repressor-bound DNA *in vivo*. *EMBO J.* 25:2596–2604.

Ptashne, M. 2004. *A genetic switch: Phage lambda revisited.* Cold Spring Harbor Laboratory Press, Cold Spring Harbor, NY.

Rao, C. V., D. M. Wolf, and A. P. Arkin. 2002. Control, exploitation and tolerance of intracellular noise. *Nature* 420:231–237.

Reinitz, J. and J. R. Vaisnys. 1990. Theoretical and experimental analysis of the phage lambda genetic switch implies missing levels of co-operativity. *J. Theor. Biol.* 145:295–318.

Rosenfeld, N., J. W. Young, U. Alon, P. S. Swain, and M. B. Elowitz. 2005. Gene regulation at the single-cell level. *Science* 307:1962–1965.

Rudolph, C. J., A. L. Upton, and R. G. Lloyd. 2007. Replication fork stalling and cell cycle arrest in UV-irradiated *Escherichia coli*. *Genes Dev.* 21:668–681.

Rupp, W. D. and P. Howard-Flanders. 1968. Discontinuities in the DNA synthesized in an excision-defective strain of *Escherichia coli* following ultraviolet irradiation. *J. Mol. Biol.* 31:291–304.

Sassanfar, M. and J. W. Roberts. 1990. Nature of the SOS-inducing signal in *Escherichia coli*. the involvement of DNA replication. *J. Mol. Biol.* 212:79–96.

Satumba, W. J. and M. C. Mossing. 2002. Folding and assembly of lambda Cro repressor dimers are kinetically limited by proline isomerization. *Biochem.* 41:14216–41224.

Savageau, M. A. and R. A. Fasani. 2009. Qualitatively distinct phenotypes in the design space of biochemical systems. *FEBS Lett.* 583:3914–3922.

Schlacher, K., M. M. Cox, R. Woodgate, and M. F. Goodman. 2006. RecA acts in *trans* to allow replication of damaged DNA by DNA polymerase V. *Nature* 442:883–887.

Senear, D. F., M. Brenowitz, M. A. Shea, and G. K. Ackers. 1986. Energetics of cooperative protein-DNA interactions: comparison between quantitative deoxyribonuclease footprint titration and filter binding. *Biochemistry* 25:7344–7354.

Shea, M. A. and G. K. Ackers. 1985. The O_R control system of bacteriophage lambda: a physical-chemical model for gene regulation. *J. Mol. Biol.* 181:211–230.

St-Pierre, F. and D. Endy. 2008. Determination of cell fate selection during phage lambda infection. *Proc. Natl. Acad. Sci. U. S. A.* 105:20705–20710.

Chapter 12

Thomas, R., A. M. Gathoye, and L. Lambert. 1976. A complex control circuit. Regulation of immunity in temperate bacteriophages. *Eur. J. Biochem.* 71:211–227.

Tomizawa, J. and T. Ogawa. 1966. Effect of ultraviolet irradiation on bacteriophage lambda immunity. *J. Mol. Biol.* 23:247–263.

Valenzuela, D. and M. Ptashne. 1989. P22 repressor mutants deficient in co-operative binding and DNA loop formation. *EMBO J.* 8:4345–4350.

Von Dassow, G., E. Meir, E. M. Munro, and G. M. Odell. 2000. The segment polarity network is a robust development module. *Nature* 406:188–192.

Weinberger, L. S., J. C. Burnett, J. E. Toettcher, A. P. Arkin, and D. V. Schaffer. 2005. Stochastic gene expression in a lentiviral positive-feedback loop: HIV-1 Tat fluctuations drive phenotypic diversity. *Cell* 122:169–182.

Weindl, J., Z. Dawy, P. Hanus, J. Zech, and J. C. Mueller. 2009. Modeling promoter search by *E. coli* RNA polymerase: one-dimensional diffusion in a sequence-dependent energy landscape. *J. Theor. Biol.* 259:628–634.

Weisberg, R. A. and J. A. Gallant. 1967. Dual function of the λ prophage repressor. *J. Mol. Biol.* 25:537–544.

Weisberg, R. A., M. E. Gottesman, R. M. Hendrix, and J. W. Little. 1999. Family values in the age of genomics: comparative analyses of temperate bacteriophage HK022. *Annu. Rev. Genet.* 33:565–602.

Whipple, F. W., N. H. Kuldell, L. A. Cheatham, and A. Hochschild. 1994. Specificity determinants for the interaction of λ repressor and P22 repressor dimers. *Genes Dev.* 8:1212–1223.

Yao, N. Y. and M. O'Donnell. 2009. Replisome structure and conformational dynamics underlie fork progression past obstacles. *Curr. Opin. Cell Biol.* 21:336–343.

Zeng, L., S. O. Skinner, C. Zong, J. Sippy, M. Feiss, and I. Golding. 2010. Decision making at a subcellular level determines the outcome of bacteriophage infection. *Cell* 141:682–691.

Zhu, X.-M., L. Yin, L. Hood, and P. Ao. 2004. Robustness, stability and efficiency of phage λ genetic switch: dynamical structure analysis. *J. Bioinform. Comput. Biol.* 2:785–817.

13. Chemotaxis

Howard C. Berg

13.1 Introduction

The literature on bacterial chemotaxis is enormous. It has been built up through the efforts of many workers with different interests and talents: geneticists, biochemists, and physicists (or, if you prefer, systems biologists). The ways bacteria swim and respond to changes in their environment is now the best understood behavioral system in biology. The organism of choice has been *Escherichia coli*, a bacterium that lives in your gut. But the subject is not as widely appreciated as one might imagine, because it has not been developed for a practical purpose (e.g., to cure a particular disease) but rather for the joy of understanding. We have not been disappointed: the tricks that *Escherichia coli* has learned in its long quest for survival are remarkable. And not all the data are in.

I am an experimentalist, and the subject is becoming more model dependent so it has not been easy to keep up. Given the size of the enterprise, I have had to pick and choose. Often, the work cited here is the tip of the iceberg, but it should get your feet wet, and you can learn more by tracking the leading players via PubMed. I will start with some history,

Chapter 13

Quantitative Biology: From Molecular to Cellular Systems edited by Michael E. Wall © 2012 CRC Press / Taylor & Francis Group, LLC. ISBN: 978-1-4398-2722-2

both recent and early, and then work my way from the output to the input end of the sensory-transduction system, discussing means of generating thrust, swimming style, flagellar rotation, coupling between flagella and receptors, adaptation, and other system properties. With luck, there will be something here for everyone.

For reviews of the mathematical approaches used to model bacterial chemotaxis at the single- or many-cell level, see [1,2].

13.2 Recent History

Escherichia coli was chosen as a model system for studying the molecular biology of behavior by Julius Adler, because it swam in a purposeful manner, was relatively simple (compared with a monkey), and was amenable to genetic and biochemical manipulation. Cells of *E. coli* are about 1 μm in diameter by 2 or 3 μm long: like microscopic cocktail sausages (Figure 13.1). They are propelled by flagella with long, thin helical filaments, each driven at its base by a reversible rotary motor. If properly prepared—one needs to proceed with care, because flagellar filaments are easily broken—cells swim some 30 diameters per second in a direction roughly parallel to their long axes, changing directions about once per second. The cell bodies are easy to see in phase-contrast microscopy, but the flagellar filaments are not. Dark-field microscopy works well for visualizing isolated filaments (e.g., [3]), but fluorescence microscopy is more effective for work with swimming cells, since one is not blinded by light scattered from the cell body [4].

Adler began by studying the migration of bacterial populations through very soft nutrient agar, either drawn into a melting-point capillary tube or poured into a Petri plate [5]. He found that bacteria placed at one end of the tube or at an isolated spot on the plate traveled in bands, consuming nutrients from the agar and then chasing the spatial gradients of chemicals generated by this consumption. In plates, these bands

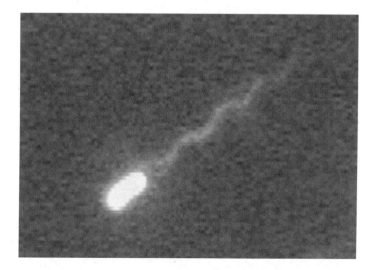

FIGURE 13.1 A wild-type *E. coli* cell (strain AW405) swimming toward 8 o'clock, propelled by a flagellar bundle that is much longer than the cell body. The cell was labeled with an amino-specific Alexa Fluor 488 dye and visualized in a fluorescence microscope with strobed laser illumination [4]. With this dye, the cell body is so bright that it is difficult to see the proximal ends of the flagellar filaments; although, individual filaments can be distinguished near the back of the cell.

appeared as concentric circular rings. For example, with a plate containing a mixture of L-amino acids (T-broth, a casein hydrolysate), cells in the outermost ring consumed serine; there was serine in front of the ring but little behind. In contrast, cells in the second ring consumed aspartate; there was aspartate in front of the second ring but little behind, and so on. This assay, known as the swarm-plate assay, is a standard tool for bacterial geneticists, because it uses materials with which they are familiar and gives results within a few hours. Mutants without flagella or with paralyzed flagella remain near the point of inoculation, as do mutants that are unable to change directions: soft agar contains blind alleys that trap such cells, preventing long-range migration [6]. Mutants with defective serine receptors fail to show the serine ring, while mutants with defective aspartate receptors fail to show the aspartate ring, and so on. The term *swarm plate* is a misnomer, because the bacteria are swimming through the agar, not swarming on its surface: for descriptions of swarming *E. coli*, see [7]. So it is better to call them swim plates.

The swim-plate assay requires uptake and metabolism of nutrients, not just the ability of cells to track gradients of these chemicals. So Adler turned to a capillary assay, improving on one introduced for studies of chemotaxis by Pfeffer [8], in which a solution of a chemical, drawn into the tip of a thin capillary tube, diffuses out into a cloud of bacteria swimming in a defined medium [9,10]. The bacteria swim up the gradient of the chemical as it diffuses from this point source, accumulate near the tip of the capillary tube, and then swim inside. By waiting 45 to 60 min, plating the contents of the tube, and then counting colonies that grew up overnight, Adler could determine the strength of the response. By repeating this procedure for tubes containing different concentrations of a given chemical, he could construct dose-response curves. In this manner, Adler showed that the response of *E. coli* to different chemicals was not a matter of material gain but rather a question of esthetics: cells responded perfectly well to chemicals that they could neither take up nor metabolize, simply because they liked the taste [11]. Surveys of responses were made to a large array of chemicals. Chemicals that cells swam toward, called attractants, included amino acids (serine, aspartate, and their analogs), dipeptides, sugars and sugar alcohols, electron acceptors (oxygen, nitrate, fumarate), salts at low concentrations, and membrane-permeant bases. Chemicals that cells swam away from, called repellents, included other amino acids (leucine, isoleucine, valine), indole, alcohols, divalent cations (cobalt and nickel), glycerol at high concentrations, other chemicals at high osmotic strength, and membrane-permeant acids. One amino acid that has played an important role in subsequent work is α-methyl-D, L-aspartate, an analog of aspartate that is not metabolized [12]. Work on repellents required a swim-plate variant, in which cells moved away from an agar plug containing the repellent [13].

The process by which cells swim toward or away from different chemicals is known as chemotaxis. This term was suggested by Pfeffer [8], who thought that a cell oriented its axis in the direction of the gradient and swam directly toward the mouth of the capillary tube. Animal physiologists use the term *taxis* for directed movement and the term *klinokinesis* for changes in the rate of random turning [14]. So, according to this scheme, *E. coli* is a master of chemoklinokinesis. At early Sensory Transduction in Microorganisms Gordon Conferences, time was spent wrangling about terminology, until the molecular biologists present asserted that they did not care what the process was called, they simply wanted to understand how it worked, that is, to discover the underlying molecular mechanisms. So the term chemotaxis has stuck. For earlier contributions by molecular biologists, see Box 2 of [15].

We still do not have an assay for bacterial chemotaxis that rivals the convenience of swim plates or the versatility of capillary assays. However, microfluidic systems are beginning to appear that might fill the bill, once system assembly, image acquisition, and data analysis are simplified; see the reviews by Englert et al. [16] and Ahmed et al. [17]. Can someone develop a stand-alone device that uses disposable cartridges?

A variation on the swim-plate story arose when Elena Budrene found conditions under which cells excrete attractants (mostly aspartate) [18,19]: cells move out from the point of inoculation in compact rings that break up into discrete spots, forming patterns with long-range symmetry. If the plate also contains nutrients toward which cells swarm, the patterns can be remarkably complex [20]. For more on this subject and a discussion of the modeling, see Brenner [21]. The model highlighted in this commentary treats spot formation as an arrested nonequilibrium phase separation, involving density-dependent motility and logistic growth, without reference to chemotaxis [22].

13.3   Early History

Bacterial motility was discovered in 1676 at the same time that bacteria were discovered, by Antony van Leeuwenhoek, as reported in his 18th letter to Henry Oldenberg, secretary of the Royal Society. It was the rapid motion of swimming cells that caught van Leeuwenhoek's attention. See [23] and Box 1 of [15]. For a description of work on bacterial chemotaxis begun in the 1880s, see [24] and Chapter 2 of [25]. Another window on this era is a review by Claes Weibull [26], cited in Adler's early papers. Weibull was a student of Arne Tiselius in Uppsala, working on preparations of polio virus isolated from feces of infected patients. These preparations were contaminated by filamentous material that Weibull was asked to identify. These proved to be filaments of bacterial flagella. Weibull showed that they were made of protein. These filaments could be salted out with ammonium sulfate, and they contained about 16% nitrogen and little if any phosphorous or sulfur. The latter was curious, because bacterial flagella were thought to be primitive muscles, and cysteine was known to play an important role in the function of actin and myosin. X-ray studies indicated that these filaments could be "regarded as primitive hairs or muscle fibers" [27]. To Weibull we owe the name for the flagellar filament protein, *flagellin* [28]. Weibull soon became involved in a debate with the South African microscopist Adrianus Pijper, who spent his entire career (1930–1964) looking at flagella in dark field. The light source described in his first paper was an arc lamp, but from 1931 onward, he used light from the sun, collected by a heliostat and directed into his laboratory through a hole in the wall. He claimed an advantage over European workers, because there were more days of sunshine in Pretoria. But in midcareer, Pijper decided that bacterial flagella were not organelles of locomotion but rather artifacts of locomotion, "viscous anisotropic polysaccharide material of the bacterial slime layer ... drawn out into tails" [29]. He held this view for the rest of his life; see the historical study by James Strick [30]. This is rather strange, given that Pijper continued to write papers that were informative and balanced, such as a 1957 review [31]. A 1938 paper contains a vivid description of what we now recognize as a tethered cell—see the discussion of the tethered-cell assay given herein—observed by Pijper upon addition of antiflagellar antibody (H-serum) to a suspension of swimming *Bacillus typhosus*, "It may happen that the end of a thickened tail becomes attached to the slide or the coverglass. The bacillus can then be seen performing a rotary movement in the plane of its long axis, revolving around a point somewhere in the middle of its body, where the tail or the two flagella are attached. ... Clockwise and anti-clockwise rotations alternate, and the whole windmill-like spectacle may last for an hour" [32].

13.4 Flagellar Propulsion

If you are a micron in size and live in an aqueous medium, you know nothing about inertia or bulk flow, only about viscosity and diffusion. What is this world like? A vivid description is given in Ed Purcell's "Life at low Reynolds number" [33]. Suppose a cell, swimming at 30 μm/s, puts in the clutch? How far does it coast? The answer turns out to be less than the diameter of a hydrogen atom! The cell is completely overwhelmed by viscosity; its Reynolds number (the ratio of the magnitude of forces due to acceleration to those due to viscous shear) is ~10^{-5}. But then the cell does not really stop, because it suffers from Brownian motion and jiggles about. Rotational Brownian motion limits a cell's ability to swim in a straight line, so even the smoothest trajectories meander. In roughly 10 s, *E. coli* can diffuse through an angle of ~90° and thus forget the direction in which it was swimming. This sets an upper limit on the time that a cell has to decide whether life is getting better or worse, because measurements made more than ~10 s ago are no longer relevant.

The only way that a cell can swim is to take advantage of viscous drag. The viscous drag on a straight segment of thin wire is about twice as large when the segment moves sideways than when it moves lengthwise. A demonstration of this is given in a film (now a DVD) by Sir Geoffrey Taylor [34]. If the wire is oriented horizontally and dropped in a viscous medium, it falls half as fast as it does when it is oriented vertically. If it is oriented slantwise, downward to the right, then it falls downward to the right. If you construct a shallow helix from segments of thin wires and rotate this assembly about its helical axis, then the sideways forces generated by each slantwise segment add, generating thrust along the helical axis; see Figure 6.3 in [35]. When the motors of *E. coli* rotate counterclockwise (CCW), their flagellar filaments, which are left-handed, form a helical bundle that pushes the cell forward. When a cell swims at constant velocity it is not accelerating, so the net force or torque acting upon it must be zero. Evidently, the thrust generated by the flagellar bundle is balanced by the drag due to translation of the cell body, and the torque generated by the flagellar bundle is balanced by the torque due to counterrotation of the cell body. Typical values for *E. coli* swimming in a dilute aqueous medium at room temperature are cell speed ~30 μm/s, bundle rotation rate ~130 Hz, and body rotation rate ~20 Hz [36]. Since the flagella and cell body rotate in opposite directions, the motor rotation rate in the reference frame of the cell body is the sum, or ~150 Hz.

When a helix is rotated, the slantwise segments are not moving one way and then moving back: the motion is not reciprocal. When swimming at low Reynolds number, reciprocal motion does not work. This was argued by Wilhelm Ludwig [37],

Chapter 13

who considered a cylindrical organism, oriented vertically, with two rigid oars (cilia), one on either side, which moved synchronously up or down. He noted that a large organism could swim this way by moving its oars rapidly downward and then slowly upward, but a microscopic organism could not; it would just return to its initial position. Purcell [33] considered a simpler case, an organism with only one oar (with only one hinge), which he drew as a scallop. He also considered an organism with two oars (two hinges) similar to Ludwig's, except that the body was thin so the organism looked like a three-link chain. He noted that this creature could swim if the oars were moved in a cyclic manner rather than in a synchronous manner. The details of this strategy turn out to be complicated [39,40]. One irony is that scallops do not move in the direction that Purcell imagined: a scallop squirts fluid out of the corners of its mouth and swims mouth first, hinge last [41]. Biology tends to be more complicated than physicists imagine.

In *E. coli* the normal flagellar helix is left-handed, with a wavelength of ~2.3 μm and a diameter of ~0.4 μm, dimensions evident in Figure 13.1, but environmental perturbations can trigger a sudden, discrete change to a new shape. These polymorphic transformations, which occur between discrete crystalline states, can be caused by a variety of means (e.g., by changing pH, salinity, or temperature) [42–44], or by applying forces or torques to the filament [3,45,46]. A geometric model due to Calladine [47], based on earlier work by Asakura [48], explains the observed spectrum of flagellar polymorphic forms. The model assumes that each individual flagellin monomer can be in one of two states, "L" or "R," that have slightly different inherent twist and length. These monomers self-segregate along 11 protofilaments that spiral down the length of the filament. The elastic energy of the filament is minimized when protofilaments of the same kind are next to one another. The trivial cases of 11L/0R and 0L/11R correspond to L-type and R-type straight filaments. The forms seen with swimming cells are left-handed normal (9L/2R), right-handed semicoiled (7L/4R), and right-handed curly 1 (6L/5R). The semicoiled form has half the normal wavelength and the normal diameter, while the curly-1 form has half the normal wavelength and half the normal diameter.

The structures of R- and L-type straight filaments are known in atomic detail, through a combination of X-ray diffraction [49,50] and electron cryomicroscopy [51,52]. Comparisons between these structures and molecular dynamics simulations [53] reveal a number of interactions important for transformations between them, but we still do not understand how the relative numbers of R- and L-type protofilaments are determined in the normal, semicoiled, and curly-1 polymorphic forms [54]. Coarse-grained models on the dynamics of polymorphic transformations are beginning to appear [55,56]. For the dynamics of flagellar bundling, see Section 7.3.2 of [57].

13.5 Runs and Tumbles

When *E. coli* swims, it executes a random walk, with relatively long intervals of smooth swimming (called runs) interrupted by relatively short intervals of erratic motion (called tumbles); these randomize the direction of motion but with a slight forward bias [58]. When cells respond to spatial gradients of attractants, favorable runs get longer; that is, tumbles are suppressed. So *E. coli* executes a biased random walk with a positive bias. Run intervals are exponentially distributed, with a mean of ~1 s. Tumble intervals also

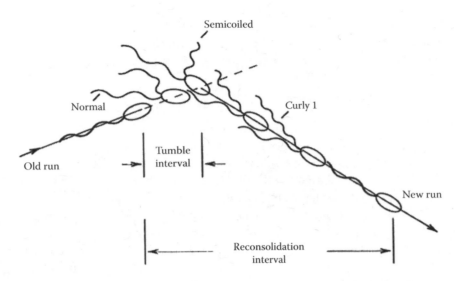

FIGURE 13.2 A tumble caused by the reversal of one flagellar motor. The filament driven by that motor comes out of the bundle and adopts the semicoiled polymorphic form, a process that alters the cell trajectory. Then the semicoiled filament relaxes to curly-1, and the cell swims in the new direction with a left-handed flagellar bundle and a right-handed curly filament. When the aberrant motor reverses again, the curly filament rejoins the bundle, and the cell swims at full speed. (From Darnton [36]. With permission.)

are exponentially distributed, with a mean of ~0.1 s. But while the means for runs vary from cell to cell (because CheY-P concentrations vary from cell to cell), the means for tumbles do not. The reason for this difference turns out to be that run lengths are a matter of motor physics, while tumble lengths are a matter of filament physics. Runs are terminated and tumbles initiated when one or more flagellar motors switch from CCW to clockwise (CW) rotation—the several motors on a given cell can do so independently. A filament driven CW comes out of the bundle and adopts the semicoiled polymorphic form, which can be seen jutting out from the side of the cell. This is shown in Figure 13.2 for a tumble involving the reversal of a single motor. By a mechanism that is not well understood (like pivoting about a fixed oar?) this generates a change in the orientation of the cell body, and the cell sets off on a new run. Thus, tumbles are terminated upon completion of the normal to semicoiled transformation. The semicoiled filament soon relaxes to curly-1: for a time, the cell is driven forward by a bundle of left-handed normal filaments turning CCW and a right-handed curly filament turning CW. When the aberrant motor switches back to CCW, the curly filament relaxes back to normal and rejoins the bundle. Only then does the cell resume its maximum run speed [36]. This process is subtler than the one commonly depicted in textbooks, where all the motors switch to CW and the bundle flies apart. It also is more complicated than just described if more than one motor turns CW at the same time; the numbers that do so have yet to be determined.

A great deal of work has been done on the hydrodynamics of flagellar propulsion, beginning with the development of resistive force theory [59]. This theory dates to an era in which it was thought that bacterial flagella propagate helical waves. This idea prevailed until 1973, when it was realized that bacterial flagella rotate rigidly [60]. However, since the dominant effects on thrust and torque are due to lateral displacements of segments of filaments rather than rotation of filaments about their local axes, the differences in hydrodynamics between helical wave propagation and rigid rotation are relatively small. For reviews on the hydrodynamics of flagellar propulsion, see, for example, [57,61,62].

13.6   Flagellar Motors

The flagellar rotary motor is a nanotechnological marvel, made of about 20 different kinds of parts (different proteins) but is only about 50 nm in diameter (1/10th of the wavelength of green light). For a schematic, see the drawings in [15,63]. The stator, embedded in the inner cell membrane but attached to the quasi-rigid framework of the cell wall (the peptidoglycan layer), is made up of a ring of the proteins MotA and MotB (for **Mot**ility), arranged in sets of four and two molecules, respectively, comprising two transmembrane proton channels. The rotor, within the ring of stator elements, includes a cytoplasmic ring (the C-ring, containing the proteins FliG,M,N), an inner-membrane ring (the M-ring, FliF), and a rod (a drive shaft, FlgB,C,F,G) that passes through the P- and L-rings, the first (FlgI) spanning the peptidoglycan layer, and the second (FlgH) embedded in the outer lipopolysaccharide membrane. The rod is connected to the flagellar filament (the propeller, FliC) via a flexible coupling called the proximal hook (FlgE). A flexible coupling (or universal joint) is required (60), because the flagella emerge from the sides of the cell, with the axes of each perpendicular to the cell surface, while the flagellar bundle tends to be aligned parallel to the long axis of the cell (Figure 13.1); hence, their orientations differ by as much as 90°. Two adapter proteins, called hook-associated proteins (FlgK,L), link the hook to the filament, while a third hook-associated protein (FliD) serves as a distal cap. Remarkably, as filaments grow, flagellin subunits pass through the filament pore and are added under the distal cap. Flagellar proteins used to be called Fla (for **Fla**gellum), but there turned out to be more than 26 (the number of letters in the alphabet) so they were renamed Flg, Flh, Fli, and Flj, depending upon the locations of their genes on the *E. coli* chromosome [64]. The synthesis, transport, and assembly of these proteins comprise an impressive feat: for reviews, [65–67]; for an animated cartoon showing the assembly, see [68]. A recent development is visualization of intact motors by cryo-electron tomography, carried out with spirochetes, because the cells are thin [69–71]. For reviews on the bacterial flageller motor, see [63,72–78].

The *E. coli* motor is driven by a proton flux, while motors of alkalophilic and marine organisms are driven by a sodium-ion flux. Some of the more innovative experiments have been done with chimaeric motors, in which all of the stator components in *E. coli* except the C-terminal end of MotB—this domain binds to the peptidoglycan—are replaced by homologous components from *Vibrio alginolyticus*, for example, an experiment with sodium ions showing that flagellar motors of *E. coli* step 26 times per revolution [79].

We don't really know how torque is generated, because we do not have atomic structures for components of the stator, except for the C-terminus of MotB. The best hypothesis, based on biochemistry, is that protonation of Asp32 of MotB, located near the cytoplasmic end of a proton channel, triggers a conformational change [80] that causes MotA to ratchet along the periphery of the rotor, interacting electrostatically with the C-terminal domain of FliG [81,82]. When the proton hops off Asp32 and enters the cytoplasm, the ratchet resets. The stator elements are thought to ratchet by an inchworm mechanism (or hand-over-hand mechanism), never letting go. If they did let go, the linkage between the rotor and the filament, which is twisted up by the motor torque, would unwind, and the motor would not work [72].

The property of the motor most relevant for chemotaxis is its ability to switch. The motor is ultrasensitive: a relatively small increase in the concentration of the signaling molecule, CheY-P, changes the motor bias from exclusively CCW to exclusively CW [83]. Remarkably, the concentration at which the bias is 0.5 (~3 μM) is nearly the same for every cell tested. We do not understand how the set point of the motors and the baseline concentration of CheY-P can be so closely matched. One suggestion is that some device in the cell monitors the motor bias and turns up the rate of CheY phosphorylation when the motors spin exclusively CCW or turns it down when they spin exclusively CW. We looked for this kind of feedback by monitoring the CheY-P concentration as motors of swimming cells were jammed by the addition of antifilament antibody, which stops flagellar rotation by cross-linking adjacent filaments in a bundle. This jamming had no effect [84].

An elegant model for switching involves a conformational spread in a ring of subunits that bind CheY-P [85]. The subunits exist in states that promote CW or CCW rotation, with the CW state stabilized by CheY-P. The energy of adjacent subunits is lower when they are in the same state. If this energy is large enough, the array switches back and forth between one in which nearly all of the subunits are in the CW state to one in which nearly all of the subunits are in the CCW state. Support for this model has been obtained from studies of the time that it takes motors, driving 0.5 μm diameter latex beads, to switch between CW and CCW conformations [86]. CheY-P is known to bind to a polypeptide at the N-terminal end of the C-ring protein FliM [87,88]. However, recent work suggests that the switch to CW rotation involves the subsequent interaction of CheY-P with FliN [89]. An NMR study of the residues of FliM perturbed by binding by CheY-P suggests a mechanism in which the binding of CheY-P displaces the C-terminal domain of FliG, altering the sign of the generated torque. A number of mutations have been made to learn more about the hinge in FliG that allows such movement; see, for example, [90].

It is important to note that motors are not required to spin CW when CheY-P binds, the binding simply increases the probability that they do so. One can enable CW events in motors of cells lacking CheY simply by lowering the temperature [91]. We still do not know whether switching is powered by ligand binding or whether it is powered by an external energy source (e.g., membrane potential) [92]. The latter possibility is considered in [93].

One of the most useful assays for studying motor function is the tethered-cell assay introduced by Silverman and Simon [94], in which a cell is attached to the substratum (usually a glass coverslip) by a single flagellar filament. Now the motor, instead of spinning the filament, spins the cell body, alternately CW and CCW. Initially, cells to be tethered were grown in a defined medium on glucose [95], which reduced flagellar number via catabolite repression [10]. Cells with single filaments were easily tethered by exposing the glass to antifilament antibody. But growth on glucose was risky, because it altered a cell's metabolic state, and it did not work for *Salmonella*. So a shearing technique was introduced to reduce the numbers of its flagellar filaments that used a Waring blender [96]. A gentler procedure was developed for *E. coli*, in which a cell

Chapter 13

suspension was forced back and forth between two syringes connected by a thin poly-ethylene tube [97]. Shearing made it possible for cells to be grown on a rich medium (e.g., T-broth), which supported higher motor speeds. A tethering medium was developed, based upon Adler's motility medium, which supported flagellar rotation for many hours [97]. Finally, a sticky-filament variant was found that tethers cells to almost any surface without requiring an antibody [98]. Sticky filaments have been particularly useful for attaching latex beads to filament stubs.

One of the things that helped end the Pijper–Weibull debate was the demonstration, in *Salmonella*, that flagellar genes could be transferred from one strain to another with phages [99]. The recipient strains, initially nonmotile, grew flagella and swam through semisolid agar. When this method was applied to *E. coli* strains that had no flagella (*flhC* strains) or to strains with paralyzed flagella (*motA* strains), both were resurrected [100]. This happened more rapidly in the latter case, because less protein synthesis was required. We tried this technique with tethered cells but failed completely, because the heads of Lambda phage adsorbed to glass and blocked cell movement. The method was a success once genes could be moved around on plasmids [101,102]: rotation was restored in a series of speed increments of equal size, implying that motors were propelled by several independent force-generating elements, each generating the same torque. We never saw more than eight steps, but recent work done at lower loads (with latex beads of diameter ≤ 1 μm) suggests that there might be as many as 11 [103]. If the resurrection experiment is done near zero load (with 60 nm gold spheres attached to hooks on cells without filaments) the resurrection occurs in a single step (from 0 to ~300 Hz) [104]. At zero load, each force generator works at its kinetic limit. It transfers protons or changes shape as rapidly as it possibly can without having to do any external work, so two force generators do not step any faster than one. As noted already, each force generator remains attached to the rotor for its entire mechanochemical cycle; its duty ratio is 1. The speed is the displacement per cycle divided by the cycle time and is the same for each torque generator [72,105].

One can remove (or inactivate) force-generating elements by lowering the proton motive force (pmf) [106]. When the pmf is restored, the force-generating elements reassemble (or reactivate) in a stepwise manner. Exchange of stator elements between the motor and the cytoplasmic membrane has been visualized by TIRF microscopy using GFP-MotB fusions [107]. Exchange of FliM has been demonstrated in a similar manner, but on a longer timescale and only in the presence of CheY-P [108]. See the commentary by Manson [109].

An important benchmark for motor models is prediction of the torque–speed relationship, the torque that the motor can exert while running at different speeds. The speed at which the motor actually runs depends upon the external load, which on a torque–speed plot appears as a straight line running from the origin with a slope equal to the load's rotational frictional drag coefficient—for a sphere of radius a, $8\pi\eta a^3$, where η is the viscosity—1,000 times smaller for 100 nm gold than for 1 μm latex. The operating point is located at the intersection of the torque–speed curve and the load line. Most measurements of torque–speed curves have been made with *cheY* mutants, which spin their motors exclusively CCW. In this case, the torque declines slightly up to an intermediate speed called the *knee speed* at ~170 Hz at room temperature, and then it falls rapidly to zero at ~300 Hz [110]. This result is consistent with a *power-stroke* mechanism [111,112]. Recently, using a mutant that spins its motors exclusively CW, we found that the CW torque-speed curve is different: the stall torque and zero-load speed are the same as before, but the torque declines linearly from 0 to ~300 Hz, without any plateau [113]. This behavior is consistent with a *Brownian ratchet*, but the biochemistry, discussed earlier, suggests otherwise [80]. In any event, the motor is asymmetric.

For a list of constraints that need to be recognized when modeling the flagellar motor, see [72]. For a catalog of the ideas envisaged before 2003, see [114]. Recent models make more quantitative predictions, including the shape of the torque–speed curves, see, for example, [115–117].

13.7 Signaling

The fraction of time that the motor spins CCW, and hence the cell's run/tumble behavior, depends upon the concentration of the phosphorylated signaling molecule CheY-P, also known as the response regulator. CheY is a small, soluble monomeric protein [118]. It is phosphorylated by a kinase, CheA [119], whose activity is controlled by a set of chemoreceptors, the best understood of which are called methyl-accepting chemotaxis proteins (MCPs) [120]. See the schematic of the signal-transduction pathway shown in Figure 13.3. The binding of attractants to these receptors reduces the kinase activity, while the binding of repellents increases it. The MCP Tsr senses serine and certain repellents (e.g., leucine); Tar senses aspartate, maltose, and certain other repellents

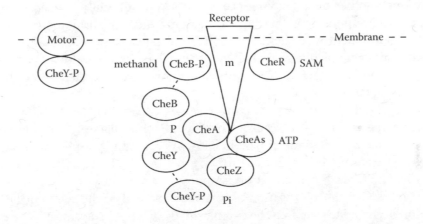

FIGURE 13.3 The *E. coli* sensory transduction pathway. A receptor trimer of homodimers (triangle) is shown traversing the inner cell membrane. The ligand binding sites are at the top (not shown). The kinase dimer, CheA/CheA or CheA/CheAs, is bound at the cytoplasmic end (bottom). Its coupling factor, CheW, is not shown. CheY-P diffuses through the cytoplasm and binds to the flagellar motors (only one is shown, at the left), increasing the probability that they spin CW. CheY is activated (phosphorylated) by CheA, a reaction indicated by the short dashed line. The phosphate is transferred from ATP by the second component of the dimer (CheAs or CheA). CheY-P is dephosphorylated by CheZ, shown bound to CheAs, liberating inorganic phosphate (Pi). CheB also is activated (phosphorylated) by CheA. It dephosphorylates spontaneously. The kinase activity is regulated by the receptor. When attractant binds to the receptor, the kinase activity goes down. Kinase activity is restored by methylation of the receptor's cytoplasmic domain. Methyl groups (m) are transferred from S-adenosylmethionine (SAM) by the methyltransferase, CheR. When attractant leaves the receptor, the kinase activity goes up. Kinase activity is restored by removal of the methyl groups by the methylesterase, CheB-P, liberating methanol.

(e.g., Co^{++} and Ni^{++}); Tap senses dipeptides; and Trg senses ribose and galactose. These are long homodimeric molecules, α-helical coiled coils, which span the cytoplasmic membrane, with their ligand binding sites exposed to the periplasmic space. Ligands of low molecular weight reach these sites after traversing channels in the outer membrane called porins. Some do so indirectly after binding periplasmic binding proteins, such as for maltose or ribose and galactose. Another receptor often included in this family, Aer, senses redox potential, but it is not methylated. Tsr and Tar number several thousand molecules per cell; the other receptors only several hundred [121]. CheA has a domain structure (P1-P5) and is homodimeric. One monomer catalyzes the transfer of inorganic phosphate from ATP to His48 of the other monomer [122]. CheY then catalyzes the transfer of this phosphate to its own Asp57 [123]. The lifetime of CheY-P is about 10 s [119,124], which is too long; it is reduced to tenths of seconds by a phosphatase, CheZ, which also is homodimeric [125]. So CheY is phosphorylated by interaction with one enzyme and dephosphorylated by another. ATP is expended, but this shortens the response time of the system, so that the motors are sensitive to sudden decrements in the concentrations of attractants.

The chemotaxis system is one of a large family of related systems called two-component signaling systems. The first component is a sensor kinase that spans the cytoplasmic membrane. The second component is a response regulator with a receiver domain coupled to an effector. When the receiver domain is phosphorylated, the effector domain becomes active and binds to a transcription factor, and this changes gene expression. For chemotaxis, the sensor and the kinase are not part of the same molecule but are linked to one another by a coupling protein, CheW [126]. Only the sensor spans the membrane. The response regulator, CheY, has only the receiver domain, which when phosphorylated binds to the flagellar motor. A second response regulator, CheB, has a receiver domain coupled to a methylesterase catalylic domain, described in the next section. CheY and CheB bind to CheA until phosphorylated, and then they diffuse freely within the cytoplasm.

The *cheA* gene has two start sites and encodes two proteins, one 97 amino-acids shorter than the other, called $CheA_S$. The CheY-P phosphatase, CheZ, binds to the truncated N-terminal domain of $CheA_S$ [127]. As a result, the phosphorylation and dephosphorylation of CheY occur at the same regions of the cell. This colocalization ensures that the CheY-P concentration is approximately constant along the length of the cell [128]. If the phosphorylation occurred at one end of the cell and the dephosphorylation occurred throughout the cytoplasm, the CheY-P concentration would drop exponentially, so that motors closer to the kinase would have a larger CW bias. For a simulation of this effect, see [129].

It was found by immuno-gold staining that chemoreceptors, and thus the kinase, tend to cluster, usually near one pole [130]. This result was confirmed by imaging fluorescent protein fusions [131]. More recent work using photoactivated localization microscopy (PALM) indicates that clusters occur in all sizes, albeit with the largest clusters at the poles [132]. This is consistent with a model in which receptors, inserted in the membrane at random sites, aggregate when colliding with other receptors or receptor clusters. The clusters are ordered arrays of receptors, arranged hexagonally as trimers of homodimers, which can be visualized by cryo-electron tomography [133]. In the absence of CheA and ChW, they exist as isolated trimers of homodimers [134].

13.8 Adaptation

If a lot of attractant is added to swimming cells (or to tethered cells), the cells swim smoothly (or spin CCW) for minutes, tumble (or spin CW) briefly, and eventually recover [135]. The cells adapt. For aspartate, the recovery is complete (the adaptation is exact), but for serine it is less so [58]. Adler found that this process involves methylation of up to four glutamic-acid residues in each receptor monomer's cytoplasmic domain. This is catalyzed by the methyltransferase, CheR, which transfers methyl groups from S-adenosylmethionine. An early hint to this process was the observation that methionine-starved cells swam smoothly and could no longer respond to gradients [10]. Methylation of the receptors increases the activity of the CheA kinase. When attractant concentrations fall, the methyl groups are removed by the activated methylesterase, CheB-P. Receptors are synthesized with two of the four glutamates glutamines. The glutamines are deaminated by CheB-P. In *cheR cheB* mutants, the glutamines remain as such, so by using site-directed mutagenesis receptors with any number of glutamines can be constructed. Glutamine approximates the function of methyl glutamate, but it does not activate the kinase to the same extent. An interesting feature of this process is that the methylation enzyme tethers to a receptor C-terminal pentapeptide at the end of a flexible arm [136], allowing the enzyme to move through a receptor cluster in a hand-over-hand fashion, like a gibbon in a tree [137].

Receptors have been studied in vitro in an ingenious way embedded in small membrane patches (nanodiscs). A single homodimer can bind ligand and activate the methyltransferase, exhibiting a simple dose-response curve (no cooperativity) [138]. However, activation or inactivation of the kinase at levels observed in native membrane vesicles requires nanodiscs averaging ~5 dimers per disk (i.e., with at least three in parallel), suggesting that coupling to the kinase occurs at the level of trimers of homodimers [139]. Conformational changes in trimers of homodimers have been observed in vivo by tagging the C-terminal ends of the receptors with yellow fluorescent protein and measuring changes in fluorescence anisotropy: addition of repellents moves the fluorophores closer together, while addition of attractants moves them farther apart [140]. Thus, coupling to the kinase and coupling between receptors might have a simple mechanical basis.

A great deal of work has been done with the aim of learning how the chemotactic signal crosses the membrane. There is a consensus that a small piston movement of one transmembrane α-helical segment coupled to a change in stability of an adjacent four-helix bundle (a HAMP domain) changes the shape of the cytoplasmic end of the homodimer, affecting both methylation and kinase activation [141–143].

13.9 System Properties

So how does all this hardware help cells respond when swimming in spatial gradients? Note first that cells respond to changes in concentration that occur with time; they do not make spatial comparisons per se. They could do so, in principle, by time-averaging information obtained at either end of the cell, but they swim too rapidly: their motion increases the diffusive influx of molecules in front compared to behind, so that any new direction would be deemed favorable [144]—for corrections to Equation (D1) relevant to this argument, see p. 123 of [145]. We proved that the measurements were temporal by

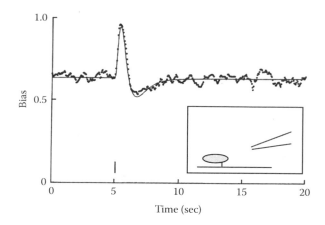

FIGURE 13.4　The response of *E. coli* to a short pulse of aspartate or a-methyl-D, L-aspartate delivered at 5.06 s. Tethered cells were stimulated by iontophoretic pipettes, as shown in the inset, and their CCW biases were measured. See [148]. (From Berg [185]. With permission.)

tracking cells swimming in homogeneous, isotropic media in which an attractant was either generated or consumed enzymatically: runs got longer over the time span that attractant was generated, but they did not get shorter when attractant was consumed [146]. The magnitude and asymmetry of the response was what was expected from what we had learned by tracking cells in spatial gradients [58].

To control the temporal stimulus, we used iontophoretic pipettes, first to measure the time it takes tethered cells to respond to large step stimuli (~0.2 s) [147] and later to gauge their response to short pulses or small steps [148]. The impulse response was biphasic, increasing the CCW bias for the first second and decreasing it for the next three seconds, with the areas of the two lobes being equal, as shown in Figure 13.4. The small-step response was the integral of the impulse response, as expected for a system that is linear, increasing the CCW bias to a higher level over the first second and bringing it back to the initial level over the next three seconds. A step of aspartate (or α-methyl-D, L-aspartate) that changed the occupancy of the Tar receptors from 600 to 601 resulted in a transient change in bias of about 0.1, so the gain of the system was prodigious. For *cheR cheB* mutants, the second lobe of the impulse response was missing and cells failed to adapt to small steps. For *cheZ* mutants, everything slowed down by a factor of ~10. All this made sense: cells measure the concentration of an attractant over the past 1 s and compare the result with measurements made over the previous 3 s. They complete the task in about 4 s, a time below the limit set by rotational Brownian motion (~10 s). But why should the impulse response have this particular shape? deGennes [149] argued that to climb a gradient, only the first lobe is necessary. Clark and Grant [150] noted that to accumulate at an attractant peak, the second lobe is needed; perhaps a composite optimization gives the right result. A more interesting suggestion has been made by Celani and Vergassola [151], who noted that bacteria must survive in complex environments. Given this complexity, what response function would ensure the highest minimum uptake of attractants over a wide range of concentration profiles? The analysis of this maximin strategy is mathematically demanding, but it gives the right result. Whether this might be the last word on the subject, I do not know.

The modern view of receptors is that they toggle rapidly between two states—one that activates the kinase and one that does not—an idea proposed in [152]. The dissociation constant for the inactive state is smaller than the dissociation constant for the active state. The probability that the system is in the active state is decreased by ligand binding (of attractant) and increased by methylation. However, changes in kinase activity are much larger than changes in ligand binding, when these quantities are measured directly in vitro, such as [153], or indirectly in vivo (with changes in binding inferred from changes in fluorescence anisotropy) [140]. Evidently, the signaling cycle is out of equilibrium.

If the methyltransferase works at saturation and the methylesterase acts only on active receptors (at a rate that depends on its level of phosphorylation, i.e., on the system's output), then adaptation is robust [154]; the kinase activity returns to its initial value regardless of enzyme concentrations. What that level of activity might be, however, does depend on these concentrations. Robust adaptation was established experimentally for aspartate in an elegant series of experiments by Alon et al. [155]. According to Yi et al. [156] this strategy is well known in control theory as integral feedback control; see also [157]. The failure of robust adaptation for serine appears to be due to full methylation beginning at ~0.1 mM serine, with subsequent loss of activity [158].

It was soon found that the steady state to which cells were thought to adapt was subject to large temporal variations, evident in the analysis of the output of single cells [159]. This had been seen earlier in studies of signal propagation in filamentous cells [160], but its significance was not fully recognized. It was not simply a question of variation from cell to cell due to the small numbers of signaling molecules [161] but rather of variations in the behavior of a single cell over time. This behavior appeared to be by design, because it could be eliminated by increasing the wild-type concentration of CheR by a factor of ~4 [159]. Further study showed that methylation and demethylation cycles are ultrasensitive: the methyltransferase and methylesterase operate near saturation, outside the region of first-order kinetics [162] (see Chapter 8). The greater the variability of the output of a nonstimulated cell, the greater its sensitivity to a chemotactic stimulus and the greater its relaxation time. Simulations of digital bacteria [163] suggest that this variability increases the spread of bacteria when swimming in the absence of gradients and increases the size of their response when exposed to gradients.

Computer simulations have played an important role in illuminating our understanding (or lack of understanding) of the chemotaxis system. The best example is the work of Dennis Bray and his colleagues on the question of chemotactic gain (the change in the fraction of time that a motor spins CCW divided by the change in the fraction of receptor occupied) [164], something I call the gain paradox [165]. The work began with representations of biochemical reactions by deterministic rate equations [166] and evolved to a stochastic analysis in which rate equations are replaced by reaction probabilities and the interactions between molecules are examined one by one; see, for example, [167]. Computer graphics have dramatized the results [168]. The models were not able to explain the missing gain, which led Bray to suggest that this gain might be supplied by interactions between adjacent receptors in clusters [169], an idea that was fleshed out with the help of Tom Duke and Yu Shi [170,171]. See also [172].

With the hope of finding the missing gain, we had developed a method for determining the CheA kinase activity that did not require the enormous labor of measuring changes

Chapter 13

in the rotational bias of tethered cells. Since at steady state the rate of phosphorylation of CheY equals the rate of dephosphorylation of CheY-P, the kinase activity can be inferred from the phosphatase activity (given that the rate of spontaneous dissociation is small). Since dephosphorylation requires the formation of an enzyme-substrate complex, a complex between CheZ and CheY-P, the process can be monitored by fluorescence resonance energy transfer (FRET) between CheZ-CFP (cyan fluorescent protein) and CheY-P-YFP (yellow fluorescent protein). CFP is excited, and the CFP and YFP fluorescence is measured: when the complex forms, the former goes down and the latter goes up. It is possible to look at several hundred cells at once, fixed to the window of a flow cell, and measure these changes as the cells are exposed to different concentrations of attractants; for a review, see [173]. Victor Sourjik and I found the missing gain at the front end of the sensory-transduction pathway: the fractional change in kinase activity was more than 30 times larger than the estimate for the fractional change in receptor occupancy [174]; see also Sourjik [175]. The fluorescence resonance energy transfer (FRET) data were sufficiently accurate to inspire an impressive series of theoretical papers (e.g., [176]), which showed how the energy-function approach of statistical mechanics could be used to extend the allosteric model of Monod et al. [177] to arrays of different kinds of membrane-bound receptors, and the more recent work of Endres et al. [178] showing that receptor signaling teams need contain only a limited number of receptor homodimers (~10). The latter result suggests that the smaller clusters seen in the PALM study [132] might be active, adding support to the argument that the most efficient way to intercept ligand diffusing in the external medium would be to distribute receptors widely over the surface of the cell [144].

Unless the receptors are saturated or fully methylated, the chemotactic response is logarithmic. This was clear in the early work showing that swimming cells migrate up exponential gradients of L-serine at the same rate whether near the bottom, middle, or top of the gradient [179]. It was shown later in experiments with tethered cells that were exposed to exponential temporal gradients of α-methyl-D, L-aspartate, which shifted the rotational bias to new steady state levels [97]. And it was shown recently from measurements of number-densities of cells swimming in a microfluidic device in linear gradients of α-methyl-D, L-aspartate [180].

Before the discovery of receptor methylation [120] and an appreciation of its role in adaptation [181], we assumed that the chemotactic response could be defined in terms of the time-rate-of-change in receptor occupancy, characterized by a single dissociation constant [146]. This was naïve, but it led to the use of exponential temporal gradients [97]. It is clear now that one should think in terms of a receptor module, whose output (the activity of CheA) depends not only on ligand occupancy but also on the level of receptor methylation, and in terms of an adaptation module that methylates the receptor at a rate that depends upon the output of the receptor module, ensuring robust adaptation via integral feedback control. In collaboration with Yuhai Tu, we developed such a model, based again upon the energy function approach of statistical mechanics [182]. The model explains responses to exponential ramps and to exponentiated sine waves, as well as to impulsive and large-step stimuli, and it explains the molecular origin for logarithmic tracking. It suggests that the chemotaxis system acts as a low-pass filter for the time-derivative of the signal, with a cutoff frequency determined by the adaptation time scale. We have calibrated the receptor module by using the FRET technique to measure responses to step stimuli with receptors in different modification states, and we have calibrated the adaptation module by using the FRET technique to measure responses to exponential ramps. From our model and the ramp results, we can predict the adaptation time and the cutoff frequencies for responses to oscillatory signals. Measurements of responses to exponentiated sine waves confirmed these predictions [183]. Results obtained earlier with small numbers of tethered cells found thresholds in ramp rates below which cells failed to respond [97]. This was not seen in our FRET data. Is it possible that flagellar motors are immune to small changes in the concentration of CheY-P? This seems unlikely, given the steep dependence of motor bias on the concentration of CheY-P measured by Cluzel et al. [83], but we do not know.

Uri Alon and his colleagues have noted that certain mammalian signaling systems display a response whose entire shape (amplitude and time course) depends only on fold-changes in input, that is, whose response sensitivity is rescaled to the ambient level of attractant [184]. That behavior is predicted by our model [182], provided that the ligand concentration stays within the range of the dissociation constants for the inactive and active receptor states and the methylation level stays in the range between 0 and 4.

13.10   Afterthought

The progress made by combining theory with experiment has been gratifying, though not based upon modern guidelines. The work has been done, not by teams of mathematicians, physicists, and biologists working together but rather by individuals with distinctive talents interested in the same behavioral system. We look carefully at what the cell does and at the machinery that it has contrived to do the job, and then we try to figure out how that machinery functions. Physics helps us recognize the constraints that the cell has had to deal with, and modeling helps us appreciate relevant mechanisms and to find out what we do not understand. Most of the credit goes to *E. coli*. Chemotaxis is a stunning achievement.

Acknowledgments

I thank Tom Shimizu and Yuhai Tu for their comments on the manuscript. Recent work in my laboratory has been supported by National Institutes of Health Grant AI016478.

References

1. Tindall, M. J., P. K. Maini, S. L. Porter, and J. P. Armitage. 2008. Overview of mathematical approaches used to model bacterial chemotaxis II: bacterial populations. *Bull. Math. Biol.* 70:1570–1607.
2. Tindall, M. J., S. L. Porter, P. K. Maini, G. Gaglia, and J. P. Armitage. 2008. Overview of mathematical approaches used to model bacterial chemotaxis I: the single cell. *Bull. Math. Biol.* 70:1525–1569.
3. Hotani, H. 1982. Micro-video study of moving bacterial flagellar filaments III. cyclic transformation induced by mechanical force. *J. Mol. Biol.* 156:791–806.
4. Turner, L., W. S. Ryu, and H. C. Berg. 2000. Real-time imaging of fluorescent flagellar filaments. *J. Bacteriol.* 182:2793–2801.
5. Adler, J. 1966. Chemotaxis in bacteria. *Science* 153:708–716.
6. Wolfe, A. J., and H. C. Berg. 1989. Migration of bacteria in semisolid agar. *Proc. Natl. Acad. Sci. USA* 86:6973–6977.

Chapter 13

7. Harshey, R. M. 1994. Bees aren't the only ones: swarming in Gram-negative bacteria. *Molec. Microbiol.* 13:389–394.

8. Pfeffer, W. 1884. Locomotorische Richtungsbewegungen durch chemische Reize. *Unters. Bot. Inst. Tübingen* 1:363–482.

9. Adler, J. 1973. A method for measuring chemotaxis and use of the method to determine optimum conditions for chemotaxis by *Escherichia coli*. *J. Gen. Microbiol.* 74:77–91.

10. Adler, J., and B. Templeton. 1967. The effect of environmental conditions on the motility of *Escherichia coli*. *J. Gen. Microbiol.* 46:175–184.

11. Adler, J. 1969. Chemoreceptors in bacteria. *Science* 166:1588–1597.

12. Mesibov, R., and J. Adler. 1972. Chemotaxis toward amino acids in *Escherichia coli*. *J. Bacteriol.* 112:315–326.

13. Tso, W. W., and J. Adler. 1974. Negative chemotaxis in *Escherichia coli*. *J. Bacteriol.* 118:560–576.

14. Fraenkel, G. S., and D. L. Gunn. 1940. *The Orientation of Animals: Kineses, Taxes and Compass Reactions.* Dover [Oxford Clarendon Press] reprinted with additional notes in 1961, New York, NY.

15. Berg, H. C. 2000. Motile behavior of bacteria. *Physics Today* 53(1):24–29.

16. Englert, D. L., A. Jayaraman, and M. D. Manson. 2009. Microfluidic techniques for the analysis of bacterial chemotaxis. *Meth. Mol. Biol.* 571:1–23.

17. Ahmed, T., T. S. Shimizu, and R. Stocker. 2010. Microfluidics for bacterial chemotaxis. *Integrative Biol.* 2:604–629.

18. Budrene, E. O., and H. C. Berg. 1991. Complex patterns formed by motile cells of *Escherichia coli*. *Nature* 349:630–633.

19. Budrene, E. O., and H. C. Berg. 1995. Dynamics of formation of symmetrical patterns by chemotactic bacteria. *Nature* 376:49–53.

20. Berg, H. C. 1996. Symmetries in bacterial motility. *Proc. Natl. Acad. Sci. USA* 93:14225–14228.

21. Brenner, M. P. 2010. Chemotactic patterns without chemotaxis. *Proc. Natl. Acad. Sci. USA* 107:11653–11654.

22. Cates, M. E., D. Marenduzzo, I. Pagonabarraga, and J. Tailleur. 2010. Arrested phase separation in reproducing bacteria creates a generic route to pattern formation. *Proc. Natl. Acad. Sci. USA* 107:11715–11720.

23. Dobell, C. 1960. *Antony van Leeuwenhoek and His "Little Animals".* Dover, New York.

24. Berg, H. C. 1975. Chemotaxis in bacteria. *Ann. Rev. Biophys. Bioeng.* 4:119–136.

25. Berg, H. C. 2004. *E. coli in Motion.* Springer-Verlag, New York, NY.

26. Weibull, C. 1960. *Movement. The Bacteria,* I.C. Gunsalus and R.Y. Stanier, Eds. (Academic Press, NY) 1:153–205.

27. Weibull, C. 1950. Investigations on bacterial flagella. *Acta Chem. Scand.* 4:268–276.

28. Astbury, W. T., E. Beighton, and C. Weibull. 1955. The structure of bacterial flagella. *Symp. Soc. Exp. Biol.* 9:282–305.

29. Pijper, A. 1946. Shape and motility of bacteria. *J. Path. Bact.* 58:325–342.

30. Strick, J. 1996. Swimming against the tide: Adrianus Pijper and the debate over bacterial flagella, 1946–1956. *Isis* 87:274–305.

31. Pijper, A. 1957. Bacterial flagella and motility. *Ergeb. Mikrobiol. Immunitätsforsch. Exptl. Therap.* 58:325–342.

32. Pijper, A. 1938. Dark ground studies of flagellar and somatic agglutination of *B. Typhosus*. *J. Path. Bact.* 47:1–17.

33. Purcell, E. M. 1977. Life at low Reynolds number. *Am. J. Phys.* 45:3–11.

34. Taylor, G. I. 1967. *Low Reynolds-Number Flows,* ENDVD 21617. Encyclopaedia Britannica Educational Corp., Chicago, IL.

35. Berg, H. C. 1993. *Random Walks in Biology.* Princeton, Princeton, NJ.

36. Darnton, N. C., L. Turner, S. Rojevsky, and H. C. Berg. 2007. On torque and tumbling in swimming *Escherichia coli*. *J. Bacteriol.* 189:1756–1764.

37. Ludwig, W. 1930. Zur Theorie der Flimmerbewegung (Dynamik, Nutzeffekt, Energiebalanz). *Z. Vgl. Physiol.* 13:397–504.

38. Glatter, O. 1977. A new method for the evaluation of small-angle scattering data. *J. Appl. Cryst.* 10:415–421.

39. Becker, L. E., S. A. Koehler, and H. A. Stone. 2003. On self-propulsion of micro-machines at low Reynolds number: Purcell's three-link swimmer. *J. Fluid Mech.* 490:15–35.

40. Tam, D., and A. E. Hosoi. 2007. Optimal stroke patterns for Purcell's three-link swimmer. *Phys. Rev. Lett.* 98:068105.

41. Vogel, S. 1997. Squirt smugly, scallop! *Nature* 385:21–22.
42. Hasegawa, E. R., R. Kamiya, and S. Asakura. 1982. Thermal transition in helical forms of *Salmonella* flagella. *J. Mol. Biol.* 160:609–621.
43. Kamiya, R., and S. Asakura. 1976. Helical transformations of *Salmonella* flagella *in vitro*. *J. Mol. Biol.* 106:167–186.
44. Kamiya, R., and S. Asakura. 1977. Flagellar transformations at alkaline pH. *J. Mol. Biol.* 108:513–518.
45. Darnton, N. C., and H. C. Berg. 2007. Force-extension measurements on bacterial flagella: triggering polymorphic transformations. *Biophys. J.* 92:2230–2236.
46. Macnab, R. M., and M. K. Ornston. 1977. Normal-to-curly flagellar transitions and their role in bacterial tumbling: stabilization of an alternative quaternary structure by mechanical force. *J. Mol. Biol.* 112:1–30.
47. Calladine, C. R. 1978. Change of waveform in bacterial flagella: the role of mechanics at the molecular level. *J. Mol. Biol.* 118:457–479.
48. Asakura, S. 1970. Polymerization of flagellin and polymorphism of flagella. *Adv. Biophys.* 1:99–155.
49. Kamiya, R., S. Asakura, K. Wakabayashi, and K. Namba. 1979. Transition of bacterial flagella from helical to straight forms with different subunit arrangements. *J. Mol. Biol.* 131:725–742.
50. Yamashita, I., K. Hasegawa, H. Suzuki, F. Vonderviszt, Y. Mimori-Kiyosue, and K. Namba. 1998. Structure and switching of bacterial flagellar filaments studied by X-ray fiber diffraction. *Nat. Struct. Biol.* 5:125–132.
51. Maki-Yonekura, S., K. Yonekura, and K. Namba. 2010. Conformational change of flagellin for polymorphic supercoiling of the flagellar filament. *Nat. Struct. Biol.* 17:417–422.
52. Yonekura, K., S. Maki-Yonekura, and K. Namba. 2003. Complete atomic model of the bacterial flagellar filament by electron cryomicroscopy. *Nature* 424:643–650.
53. Kitao, A., K. Yonekura, S. Maki-Yonekura, F. A. Samatey, K. Imada, and K. Namba. 2006. Switch interactions control energy frustration and multiple flagellar filament structures. *Proc. Natl. Acad. Sci. USA* 103:4894–4899.
54. Calladine, C. R. 2010. New twists for bacterial flagella. *Nat. Struct. Biol.* 17:395–396.
55. Friedrich, B. 2006. A mesoscopic model for helical bacterial flagella. *J. Math. Biol.* 53:162–178.
56. Srigiriraju, S. V., and T. R. Powers. 2006. Model for polymorphic transitions in bacterial flagella. *Phys. Rev. E* 73:011902.
57. Lauga, E., and T. R. Powers. 2009. The hydrodynamics of swimming microorganisms. *Rep. Prog. Phys.* 72:1–36.
58. Berg, H. C., and D. A. Brown. 1972. Chemotaxis in *Escherichia coli* analysed by three-dimensional tracking. *Nature* 239:500–504.
59. Holwill, M. E. J., and R. E. Burge. 1963. A hydrodynamic study of the motility of flagellated bacteria. *Arch. Biochem. Biophys.* 101:249–260.
60. Berg, H. C., and R. A. Anderson. 1973. Bacteria swim by rotating their flagellar filaments. *Nature* 245:380–382.
61. Lighthill, J. 1976. Flagellar hydrodynamics: the John von Neumann lecture. *SIAM Rev.* 18:161–230.
62. Brennen, C., and H. Winet. 1977. Fluid mechanics of propulsion by cilia and flagella. *Annu. Rev. Fluid Mech.* 9:339–398.
63. Berg, H. C. 2008. Bacterial flagellar motor. *Curr. Biol.* 18:R689–R691.
64. Iino, T., Y. Komeda, K. Kutsukake, R. M. Macnab, P. Matsumura, J. S. Parkinson, M. I. Simon, and S. Yamaguchi. 1988. New unified nomenclature for the flagellar genes of *Escherichia coli* and *Salmonella typhimurium*. *Microbiol. Rev.* 52:533–535.
65. Macnab, R. M. 2003. How bacteria assemble flagella. *Annu. Rev. Microbiol.* 57:77–100.
66. Apel, D., and M. G. Surette. 2008. Bringing order to a complex molecular machine: the assembly of the bacterial flagella. *Biochim. Biophys. Acta* 1778:1851–1858.
67. Chevance, F. F. V., and K. T. Hughes. 2008. Coordinating assembly of a bacterial macromolecular machine. *Nat. Rev. Micro.* 6:455–465.
68. Namba, K. 2010. Movies (5): assembly process for the bacterial flagellum. http://www.fbs.osaka-u.ac.jp/labs/namba/npn/index.html.
69. Kudryashev, M., M. Cyrklaff, R. Wallich, W. Baumeister, and F. Frischknecht. 2010. Distinct *in situ* structures of the *Borrelia* flagellar motor. *J. Struct. Biol.* 169:54–61.
70. Liu, J., T. Lin, D. J. Botkin, E. McCrum, H. Winkler, and S. J. Norris. 2009. Intact flagellar motor of *Borrelia burgdorferi* revealed by cryo-electron tomography: evidence for stator ring curvature and rotor/C-ring assembly flexion. *J. Bacteriol.* 191:5026–5036.

Chapter 13

71. Murphy, G. E., J. R. Leadbetter, and G. J. Jensen. 2006. *In situ* structure of the complete *Treponema primitia* flagellar motor. *Nature* 442:1062–1064.

72. Berg, H. C. 2003. The rotary motor of bacterial flagella. *Annu. Rev. Biochem.* 72:19–54.

73. Blair, D. F. 2003. Flagellar movement driven by proton translocation. *FEBS Lett.* 545:86–95.

74. Blair, D. F. 2009. Structure and mechanism of the flagellar rotary motor. In Pili and Flagella. K. F. Jarrell, editor. Caister, Norwich, UK. 121–136.

75. Kojima, S., and D. F. Blair. 2004. The bacterial flagellar motor: structure and function of a complex molecular machine. *Int. Rev. Cytol.* 233:93–134.

76. Minamino, T., K. Imada, and K. Namba. 2008. Molecular Motors of the bacterial flagella. *Curr. Opin. Struct. Biol.* 18:693–701.

77. Sowa, Y., and R. M. Berry. 2008. The bacterial flagellar motor. *Quart. Rev. Biophys.* 41:103–132.

78. Baker, M. A. B., and R. M. Berry. 2009. An introduction to the physics of the bacterial flagellar motor: a nanoscale rotary electric motor. *Contemp. Physics* 50:617–632.

79. Sowa, Y., A. D. Rowe, M. C. Leake, T. Yakushi, M. Homma, A. Ishijima, and R. M. Berry. 2005. Direct observation of steps in rotation of the bacterial flagellar motor. *Nature* 437:916–919.

80. Kojima, S., and D. F. Blair. 2001. Conformational change in the stator of the bacterial flagellar motor. *Biochemistry* 40:13041–13050.

81. Yakushi, T., J. Yang, H. Fukuoka, M. Homma, and D. F. Blair. 2006. Roles of charged residues of rotor and stator in flagellar rotation: comparative study using H^+-driven and Na^+-driven motors in *Escherichia coli*. *J. Bacteriol.* 188:1466–1472.

82. Zhou, J., S. A. Lloyd, and D. F. Blair. 1998. Electrostatic interactions between rotor and stator in the bacterial flagellar motor. *Proc. Natl. Acad. Sci. USA* 95:6436–6441.

83. Cluzel, P., M. Surette, and S. Leibler. 2000. An ultrasensitive bacterial motor revealed by monitoring signaling proteins in single cells. *Nature* 287:1652–1655.

84. Shimizu, T. S., N. Delalez, K. Pichler, and H. C. Berg. 2006. Monitoring bacterial chemotaxis by using bioluminescence resonance energy transfer: absence of feedback from the flagellar motors. *Proc. Natl. Acad. Sci. USA* 103:2093–2097.

85. Duke, T. A. J., N. LeNovère, and D. Bray. 2001. Conformational spread in a ring of proteins: a stochastic approach to allostery. *J. Mol. Biol.* 308:541–553.

86. Bai, F., R. W. Branch, D. V. Nicolau, Jr., T. Pilizota, B. C. Steel, P. K. Maini, and R. M. Berry. 2010. Conformational spread as a mechanism for cooperativity in the bacterial flagellar switch. *Science* 327:685–689.

87. Bren, A., and M. Eisenbach. 1998. The N terminus of the flagellar switch protein, FliM, is the binding domain for the chemotactic response regulator, CheY. *J. Mol. Biol.* 278:507–514.

88. Welch, M., K. Oosawa, S.-I. Aizawa, and M. Eisenbach. 1993. Phosphorylation-dependent binding of a signal molecule to the flagellar switch of bacteria. *Proc. Natl. Acad. Sci. USA* 90:8787–8791.

89. Sarkar, M. K., K. Paul, and D. Blair. 2010. Chemotaxis signaling protein CheY binds to the rotor protein FliN to control the direction of flagellar rotation in *Escherichia coli*. *Proc. Natl. Acad. Sci. USA* 107:9370–9375.

90. Van Way, S. M., S. G. Millas, A. H. Lee, and M. D. Manson. 2004. Rusty, jammed, and well-oiled hinges: mutations affecting the interdomain region of FliG, a rotor element of the *Escherichia coli* flagellar motor. *J. Bacteriol.* 186:3173–3181.

91. Turner, L., S. R. Caplan, and H. C. Berg. 1996. Temperature-induced switching of the bacterial flagellar motor. *Biophys. J.* 71:2227–2233.

92. Turner, L., A. D. T. Samuel, A. S. Stern, and H. C. Berg. 1999. Temperature dependence of switching of the bacterial flagellar motor by the protein CheY[13DK106YW]. *Biophys. J.* 77:597–603.

93. Tu, Y. 2008. The nonequilibrium mechanism for ultrasensitivity in a biological switch: sensing by Maxwell's demons. *Proc. Natl. Acad. Sci. USA* 2008:11737–11741.

94. Silverman, M., and M. Simon. 1974. Flagellar rotation and the mechanism of bacterial motility. *Nature* 249:73–74.

95. Larsen, S. H., R. W. Reader, E. N. Kort, W. Tso, and J. Adler. 1974. Change in direction of flagellar rotation is the basis of the chemotactic response in *Escherichia coli*. *Nature* 249:74–77.

96. Khan, S., R. M. Macnab, A. L. DeFranco, and D. E. Koshland. 1978. Inversion of a behavioral response in bacterial chemotaxis: explanation at the molecular level. *Proc. Natl. Acad. Sci. USA* 75:4150–4154.

97. Block, S. M., J. E. Segall, and H. C. Berg. 1983. Adaptation kinetics in bacterial chemotaxis. *J. Bacteriol.* 154:312–323.

98. Scharf, B. E., K. A. Fahrner, L. Turner, and H. C. Berg. 1998. Control of direction of flagellar rotation in bacterial chemotaxis. *Proc. Natl. Acad. Sci. USA* 95:201–206.

99. Stocker, B. A. D., N. D. Zinder, and J. Lederberg. 1953. Transduction of flagellar characters in *Salmonella*. *J. Gen. Microbiol.* 9:410–433.

100. Silverman, M., P. Matsumura, and M. Simon. 1976. The identification of the *mot* gene product with *Escherichia coli*-lambda hybrids. *Proc. Natl. Acad. Sci. USA* 73:3126–3130.

101. Blair, D. F., and H. C. Berg. 1988. Restoration of torque in defective flagellar motors. *Science* 242:1678–1681.

102. Block, S. M., and H. C. Berg. 1984. Successive incorporation of force-generating units in the bacterial rotary motor. *Nature* 309:470–472.

103. Reid, S. W., M. C. Leake, J. H. Chandler, C.-J. Lo, J. P. Armitage, and R. M. Berry. 2006. The maximum number of torque-generating units in the flagellar motor of *Escherichia coli* is at least 11. *Proc. Natl. Acad. Sci. USA* 103:8066–8071.

104. Yuan, J., and H. C. Berg. 2008. Resurrection of the flagellar rotary motor near zero load. *Proc. Natl. Acad. Sci. USA* 105:1182–1185.

105. Ryu, W. S., R. M. Berry, and H. C. Berg. 2000. Torque-generating units of the flagellar motor of *Escherichia coli* have a high duty ratio. *Nature* 403:444–447.

106. Fung, D. C., and H. C. Berg. 1995. Powering the flagellar motor of *Escherichia coli* with an external voltage source. *Nature* 375:809–812.

107. Leake, M. C., J. H. Chandler, G. H. Wadhams, F. Bai, R. M. Berry, and J. P. Armitage. 2006. Stoichiometry and turnover in single, functioning membrane protein complexes. *Nature* 443:355–358.

108. Delalez, N. J., G. H. Wadhams, G. Rosser, Q. Xue, M. T. Brown, I. M. Dobbie, R. M. Berry, M. C. Leake, and J. P. Armitage. 2010. Signal-dependent turnover of the bacterial flagellar switch protein FliM. *Proc. Natl. Acad. Sci. USA* 107:11347–11351.

109. Manson, M. D. 2010. Dynamic motors for bacterial flagella. *Proc. Natl. Acad. Sci. USA* 107:11151–11152.

110. Chen, X., and H. C. Berg. 2000. Torque-speed relationship of the flagellar rotary motor. *Biophys. J.* 78:1036–1041.

111. Berry, R. M., and H. C. Berg. 1999. Torque generated by the flagellar motor of *Escherichia coli* while driven backward. *Biophys. J.* 76:580–587.

112. Howard, J. 2006. Protein power strokes. *Curr. Biol.* 16:R517–R519.

113. Yuan, J., K. A. Fahrner, L. Turner, and H. C. Berg. 2010. Asymmetry in the clockwise and counterclockwise rotation of the bacterial flagellar motor. *Proc. Natl. Acad. Sci. USA* 107:12846–12849.

114. Berg, H. C. 2003. The bacterial rotary motor. *The Enzymes* 23:143–202.

115. Bai, F., C.-J. Lo, R. M. Berry, and J. Xing. 2009. Model studies of the dynamics of bacterial flagellar motors. *Biophys. J.* 96:3154–3167.

116. Meacci, G., and Y. Tu. 2009. Dynamics of the bacterial flagellar motor with multiple stators. *Proc. Natl. Acad. Sci. USA* 106:3746–3751.

117. Mora, T., H. Yu, and N. S. Wingreen. 2009. Modeling torque versus speed, shot noise, and rotational diffusion of the bacterial flagellar motor. *Phys. Rev. Lett.* 103:248102.

118. Lee, S. Y., H. S. Cho, J. G. Pelton, D. Yan, E. A. Berry, and D. E. Wemmer. 2001. Crystal structure of activated CheY: comparison with other activated receiver domains. *J. Biol. Chem.* 276:16425–16431.

119. Hess, J. F., K. Oosawa, N. Kaplan, and M. I. Simon. 1988. Phosphorylation of three proteins in the signaling pathway of bacterial chemotaxis. *Cell* 53:79–87.

120. Kort, E. N., M. F. Goy, S. H. Larsen, and J. Adler. 1975. Methylation of a membrane protein involved in bacterial chemotaxis. *Proc. Natl. Acad. Sci. USA* 72:3939–3943.

121. Li, M., and G. L. Hazelbauer. 2004. Cellular stoichiometry of the components of the chemotaxis signaling complex. *J. Bacteriol.* 186:3687–3694.

122. Bilwes, A. M., L. A. Alex, B. R. Crane, and M. I. Simon. 1999. Structure of CheA, a signal-transducing histidine kinase. *Cell* 96:131–141.

123. Lukat, G. S., W. R. McCleary, A. M. Stock, and J. B. Stock. 1992. Phosphorylation of bacterial response regulator proteins by low molecular weight phospho-donors. *Proc. Natl. Acad. Sci. USA* 89:718–722.

124. Wylie, D., A. Stock, C.-Y. Wong, and J. Stock. 1988. Sensory transduction in bacterial chemotaxis involves phosphotransfer between Che proteins. *Biochem. Biophys. Res. Commun.* 151:891–896.

125. Zhao, R., E. J. Collins, R. B. Bourret, and R. E. Silversmith. 2002. Structure and catalytic mechanism of the *E. coli* chemotaxis phosphatase CheZ. *Nat. Struct. Biol.* 9:570–575.

126. Kirby, J. R. 2009. Chemotaxis-like regulatory systems: unique roles in diverse bacteria. *Annu. Rev. Microbiol.* 63:45–59.

Chapter 13

127. Cantwell, B. J., and M. D. Manson. 2009. Protein domains and residues involved in the CheZ/CheAs interaction. *J. Bacteriol.* 191:5838–5841.

128. Vaknin, A., and H. C. Berg. 2004. Single-cell FRET imaging of phosphatase activity in the *Escherichia coli* chemotaxis system. *Proc. Natl. Acad. Sci. USA* 101:17072–17077.

129. Lipkow, K., S. S. Andrews, and D. Bray. 2005. Simulated diffusion of phosphorylated CheY through the cytoplasm of *Escherichia coli*. *J. Bacteriol.* 187:45–53.

130. Maddock, J. R., and L. Shapiro. 1993. Polar location of the chemoreceptor complex in the *Escherichia coli* cell. *Science* 259:1717–1723.

131. Sourjik, V., and H. C. Berg. 2000. Location of components of the chemotaxis machinery of *Escherichia coli* using fluorescent-protein fusions. *Molec. Microbiol.* 37:740–751.

132. Greenfield, D., A. L. McEvoy, H. Shroff, G. E. Crooks, N. S. Wingreen, E. Betzig, and J. Liphardt. 2009. Self-organization of the *E. coli* chemotaxis system imaged with super-resolution light-microscopy. *PLoS Biol.* 7:e1000137.

133. Briegel, A., D. R. Ortega, E. I. Tocheva, K. Wuichet, Z. Li, S. Chen, A. Müller, C. V. Iancu, G. E. Murphy, M. J. Dobro, I. B. Zhulin, and G. J. Jensen. 2009. Universal architecture of bacterial chemoreceptor arrays. *Proc. Natl. Acad. Sci. USA* 106:17181–17186.

134. Parkinson, J. S., P. Ames, and C. A. Studdert. 2005. Collaborative signaling by bacterial chemoreceptors. *Curr. Opin. Microbiol.* 8:1–6.

135. Berg, H. C., and P. M. Tedesco. 1975. Transient response to chemotactic stimuli in *Escherichia coli*. *Proc. Natl. Acad. Sci. USA* 72:3235–3239.

136. Muppirala, U. K., S. Desensi, T. P. Lybrand, G. L. Hazelbauer, and Z. Li. 2009. Molecular modeling of flexible arm-mediated interactions between bacterial chemoreceptors and their modification enzyme. *Protein Sci.* 18:1702–1714.

137. Levin, M. D., T. S. Shimizu, and D. Bray. 2002. Binding and diffusion of CheR molecules within a cluster of membrane receptors. *Biophys. J.* 82:1809–1817.

138. Amin, D. N., and G. L. Hazelbauer. 2010. The chemoreceptor dimer is the unit of conformational coupling and transmembrane signaling. *J. Bacteriol.* 192:1193–1200.

139. Hazelbauer, G. L., and W.-C. Lai. 2010. Bacterial chemoreceptors: providing enhanced features to two-component signaling. *Curr. Opin. Microbiol.* 13:124–132.

140. Vaknin, A., and H. C. Berg. 2007. Physical responses of bacterial chemoreceptors. *J. Mol. Biol.* 366:1416–1423.

141. Falke, J. J., and G. L. Hazelbauer. 2001. Transmembrane signaling in bacterial chemoreceptors. *Trends Biochem. Sci.* 26:257–265.

142. Zhou, Q., P. Ames, and J. S. Parkinson. 2009. Mutational analyses of HAMP helices suggest a dynamic bundle model of input-output signalling in chemoreceptors. *Molec. Microbiol.* 2009:801–814.

143. Khursigara, C. M., X. Wu, P. Zhang, J. Lefman, and S. Subramaniam. 2008. Role of HAMP domains in chemotaxis signaling by bacterial chemoreceptors. *Proc. Natl. Acad. Sci. USA* 105:16555–16560.

144. Berg, H. C., and E. M. Purcell. 1977. Physics of chemoreception. *Biophys. J.* 20:193–219.

145. Mielczarek, E. V., E. Greenbaum, and R. S. Knox. 1993. *Biological Physics*. Am. Inst. Physics, New York, NY.

146. Brown, D. A., and H. C. Berg. 1974. Temporal stimulation of chemotaxis in *Escherichia coli*. *Proc. Natl. Acad. Sci. USA* 71:1388–1392.

147. Segall, J., M. Manson, and H. C. Berg. 1982. Signal processing times in bacterial chemotaxis. *Nature* 296:855–857.

148. Segall, J. E., S. M. Block, and H. C. Berg. 1986. Temporal comparisons in bacterial chemotaxis. *Proc. Natl. Acad. Sci. USA* 83:8987–8991.

149. de Gennes, P. G. 2004. Chemotaxis: the role of internal delays. *Eur. Biophys. J.* 33:691–693.

150. Clark, D. A., and L. C. Grant. 2005. The bacterial chemotactic response reflects a compromise between transient and steady-state behavior. *Proc. Natl. Acad. Sci. USA* 102:9150–9155.

151. Celani, A., and M. Vergassola. 2010. Bacterial strategies for chemotaxis response. *Proc. Natl. Acad. Sci. USA* 107:1391–1396.

152. Asakura, S., and H. Honda. 1984. Two-state model for bacterial chemoreceptor proteins. *J. Mol. Biol.* 176:349–367.

153. Levit, M. N., and J. B. Stock. 2002. Receptor methylation controls the magnitude of stimulus-response coupling in bacterial chemotaxis. *J. Biol. Chem.* 277:36760–36765.

154. Barkai, N., and S. Leibler. 1997. Robustness in simple biochemical networks. *Nature* 387:913–917.

155. Alon, U., M. G. Surette, N. Barkai, and S. Leibler. 1999. Robustness in bacterial chemotaxis. *Nature* 397:168–171.

156. Yi, T. M., Y. Huang, M. I. Simon, and J. Doyle. 2000. Robust perfect adaptation in bacterial chemotaxis through integral feedback control. *Proc. Natl. Acad. Sci. USA* 97:4649–4653.

157. Yi, T.-M. 2001. Constructing mathematical models of biological signal transduction pathways: an analysis of robustness. In *Foundations of Systems Biology*, H. Kitano, editor. MIT Press, Cambridge, MA. 151–162.

158. Hansen, C. H., R. G. Endres, and N. S. Wingreen. 2008. Chemotaxis in *Escherichia coli*: a molecular model for robust precise adaptation. *PLoS Comp. Biol.* 4:14–27.

159. Korobkova, E., T. Emonet, J. M. G. Vilar, T. S. Shimizu, and P. Cluzel. 2004. From molecular noise to behavioral variability in a single bacterium. *Nature* 428:574–578.

160. Ishihara, A., J. E. Segall, S. M. Block, and H. C. Berg. 1983. Coordination of flagella on filamentous cells of *Escherichia coli*. *J. Bacteriol.* 155:228–237.

161. Levin, M. D., C. J. Morton-Firth, W. N. Abouhamad, R. B. Bourret, and D. Bray. 1998. Origins of individual swimming behavior in bacteria. *Biophys. J.* 74:175–181.

162. Emonet, T., and P. Cluzel. 2008. Relationship between cellular response and behavioral variability in bacterial chemotaxis. *Proc. Natl. Acad. Sci. USA* 105:3304–3309.

163. Emonet, T., C. M. Macal, M. J. North, C. E. Wickersham, and P. Cluzel. 2005. AgentCell: a digital single-cell assay for bacterial chemotaxis. *Bioinformatics* 21:2714–2721.

164. Bray, D. 2002. Bacterial chemotaxis and the question of gain. *Proc. Natl. Acad. Sci. USA* 99:7–9.

165. Berg, H. 2009. The gain paradox. *Prog. Biophys. Mol. Biol.* 100:2–3.

166. Bray, D., R. B. Bourret, and M. I. Simon. 1993. Computer simulation of the phosphorylation cascade controlling bacterial chemotaxis. *Mol. Biol. Cell* 4:469–482.

167. Morton-Firth, C. J., T. S. Shimizu, and D. Bray. 1999. A free-energy-based stochastic simulation of the Tar receptor complex. *J. Mol. Biol.* 286:1059–1074.

168. Bray, D., M. D. Levin, and K. Lipkow. 2007. The chemotactic behavior of computer-based surrogate bacteria. *Curr. Biol.* 17:12–19.

169. Bray, D., M. D. Levin, and C. J. Morton-Firth. 1998. Receptor clustering as a cellular mechanism to control sensitivity. *Nature* 393:85–88.

170. Duke, T. A. J., and D. Bray. 1999. Heightened sensitivity of a lattice of membrane receptors. *Proc. Natl. Acad. Sci. USA* 96:10104–10108.

171. Shi, Y., and T. Duke. 1998. Cooperative model of bacterial sensing. *Phys. Rev. E* 58:6399–6405.

172. Shimizu, T. S., S. V. Aksenov, and D. Bray. 2003. A spatially extended stochastic model of the bacterial chemotaxis signalling pathway. *J. Mol. Biol.* 329:291–309.

173. Sourjik, V., A. Vaknin, T. S. Shimizu, and H. C. Berg. 2007. In vivo measurement by FRET of pathway activity in bacterial chemotaxis. *Meth. Enzymol.* 423:365–391.

174. Sourjik, V., and H. C. Berg. 2002. Receptor sensitivity in bacterial chemotaxis. *Proc. Natl. Acad. Sci. USA* 99:123–127.

175. Sourjik, V. 2004. Receptor clustering and signal processing in *E. coli* chemotaxis. *Trends Microbiol.* 12:569–576.

176. Mello, B. A., and Y. Tu. 2005. An allosteric model for heterogeneous receptor complexes: understanding bacterial chemotaxis responses to multiple stimuli. *Proc. Natl. Acad. Sci. USA* 102:17354–17359.

177. Monod, J., J. Wyman, and J.-P. Changeux. 1965. On the nature of allosteric transitions: a plausible model. *J. Mol. Biol.* 12:88–118.

178. Endres, R. G., O. Oleksiuk, C. H. Hansen, Y. Meir, V. Sourjik, and N. S. Wingreen. 2008. Variable sizes of *Escherichia coli* chemoreceptor signaling teams. *Mol. Sys. Biol.* 4:211.

179. Dahlquist, F. W., P. Lovely, and D. E. Koshland, Jr. 1972. Quantitative analysis of bacterial migration in gradients. *Nat. New Biol.* 236:120–123.

180. Kalinin, Y. V., L. Jiang, Y. Tu, and M. Wu. 2009. Logarithmic sensing in *Escherichia coli* bacterial chemotaxis. *Biophys. J.* 96:2439–2448.

181. Springer, M. S., M. F. Goy, and J. Adler. 1979. Protein methylation in behavioral control mechanisms and in signal transduction. *Nature* 280:279–284.

182. Tu, Y., T. S. Shimizu, and H. C. Berg. 2008. Modeling the chemotactic response of *Escherichia coli* to time-varying stimuli. *Proc. Natl. Acad. Sci. USA* 105:14855–14860.

183. Shimizu, T. S., Y. Tu, and H. C. Berg. 2010. A modular gradient-sensing network for chemotaxis in *Escherichia coli* revealed by responses to time-varying stimuli. *Mol. Sys. Biol.* 6:382.

184. Shoval, O., L. Goentoro, Y. Hart, A. Mayo, E. Sontag, and U. Alon. 2010. Fold change detection and scalar symmetry of sensory input fields. *Proc. Natl. Acad. Sci. USA* 107:15995–16000.

185. Berg, H. C. 2006. Marvels of bacterial behavior. *Proc. Am. Philosoph. Soc.* 150:428–442.

Chapter 13

Index